Advances in Japanese Business and Economics

Volume 31

Advances in Japanese Business and Economics (AJBE) showcases the work of Japanese and non-Japanese scholars researching the Japanese economy and Japanese businesses. Published in English, the series highlights for a global readership the unique perspectives of Japan's most distinguished and emerging scholars of business and economics. It covers research of either theoretical or empirical nature, in both authored and edited volumes, regardless of the sub-discipline or geographical coverage, including, but not limited to, such topics as macroeconomics, microeconomics, industrial relations, innovation, regional development, entrepreneurship, international trade, globalization, financial markets, technology management, and business strategy. At the same time, as a series of volumes written by Japanese and non-Japanese scholars studying Japan, it includes research on the issues of the Japanese economy, industry, management practice, and policy, such as the economic policies and business innovations before and after the Japanese "bubble" burst in the 1990s.

AJBE endeavors to overcome a historical deficit in the dissemination of Japanese economic theory, research methodology, and analysis. The volumes in the series contribute not only to a deeper understanding of Japanese business and economics but to revealing underlying universal principles.

Overseen by a panel of renowned scholars led by Editor-in-Chief Professor Ryuzo Sato, AJBE employs a single-blind review process in which the Editor-in-Chief, together with the Managing Editors and specialized scholars designated by the Editor-in-Chief or Managing Editors, rigorously reviews each proposal and manuscript to ensure that every submission is a valuable contribution to the global scholarly readership.

All books and chapters in AJBE are indexed in Scopus.

Masatoshi Fujiwara · Yaichi Aoshima

Mechanisms for Long-Term Innovation

Technology and Business Development of Reverse Osmosis Membranes

Springer

Masatoshi Fujiwara
Graduate School of Business Administration
Hitotsubashi University
Tokyo, Japan

Yaichi Aoshima
Graduate School of Business Administration
Hitotsubashi University
Tokyo, Japan

ISSN 2197-8859 ISSN 2197-8867 (electronic)
Advances in Japanese Business and Economics
ISBN 978-981-19-4895-4 ISBN 978-981-19-4896-1 (eBook)
https://doi.org/10.1007/978-981-19-4896-1

This Springer imprint is published by the registered company Springer Nature Singapore Pte Ltd.
The registered company address is: 152 Beach Road, #21-01/04 Gateway East, Singapore 189721, Singapore

Preface

By exploring the history of reverse osmosis membranes (RO membranes) development, this book aims to identify the mechanisms that enable companies to continue long-term development under a high level of uncertainty.

It was in 2010 that the authors began researching the development of RO membranes. At that time, the B2C business, represented by the electronics industry, was in a severe predicament in Japan. Thus there was a conspicuous debate about the need to shift to the B2B business, including the upstream materials industry, to maintain Japan's industrial competitiveness. In the midst of this debate, the RO membranes, in which Japanese companies have a large market share, attracted attention as a leading material business that could revitalize the stagnant Japanese economy.

In January 2008, the Ministry of Economy, Trade and Industry (METI) established a study group on water resource policy, where RO membranes were discussed. Following this trend, we became interested in RO membranes as a representative example of internationally competitive components and decided to conduct research to analyze not only the companies that developed RO membranes but also the industry as a whole.

As soon as we started conducting research, a simple question arose in our minds. Why did the development of RO membranes continue for so long in Japan from the 1960s onward, although dams were actively built and water resources were not severely depleted? We thought that we could deepen our understanding of the innovation processes by uncovering mechanisms that enabled the continuation of a new business development under technological and market uncertainties for a long time. We also felt we could make some suggestions for Japanese companies whose development activities were becoming short-lived. In retrospect, this research, which began with such a simple question, has taken quite a long time.

A conclusion drawn from analyses in this book is the importance of shaping expectations in the face of diverse uncertainties about technology, customers, competition, organizations, and society. For more details, please read the book. By analyzing long-term development cases, this book clarifies the mechanisms in which internal and external expectations give a future prospect for development activities at any given time and draw the resources needed to keep them going.

The RO membrane discussed in this book was developed as a component to solve the serious social problem of securing drinking water. In the midst of global population growth and environmental pollution, there are many social issues, not limited to water problems, that we need to solve. It would be a great pleasure if the analysis in this book can be of some help in the development of new technologies that lead to the solution of such social problems.

Tokyo, Japan

Masatoshi Fujiwara
Yaichi Aoshima

Acknowledgments

Without the cooperation and generosity of the many people involved in the development of reverse osmosis membranes (RO membranes), it would not have been possible to carefully trace the long-term development process and clarify its history. We are very grateful for the generosity of the people involved in the industry who have given us the opportunity to conduct this research. This book is wholly the result of their kindness.

The reason we started this research in 2010 was to write a case study on Toray Industries, Inc. which was awarded the Okochi Prize for the development of the RO membrane. At the time, Hitotsubashi University, to which the authors belonged, was conducting the Okochi Prize Case Study Project to elucidate the mechanisms of innovation created by Japanese companies with the cooperation of the Okochi Memorial Foundation. The case study writing that led to this book was conducted as part of this project. We would like to express our gratitude again to the members of the Okochi Memorial Foundation.

Afterward, we were able to interview the main related Japanese companies, Toyobo and Nitto Denko. It is a rare opportunity to be able to provide a comprehensive picture of the long process of innovation, from the start of development to the present day, for major companies in the industry. We are truly grateful for the opportunity. It would be our great pleasure if the knowledge we have gained through the cooperation of everyone involved could be repaid in some small way in the form of this book.

We have tried to describe and analyze the facts as accurately as possible by carefully interviewing people and consulting many secondary sources. The names of organizations and institutions whose English names were not available have been translated into English by the authors in accordance with the original intent. The authors have also translated the statements in Japanese obtained from the interviews and articles.

However, differing views and interpretations of development history could always be possible depending on one's position even if the facts are the same. Our descriptions may not be enough to include such diverse perspectives and interpretations and may also have some errors, which we are responsible for.

It is not limited to individuals in the industry who kindly supported us. We would like to express our gratitude to Hideki Nakayama of Toyo Keizai Inc. for his generous support, four students (Mitsuki Ohashi, Yuka Ohashi, Atsuhiro Ozaki, and Naoki Takada) of our university for data collection, Vincent Coleman for proofreading our English manuscript, and Mayuka Fujiwara for editing the reference list. We also received support from academics: Tomono Miki (Chuo University), Takushi Hara (Kansai University), Akira Takeishi (Gakushuin University), and our colleagues at Hitotsubashi University. They gave us insightful advice and support. It is not possible to mention everyone's name here, but we would like to express our gratitude to all of them.

Travel within Japan and abroad was necessary for this research project. Okinawa and Fukuoka are the only two cities in Japan that have large desalination plants. The three leading Japanese companies all have their business bases in the Kansai region. The birthplace of desalination research is the United States, and Saudi Arabia has many large-scale desalination plants. Financial support was essential for the Tokyo-based authors to carry out these investigations, and we were able to use a portion of the JSPS Leading-edge Research Promotion Fund (Funding Program for Next Generation World-Leading Researchers) GZ002 and JSPS Grants-in-Aid for Scientific Research 26285081, 24243046, 21730329, 20K20754, and 16K03855. We are very grateful for this support.

We have been substantially away from home because of frequent travels. We would like to express our heartfelt gratitude to our family for their unwavering support.

Masatoshi Fujiwara
Yaichi Aoshima

Contents

About the Authors

Masatoshi Fujiwara is a Professor of Graduate School of Business Administration at Hitotsubashi University. He received his B.A., M.A., and Ph.D. (commerce and management) from Hitotsubashi University. His research interest is in strategic management and innovation management. He has been doing qualitative research on diversification strategy and the dynamic aspects of knowledge accumulation and utilization within large established firms. He is currently focusing on the dynamic interactions among established firms and start-ups.

Yaichi Aoshima is currently a Professor of Institute of Innovation Research at Hitotsubashi University. His research interests include the management issues related to innovation and new product development, focusing on justification processes of mobilizing resources for innovation, and co-evolution of industry, technology, and firm capabilities. His research fields ranged from the electronics industry, the advanced material industries, and the environment and energy industries. He has recently extended his research to explore possibilities of new business creation by collaboration between large established firms and start-ups. He holds a Master's degree in Commerce and Management from Hitotsubashi University and Ph.D. in Management from MIT Sloan School.

Chapter 1
Introduction: The Objectives and Research Questions of This Book

Abstract In this chapter, we first clarify the research question addressed in this book, "Why is development of new technologies and new businesses continued for a long period of time under a high level of uncertainty?" and explain a background that we selected the history of reverse osmosis membrane development as our research target. We next propose the framework for our analysis, which focuses on four types of uncertainty associated with innovation processes: technological uncertainty, customer uncertainty, competition, uncertainty, and social and organizational uncertainty. We also discuss that breakthrough technologies, initial market, policy support, and company-specific rationales play important roles in dealing with these uncertainties. At the end of this chapter, we explain the overall structure of this book.

1.1 The Research Questions of This Book

Under the banner of innovation, we are increasingly hearing calls for the creation of new industries and new business fields. As existing industries and business fields mature, nations aim to develop new industries that will drive economic growth, and companies attempt to place the creation of new business fields at the center of their medium- and long-term strategies.

However, it is clear that new industries and new business fields are not created overnight, as we can see from history. It usually takes many years from the birth of a technology or idea that forms the core of a new business until it is brought to fruition as a product or service, introduced to the market, and widely accepted by customers. When the seeds of technologies and ideas are still new, it is not clear whether they can be materialized into products and services in the future. Even if they do become products or services, it is even more difficult to predict in advance whether they will be widely accepted by customers. However, even in the face of such uncertainty, companies must continue commercializing their new technologies and ideas if they want to be able to shape new industries. Therefore, to explore the mechanisms of new industry creation, we need to understand why firms can continuously invest

M. Fujiwara and Y. Aoshima, *Mechanisms for Long-Term Innovation*,
Advances in Japanese Business and Economics 31,
https://doi.org/10.1007/978-981-19-4896-1_1

resources in developing new technologies and exploring new business fields despite high levels of uncertainty.

Of course, there is a profit motive behind the development of new technology and new businesses by commercial firms. Companies should try to invest in them based on economically rational judgments. However, in the early stage when the technology is not yet mature and the market is not yet clear, it is difficult not only to expect short-term profits but also to reasonably estimate the future profits from the investment. Even if a company successfully brings products or services to the market, it is not uncommon for the businesses to suffer because they are not widely accepted by customers. Nevertheless, if a company does not continue to innovate and improve its products and services to overcome such a situation, it will not be able to develop them to a level where they will eventually be widely accepted by customers and recognized as a business or an industry.

It seems impossible to explain the long and winding innovation process that leads to the creation of new industries by reducing this process to the profit motive of firms and the function of market competition. It seems difficult to explain in advance the economic rationale for allocating resources for highly uncertain innovations in a way that is transparent and acceptable to a wide range of people. If that is the case, how can companies, being exposed to profit pressure, continue development activities, for which expected profits cannot be calculated, until the new industry is finally formed? Why is development of new technologies and new businesses continued for a long period of time under a high level of uncertainty? This question is the starting point of this book. To elucidate industrial formation through innovation, this question must be answered.

Based on this question, this book analyzes the development history of RO membranes, which are semipermeable membranes used to produce fresh water from various types of raw water. The development of RO membranes began with the aim of solving a clear social problem: water shortage. As will be explained in the next section, the development of technologies to solve social problems is often accompanied by a particularly high degree of uncertainty. Companies inevitably face a variety of difficulties in the process of sustaining that development over a long period of time. Therefore, by focusing on the development of social problem solving technologies, the process of overcoming the many difficulties that stand in the way of innovation can be more clearly depicted.

1.2 Resolution of Social Problems and Innovation

In recent years, with the economic development of emerging countries, various social problems have been exposed, such as environmental pollution and global warming caused by industrialization, food and water shortages caused by population growth, and the widening gap between the rich and the poor. It is already common understanding that to achieve sustainable growth in the global economy, it is necessary to focus on solving these social issues in each country as well as to promote free competition. In this environment, corporate social responsibility (CSR) is being strongly emphasized, and the market is beginning to evaluate companies not only on the basis of profitability but also from the perspective of environmental, social, and governance (ESG) criteria.

In line with this trend, there are growing expectations for innovation to solve various social problems (e.g., Nidumolu et al., 2009). For example, a variety of new technologies are being developed with anticipation from society, such as renewable energy technologies and electric vehicles to address global environmental problems, materials of natural origin, genetic modification technology, and plant factories to address food problems, robot technology to address aging populations, and desalination technology to address water shortages.

However, it is by no means easy for these new technologies to be developed and spread widely throughout society to solve the social issues of concern. First, budding technologies are often far behind old technologies in terms of economic efficiency. In Japan, for example, the cost of power generation using renewable energies such as solar, wind, and geothermal energy is still higher than the cost of thermal power generation using coal or natural gas. People with a strong interest in global environmental issues may be willing to pay an additional cost burden to embrace renewable energy, but not all customers are willing to pay a high premium. Electric vehicles are also difficult to purchase without government subsidies because of their expensive storage batteries. Plant-derived biodegradable plastics also have a low environmental impact, but their cost is much higher than that of conventional petroleum-based plastics, and it will be some time before they are widely accepted by the market as economically viable.

Therefore, the development and diffusion of these new technologies are often not possible without some kind of policy support, such as subsidies. However, it cannot be said that a new technology has successfully solved a social problem as long as it relies on policy support. Unless the new technology establishes economic efficiency that is at least comparable to that of the old technology, it will not take root in society and will be used continuously.

Of course, if the price-performance ratio improves by technological progress and the market grows as a result, sufficient economic efficiency will be actualized through the effect of scale, and the profits earned from this will be used to further advance the next stage of progress. In this case, a virtuous cycle can be expected. However, technological progress does not occur spontaneously. It requires active and continuous investment in technological development by a variety of actors and the commitment of people for progress to be actualized. How can this be achieved?

Providing incentives through policy support is one way, but barring exceptions such as pharmaceuticals, such support cannot continue forever. Reliance on policy support even risks discouraging the private sector from pursuing economic efficiency. Therefore, the continuous technological development that drives the growth of industry must eventually be carried out by the private sector. However, in the early stages of development, new technology is often inferior to old technology in terms of economic efficiency. A scenario in which the new technology outperforms and replaces the old technology in the future is accompanied by a substantial amount of uncertainty. Under such uncertainty, how can profit-seeking private companies justify their investment in the development of new technologies?

This problem of substituting old technologies with new technologies is not limited to technologies for solving social problems. If taking even the briefest review of history, we can observe a broad panorama of this phenomenon of substituting new technologies for old technologies: from vacuum tubes to transistors, from cathode-ray tube televisions to liquid crystal televisions, and from internal combustion engines to electric motors. Even in these cases, the price-performance ratio of the new technology at the initial stage was overwhelmingly inferior to that of the old technology. There is usually a high degree of uncertainty as to whether technological development aimed at improving the price-performance ratio will actually bear fruit as expected. Despite this high level of uncertainty, the need to continue to invest in new developments is a challenge common to all types of technology.

However, in the case of technology that solves a social issue, the substitution problem is considered to be more serious. This is because the customers who truly need these new technologies often do not have sufficient ability or willingness to pay for them. For example, people living in South Asia and Africa, where food and water shortages are currently worsening, unfortunately lack the ability to pay for new technologies. Most of the countries without access to safe water are developing countries with a low GDP per capita. These countries are projected to experience population growth, and there are concerns that they will experience even more severe water shortages in the future. Furthermore, when developing countries are in the process of accelerating industrialization, they may be slow to invest in environmental technologies that would undermine industrial development.

Thus, new technologies for solving social problems are subject not only to uncertainty in technological development but also to another uncertainty: even if the technology is successfully developed, will customers have the ability and willingness to pay for investment in that technological development? In other words, there is a fundamental dilemma: where there is demand, there is no willingness or ability to pay. From the standpoint of the private sector, which seeks profit, it would not be easy to invest aggressively in such an uncertain technology. It is quite possible that investment in this kind of technology is made as part of CSR efforts, without regard to profitability. However, it is not possible to continue to invest the enormous amount of management resources needed to commercialize new technologies solely from the perspective of CSR. As long as the company is a profit-making enterprise, it cannot continue to allocate resources without expectation of profit.

In this light, the phenomenon in which profit-motivated commercial companies continue long-term efforts to develop technologies to solve social problems in the midst of an uncertain situation in which they cannot earn profits seems strange at first glance.

1.3 Perspective of Analysis

1.3.1 The Four Types of Uncertainty

Why do new technologies and business fields continue to be developed for long periods of time when they are subject to a high level of uncertainty? This is a valid research question because at first glance, it seems strange that private companies, which are supposed to make investment decisions based on profit expectations under high profit pressures, have continued to develop technologies and businesses even when the expected markets have not grown significantly for many years and the companies have been unable to earn profits.

To solve this mystery, we need to clarify the factors and mechanisms that influence the decision-making of companies under conditions of uncertainty. To this end, we would first like to clarify the nature of uncertainty upon which this book focuses. Uncertainty generally refers to a situation in which it is impossible to predict, even probabilistically, that a phenomenon will occur (Knight, 1921). For example, corporate finance textbooks teach us to determine the level of investment in light of NPV (Net present value), but when uncertainty is high, we cannot even calculate NPV of the investment because we cannot know the future profits even probabilistically. When this happens, decision makers are unable to determine the appropriate investment level, and they are likely to underinvest to avoid ambiguity as existing research has indicated (e.g., Kellogg, 2014; Leahy & Whited, 1996).

In the case of investment in technological development, a rational company should decide the level of resource input after envisioning a series of cause-and-effect relationships in which the additional input of resources improves the technological level, resulting in the development of a new market, and the company earns profits from it. However, in the early stage of innovation, it is often impossible to predict how this causal chain will occur even in terms of probability, and hence, the investment in technological development tends to be underestimated. This situation is surrounded by four different types of uncertainty, which are described below.

The first type of uncertainty is "technological uncertainty". This is a situation in which it is not possible to predict the extent to which investment or effort in technological development will result in improvements in the level of technology. It can be said that this type of uncertainty arises from a lack of knowledge about the laws of nature. The second type of uncertainty is "market uncertainty". This indicates a situation in which it is impossible to predict how many markets will be opened up by improvements in the level of technology and how much profit will be generated for the company as a result. This uncertainty arises from a lack of knowledge about the market value of technology. The types of uncertainty that have been frequently addressed by the existing studies are these technological uncertainty and market uncertainty (Jalonen, 2012; McGrath & MacMillan, 2000).

Market uncertainty can be understood by breaking itself down into uncertainty about the acceptance of technology by customers and uncertainty about the impact of market competition on corporate profits. The former, which can be called "customer uncertainty", refers to a situation in which it is impossible to predict how much of the market will be opened up by improvements in the level of technology. The latter, on the other hand, can be called "competition uncertainty". Even if a large market is created, if competition becomes fierce, companies may not be able to earn enough profit to justify their investment. The profits earned by firms depend on market competition. The case in which the competitive situation cannot be ascertained in advance can be called competition uncertainty.

These three types of uncertainty can be bundled together as uncertainties related to the economic calculations that developers in companies make when they promote innovation. Uncertainty about the economic rationale for development due to uncertainty about the technical outcomes of development investments and the customer value and future revenues that could be realized from those outcomes is a major obstacle to development.

In addition to these three types of uncertainty, another type of uncertainty that affects the innovation of firms is "social and organizational uncertainty". Companies exist in a society consisting of people with various interests. There are also social relationships consisting of diverse people within the corporate organization. These factors affect the input of resources for innovation by firms according to a logic different from simple economic rationality. For example, there are cases, such as genetic engineering technology, where companies are forced to hesitate to invest in development from an ethical point of view, no matter how high the profitability is expected. On the other hand, there are cases where development is socially recommended even if profitability is low, as in the case of environment-related technologies. In addition, it is not uncommon for the commercialization of a profitable technology to be abandoned due to a political struggle within an organization or for an unprofitable technology to be commercialized and introduced into the market for that same reason. How much resource to be invested for the expected future profit depends on the social factors inside and outside the organization. The situation in which the effects of these factors cannot be grasped in advance is the "social and organizational uncertainty". Figure 1.1 shows the above four types of uncertainty that affect the development of new technologies and businesses.

1.3.2 Four Analytical Perspectives: Factors that Influence the Formation of Expectation

The process of realizing innovation and creating new businesses is subject to these four types of uncertainty. Normally, these uncertainties are gradually reduced as each company's technological development efforts and interactions with customers in the market progress and as the industry develops. The problem for companies, however, is the speed with which this happens. If the speed of uncertainty reduction is slow and the future is uncertain for a substantial length of time, companies will not be able to continue to take development risks and will not be able to justify

Fig. 1.1 Four types of uncertainty. *Source* Prepared by the authors

the allocation of resources to the creation of new businesses. For technology and business development to continue even in such a situation, the action of forces that make people expect good future prospects is necessary. We explore the forces that shape such expectations from the following four perspectives, corresponding to the four types of uncertainty mentioned above.

1.3.2.1 Emergence of Breakthrough Technologies and Convergence of Technological Approaches

First, this book focuses on the "emergence of breakthrough technologies" and the resultant "convergence of technological approaches" as forces that raise expectations for the outcome of technological development. Normally, in the early stage of industrial development, each company explores various technological approaches, and there are often a variety of technologies in an industry. In this juxtaposing situation, a technological race is waged to become mainstream in the industry (Suarez, 2004). This situation of technological diversity, when viewed at the industry level, means the dispersion of technological development efforts and investments, which thus slows down the speed of technological progress. It is the dominant design (Abernathy, 1978; Utterback & Abernathy, 1975), which is created by breakthrough technologies, that overcome this situation and raises expectations toward the technology.

A breakthrough technology is one that presents a promising "set of solutions" that everyone in the industry agrees on. It provides a clear new solution to a problem that has been recognized in the industry for many years. At the time of its emergence, it does not necessarily guarantee practicality or economic improvement. However, its timeliness and the resulting dramatic improvement in technological results will attract the attention of many players, and diverse technological approaches will converge at once. The investment in technological development by each company is concentrated in a particular direction, and competition develops on the same technological trajectory (Dosi, 1982). As a result, the economic potential of the technology will improve at an accelerated rate, and it will spark the beginning of a diffusion that will replace the old technology.

1.3.2.2 Exploration of Initial Market

Second, this book focuses on the "exploration of initial market" as a force that directly stimulates expectations for customer acquisition in a situation of high customer uncertainty. In the early stage, when the level of technology is low and immature, it is difficult to persuasively explain the expansion of the target market and the existence of future profits to the management and investors of a company. Nevertheless, it would be extremely difficult for a commercial company to continue development for decades without any market visibility. An initial commercial market can overcome this situation and allow companies to continue developing. This may be a byproduct of the technology development process, or it could be an "unexpected" small niche market discovered as a result of an active search for a commercial market that is different from what was originally envisioned. Even if it is not a large market, the fact that such a niche market does indeed exist, albeit in a limited way, and that customers are buying, will raise expectations for future market expansion and form a sufficient reason to continue development.

In particular, the existence of such niche markets is considered to play an important role in promoting innovation that solves social problems (Kemp et al., 1998; Weber et al., 1999). As already mentioned, in order for a technology developed for the purpose of solving a social problem to achieve sufficient economic viability, more difficulties must be overcome compared to the technology development for normal commercial purposes. For this to occur, a grace period is necessary. In such a situation, an initial niche market allows for the immediate continuation of technological development activities, even if it does not directly serve the purpose of solving social problems. In addition, learning from the initial niche market can lead to the discovery of new possibilities for the technology, which may lead to the expansion of the market beyond the initial expectation. The existing research has clarified the importance of "strategic niche management" to expand such niche markets, and pointed out the significance of clearly stating expectations and visions for the future (Schot & Geels, 2008).

1.3.2.3 Policy Support

Third, we will focus on "policy support" as a force that raises expectations toward future profits in the face of high levels of competition uncertainty and customer uncertainty. If highly external technological development is left to the market mechanism alone, there is a risk of under-investment in technology that is necessary for society. This may be especially true in the case of technological development aimed at solving social problems (Jaffe et al., 2005). This is not only because future benefits are unpredictable due to a high level of uncertainty but also because technology spillovers are predicted due to high public interest.

This is where the government plays an important role. When an entity with a different objectives from that of the private sector is involved in the decision-making process for the allocation of resources in society, it can encourage the development of new technologies.[1] Since firms are sensitive to domestic and international policy and regulatory uncertainty (Marcus, 1981; Engau & Hoffmann, 2011; Porter & van der Linde, 1995), the policy attitudes the government takes and which R&D strategies it supports or regulates greatly influence the R&D strategies of private firms.[2]

Government policy support directly and indirectly influences the development activities of firms (Hall & Van Reenen, 2000; Leyden & Link, 1991). Directly, the government subsidizes the cost of developing new technologies that are budding and whose future prospects are uncertain, thereby enabling firms to take on the challenge of developing new technologies that would not be profitable without assistance. On the other hand, as an indirect effect, government policy has a kind of priming effect that induces private firms to invest by drawing the world's attention (Anderson, 1944). It is conceivable that the attention of society is drawn to the technological field that has received policy support, and the increased expectations for that technology and the expansion of its market stimulate the entry of firms. In fact, from this perspective, policy support has been provided in various industries in Japan, including the semiconductor industry and the energy industry.

Firms selected for public support may believe that they have a competitive advantage over firms that are not selected. On the other hand, firms that are not selected may recognize their competitive disadvantage and withdraw from the market. Both of these factors will work to raise the profit expectations of publicly supported firms.

Thus, public support at the early stage of an industry acts as a "hiding hand" for uncertainty (Hirschman, 1967). By diverting attention temporarily from the difficulties faced in the future, this allows firms to develop new technologies.[3]

---

[1] An example of public support for private sector development in Japan is the large-scale project program that the Ministry of International Trade and Industry (now the Ministry of Economy, Trade and Industry) promoted from the 1960s onward. Some of these large projects were particularly successful, such as the Super LSI Technology Research Association, which was started in 1976 for semiconductors (Tatsumoto, 2008), while others, such as the Sunshine Project, did not necessarily produce the results that were initially expected (Shimamoto, 2014). Although there is a wide range in terms of results, it is clear that this program had a strong driving force to make companies start R&D. The development of RO membranes, which is the subject of this book, was also supported by the Large Project Program under the title "Research on Desalination and Byproduct Utilization".

[2] The impact of public institutions on the innovation of private firms has been actively discussed in the framework of national innovation systems since the late 1980s (Freeman, 1987; Nelson, 1992, 1993; Mowery, 1998; Lundvall, 1992; Guan & Chen, 2012; Intarakumnerd & Goto, 2018).

[3] The "hiding hand" is sometimes viewed negatively as a cause of rather unfavorable outcomes (Flyvbjerg & Sunstein, 2016), whereas Hirschman views it favorably. However, both positions share the point that the "hiding hand" causes projects to be initiated with a high degree of uncertainty, and we believe that this way of thinking can be applied to development activities in the private sector.

1.3.2.4 Company-Specific Rationales

Fourth, in the presence of social and organizational uncertainty, we focus on a "company-specific rationales" as a force that makes possible the long-term development of new technologies and new businesses. Such rationales are provided from both inside and outside the organization.

First, "company-specific rationales" refers to the logic specific to each company that guides toward a decision to continuously invest efforts and resources in technological development even though the future earnings are not sufficiently estimated under a high level of uncertainty. In general, as companies engage in business activities for a long time, they tend to form their own specific values, which can be called organizational culture. The beliefs and values unique to the organization, which are shared by its members as invisible basic assumptions (Schein, 1990), often allow unique behavioral patterns that are not recognized by simple economic rationality standards. This means that there are unique reasons in business organizations that support innovation, in addition to the generic reasons that are accepted by all, such as economic rationality.

When developing a new technology, it is difficult to specify in advance the economic rationality that lies ahead. In the actual innovation process, resources are allocated to a new scheme that seems irrational in advance, and technological development continues. This eventually becomes widely accepted in the market and, as a result, acquires rationality. In such a long-term development process, the existence of non-economic rational values unique to the organization becomes a critically important base for developers. Through the creative justification process (Takeishi et al., 2012) that appeals to such organization-specific values by using various measures, it is possible to progressively mobilize resources and move toward the goal of innovation.

Second, "social logic provided from outside the organization" refers to a social reason that is different from the profit motive. Although commercial companies act mainly out of a profit motive, not all of the stakeholders surrounding a company are directed by profit. Corporations exist within the larger society and cannot exist without paying attention to social rules and social demands. In recent years, CSR and CSV (creating shared value) have attracted attention because of the concern that corporate activities damage social life and the natural environment. Companies are increasingly required to create social value as well as private profit (McWilliams & Siegel, 2001; McWilliams et al., 2006; Porter & Kramer, 2011; Waddock & Graves, 1997).

Therefore, if an innovation that leads to a new business is strongly demanded by society, companies will be able to continue to invest in it even if doing so does not lead to large profits. Of course, a company cannot continue to invest in a situation that conflicts with its profit motive just because it is a social demand. At least, however, it will act as a catalyst for the development of technology and the start of business activity.

For funds and people's energy to be continuously invested in the development of new technologies that are clearly less economical than existing technologies, it may

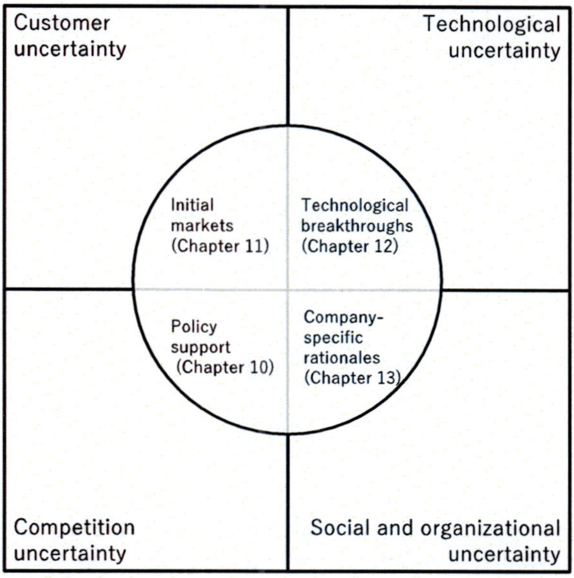

Fig. 1.2 Analytical perspective of this book. *Source* Prepared by the authors

not be enough to appeal to the profit motive of private companies. When it takes a very long time for a new technology to flourish in the hands of the private sector, it is difficult even to calculate the present value of the investment in the early stages. In such a situation, the existence of environmental problems and social issues may serve as a social demand and pressure to induce various actors to initiate development actions.[4]

Based on the above discussion, Fig. 1.2 illustrates the four forces that affect the formation of corporate expectations under the four types of uncertainty. This is the basic analytical perspective of this study.

1.4 Research Methods

1.4.1 Single Case Study

To answer the question, "Why do new technologies and business fields continue to be developed for a long period of time under a high level of uncertainty?", the research method we adopted is the single case study method.

[4] Such a rationale given by society is an important force in promoting innovation, but the analysis in this book does not include a specific analysis of this point. The reason for this is that the development of RO membranes is aimed at solving a clear social problem, namely, coping with future water shortages, and its existence is self-evident.

Qualitative research on single cases is an effective means of exploring and constructing new hypotheses that may lead to the replacement or revision of existing theories (Gerring, 2006; Levy, 2008; Yin, 1984). In particular, it is desirable to extract a single deviant case that violates the general trend to attempt a new hypothesis search. Therefore, in this book, we decided to take up cases in which technology and business development, which at first glance seem to be difficult to continue in a commercial company, actually continued for a long period of time and finally came to fruition as innovation.

In addition, this book aims to comprehensively depict the causal mechanisms that lead to the creation of new technologies and businesses aimed at solving social problems and their formation into industries. To do this, it is necessary to look at a wide range of influencing factors, including the detailed content of technological progress, internal organizational circumstances, policy intentions, and social pressures, in addition to the strategic behavior of companies and market competition. It is then necessary to unravel the relationships among them. To understand the long causal chain that includes the series of processes from the beginning of technological development to its continuation, the commercialization of the technology, and the realization of business expansion, it is necessary to provide a robust description by painstakingly retracing its history and carefully following events along the time lapse while accessing primary information as much as possible (Numagami, 1999, 2000). From this point of view, qualitative single case analysis is suitable.

The methodological position of this book is to explore logical explanations that can lead to new hypotheses or theories by qualitatively analyzing cases that go against the general trend in a broad and deep way. In proceeding with the analysis from this standpoint, it is important to clarify the underlying mechanisms of the event by asking the "how" in depth, rather than simply seeking the cause in response to the "why" question (Anteby et al., 2014). Therefore, this book aims to clarify the mechanisms that have supported the long-term development of new technologies and businesses by asking the "how" as deeply as possible while starting from the "why" question.

1.4.2 Focus on RO Membrane Technology

If the purpose of a single case study is to search for a new hypothesis or mechanism, it is desirable to select a characteristic case in which the research question is prominent. In this respect, the development process of RO membranes is a perfect case study.

RO membranes are special semi-permeable membranes developed mainly to separate salt and other substances from seawater to secure drinking water. The development of RO membranes began with the aim of solving social problems. However, the market did not expand as companies had initially envisioned, and companies were unable to earn profits for a long time. Nonetheless, the main companies continued to develop their technologies and business fields, and eventually, industrial formation was realized. The purpose of this book is to clarify this process.

The development of RO membranes started in the 1940s at the University of California, Los Angeles (UCLA) in response to the water shortage caused by urbanization on the West Coast of the United States. In 1952, the U.S. government enacted the Saline Water Conversion Act and policy support became stronger. While U.S. companies proceeded with the development toward commercialization, Japanese companies followed the U.S. companies and entered the development race in the late 1960s.

However, it took many years for RO membranes to fulfill the purpose of producing drinking water from seawater and to gain a market of the current scale. For a long time, RO was not accepted by the market as expected because it could not exceed the cost performance of the old evaporation method amidst the necessity of a cost performance level appropriate for a water supply social infrastructure project. In the meantime, companies had to carry out their development activities with a high degree of uncertainty, both technologically and in terms of the market.

Among them, at least four companies, Dow Chemical (Dow), Toray Industries (Toray), Nitto Denko, and Toyobo, have been developing RO membranes without interruption for a long period. As a result, the RO membrane market has become a mainstream technology in the seawater desalination market, which was the original purpose of the development, after more than 40 years of development by private companies. The majority of new seawater desalination plants now use the RO method, and the global market for RO membranes has grown to over 100 billion yen.

One of the characteristics of this market is that Japanese companies maintain a high level of competitiveness. Dow is the leader in the global market, but three Japanese companies, Toray, Nitto Denko, and Toyobo, hold more than 50% of the market share.

Japan is a country rich in water resources. The annual precipitation is 1,668 mm, which is about 1.5 times as much as the world average of 1,065 mm.[5] Although some people believe that the amount of precipitation per capita is low, Japan has been actively building dams since ancient times and has never suffered from chronic water shortages in many parts of the country. Although high domestic demand was not always expected, Japanese companies have been actively involved in the development of RO membranes since the 1960s and as a result have succeeded in building a strong position in the market.

Why were three companies based in Japan, a country relatively rich in water resources and without strong domestic social needs, able to continue development over the long term in the face of technological and market uncertainty? This is the more specific analytical question that this book pursues.

[5] Ministry of Land, Infrastructure, Transport and Tourism (2019).

1.4.3 Analytical Advantages of RO Membranes

For the purpose of this book, which is to elucidate the process and mechanisms by which new technologies with a high level of uncertainty are developed and diffused in society over a long period of time, RO membrane technology and business are excellent subjects in terms of the following three points.

First, RO membranes were first developed for the primary purpose of solving a very clear social problem, water shortages. Desalination by the RO method is still attracting attention as a promising method to contribute to solving water shortages in drought areas around the world.

Second, the industry's engineering community has been relatively stable, allowing access to key developers and to the entire history of the industry from early development to market penetration. In the U.S., there have been structural changes such as acquisitions, mergers, and divestitures during the development of the industry. However, the key people who developed the important early technologies are still in the industry, especially in California. Although the overall market size of the desalination business is large, if we focus on the membrane business, it is still about 100 billion yen worldwide, which is why a relatively small and cohesive community of engineers has formed in this industry. This is in line with the purpose of this study, which aims to understand the entire life cycle of the industry from initial development to commercialization and diffusion.

Third, three of the industry's leading firms are Japanese, and all of them have been in business continuously for more than 40 years, from the technology development stage to the present. For Japan-based authors, it is possible to conduct qualitative research inside companies and obtain detailed information on their entire business history, including the process of overcoming technological and economic challenges and a series of investment decisions. The case has high research potential. In fact, as described below, we were blessed with a number of primary sources for our analysis.

1.4.4 Analysis Method

This book relies on the following primary and secondary sources for its analysis. Primary sources are the records of interviews conducted by the authors. The interviews and field surveys were conducted for a total of 105 people in Japan and abroad. Most of the interviews were conducted by the authors, and additional interviews were conducted by e-mail to supplement the information. A list of these interviews is provided at the end of this book. We interviewed Toray, Nitto Denko, and Toyobo in that order. The interviews took place both in Japan and overseas and summarized the development history of RO membranes in each company. Some of the developers became politicians or university professors or worked for research institutes after their retirement. In addition, since membrane development itself started in the United States, we also interviewed and surveyed people involved in ROGA and private companies in the United States, such as Dow. As mentioned earlier, the RO membrane industry is unique in that many of the early development members still remain in the industry and continue development. This is true not only in Japan but also in the United States.

Public support played an important role in the development of RO membranes. Therefore, to confirm the viewpoints of public organizations, we interviewed the Water Reuse Promotion Center, which supported the practical application of desalination plants in Japan, and university officials who were familiar with the history of development with public support.

In addition, since RO membranes demonstrate their performance through downstream plant operations, we visited various plants, including desalination plants, to investigate the current status and potential of RO membrane technology from the operator's point of view. In Japan, we visited and conducted interviews at desalination plants in Okinawa and Fukuoka. Overseas, we visited and surveyed various plants in Saudi Arabia and Singapore.

For secondary materials, we were blessed with an abundant supply. First, both in Japan and the United States, specialized journals were published with desalination in mind, and by tracing these journals, we were able to corroborate the views of the parties involved in the early stages of industrial development. The digital library of the *American Membrane Technology Association* [AMTA] was also very valuable.

These secondary sources were very useful because they could be compared and contrasted with the information obtained from the primary sources. While interview transcripts have the advantage of providing the views and ideas of the people involved and information that has not been made available in secondary sources, they are also subject to "recall bias" (Raphael, 1987) because they are conducted after the fact. In contrast, research papers are free from such bias and contain many descriptions based on scientific evidence. Therefore, we were able to check for possible biases contained in the primary information by comparing it with the secondary information. The results confirmed that there was no serious bias in the primary information obtained by the authors. Rather, the information obtained from the interviews was treated as complementary to support the content of the paper. Furthermore, it was an unexpected harvest that the authors' individual views on the future potential of RO membranes were confirmed in the various papers.

In the U.S. government agencies, there were many official records of political discussions on the actual state of government support for the development of desalination technology and the establishment of support systems. This was not limited to the records of congressional hearings. At the Kennedy Library, an effort to preserve oral history records of people involved in the Kennedy administration has been underway since 1964, and more than 1,600 interviews have been recorded. It was the Kennedy administration that actively encouraged desalination and substantially increased the subsidies. For this reason, it was very useful that the records of the people involved at that time were made available as oral histories.

In Japan, however, the records of individuals involved in public research activities have not been sufficiently accumulated. In recent years, the former employees of the former Agency of Industrial Science and Technology and the current National

Institute of Advanced Industrial Science and Technology (both abbreviated as AIST) have been promoting a project to record their old research activities at the Sankoukai (Industrial Technology Association), the alumni association of AIST. Thanks to this valuable project to preserve history for posterity, we were able to refer to the records of those involved in the desalination research conducted by the former AIST.

Another major source of information that the authors relied on as a secondary source was the patent database provided by ULTRA Patent. In the analysis, the authors extracted a maximum of 9,296 patents related to RO membranes from the database and conducted a full search. In particular, the "F-term" classification was used to analyze Japanese patents. Although F-term, which is a unique Japanese classification, has the disadvantage that it is not internationally comparable, it has the advantage that it classifies patents according to their expected uses. The existence of the F-term, which enables us to classify patents by use, is very useful for us to analyze the materials business, which has a wide range of uses. Although there was a concern about the low accuracy of assigning F-terms, in reality, they were mostly consistent with the information obtained from the interviews. The accuracy of assigning F-terms was not a problem in the RO membrane analysis.

On the other hand, it was extremely difficult to analyze the management aspects of the RO membrane business. This is because quantitative data on management performance, such as sales and profit margins, are not available for the materials business, which is located upstream in the value chain. This was a major constraint in conducting corporate analysis. As a result, we have only been able to indirectly analyze business performance and market information based on the business report information in each company"s annual report and information from leading third-party research organizations such as Fuji Keizai.

Finally, it should be noted that since our description covers only periods up to mid-2010s, some important events, such as the merger between DuPont and Dow Chemical in 2017, are not included for the analyses.

1.5 The Structure of This Book

RO membranes are rare materials that have been developed for a very long time under a relatively high level of uncertainty in terms of the difficulty in achieving the original goal of seawater desalination. Therefore, we may contribute to policy makers for science and technology policies and R&D engineers in private companies, firstly by preserving the historical facts of the RO membrane development. For this reason, we decided to describe the historical background in the first three parts of this book. Based on these historical descriptions, in Part 4, we explored the mechanisms to answer the question of this book, "Why is it that the development of new technologies and new business fields continues for a long time under a high level of uncertainty?" The outline of each part is as follows.

Part 1, the overview section, provides an overview of RO membrane technology and outlines the current status and development of the market. We explain the principle of RO and summarize the different types of RO membranes according to material and shape. In addition, the main development issues for different types of RO membranes are identified.

In Part 2, the history of the development and commercialization of RO membranes in the United States is reviewed. We describe the history of the development of RO membranes in the United States, policy support and intentions, early research results mainly by universities, commercialization and business development by private companies, the actions of each company, and market competition.

In Part 3, the history of development and commercialization in Japan is reviewed. First, we summarize the history of RO membrane development in Japan and the actions of the early players. Then, the histories of the three main companies, Toray, Nitto Denko, and Toyobo, are described in detail. This ranges from the start of development to commercialization and business development.

In the analysis section of Part 4, we proceed with the analysis to answer the questions presented in this book based on the descriptions of industry transition and corporate behavior seen in Parts 1 to 3. The chapters in Part 4 are positioned as follows, in line with the framework shown in Figs. 1.1 and 1.2.

Focusing on the role of "policy support", Chap. 10 analyzes how development was launched in the United States and Japan. It also analyzes the role played by the Japanese and United States governments. It will be argued that in the United States, a huge amount of public support was generated by the confluence of different policy intentions, which supported early development. On the other hand, in Japan, it is argued that the development may have been triggered as a follow-up action for the U.S. predecessors such as DuPont and Dow. There, it becomes clear that U.S. government policy indirectly influenced Japanese companies through the actions of U.S. public institutions and private companies.

Chapter 11 focuses on the role of the "initial markets". We discuss the development of the commercial markets by the companies and their significance. The emergence of commercial markets played a role in supporting business continuity by raising expectations of future market expansion. Of particular importance was the emergence of ultra-pure water production applications for semiconductors. This market cropped up like an amazing miracle in the late 1970s, when the miniaturization of semiconductors was advancing and high cleanliness was required in the manufacturing process. The market made it possible for the companies to continue development and led to the establishment of technology that could be applied to seawater desalination, which was the original goal.

In Chap. 12, we focus on the role of "breakthrough technologies". In particular, we will focus on the breakthrough technology invented by John E. Cadotte (1925–2005) of FilmTec. Cadotte's patent for polyamide composite membranes, filed in 1979, is known as the '344 patent, from its patent number 4,277,344. The '344 patent was an innovative technology that achieved a high level of compatibility between various contradictory performances through the use of membrane composition and interfacial polycondensation. This led to the convergence of diversified technological approaches.

In Chap. 13, we focus on the role of "company-specific rationales" and explore the reasons why development has continued to be justified in three major Japanese companies. The chapter reveals that each of the three companies has continued to support development based on their own internal reasons and their own value standards. At Toray, development was carried out in line with the concept of technological superiority, which was a response to the highly challenging application of seawater desalination with advanced technological capabilities. The developers at Toyobo positioned their business within the company-wide policy of diversification, and when faced with a company-wide crisis, they took the path of acting as inconspicuously as possible. At Nitto Denko, the business was justified by the company-wide philosophy of making the membrane business one of the pillars of corporate growth.

In Chap. 14, based on the results of the analysis in the previous chapters, we present a descriptive model to conceptually understand the process of long-term continuation of new technology and new business development under the four uncertainties. The chapter develops an explanation that the four factors, government policy support, emergence of breakthrough technologies, initial markets, and company-specific rationales, acted in a complementary manner to each other to enable the continuation of the development of new technologies and new business fields that could not be realized with a simple profit motive. We develop an explanation of how this has been achieved.

In Chap. 15, we identify the contributions and implications of the book and then constructs arguments for future research. One of the arguments concerns the contradiction between industry development and company profitability. Groundbreaking breakthroughs that led to the emergence of dominant designs, the convergence of technological approaches in the industry, and the creation of de facto standards for element shapes have certainly contributed to industrial development. However, they also led to fierce competition for homogeneity, which squeezed the profitability of companies. The question of how the balance between the convergence and standardization of technological approaches and the profitability of individual firms affects industrial development is beyond the scope of this book, but it is an important theme. In addition, the role of the commercial market as an experimentation ground and the possibility that the specific reasons justifying development within a company influence the direction of technological development are also presented as a possible future research theme.

The interrelationship of the parts described thus far and the position of each chapter are shown in Fig. 1.3. Readers with a strong interest in analysis can go through the outline of Part 1, then jump to the analysis section of Part 4, and refer to Parts 2 and 3 as necessary to obtain background information corresponding to each analysis. Readers who are also interested in historical facts should read directly from Part 1 to Part 2.

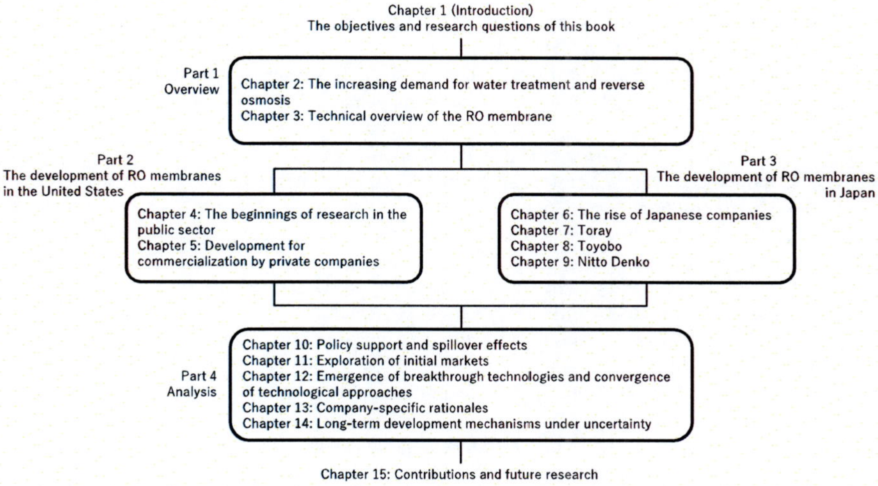

Fig. 1.3 The structure of this book. *Source* Prepared by the authors

References

Abernathy, W. J. (1978). *The productivity dilemma: Roadblock to innovation in the automobile industry*. Johns Hopkins University Press.

Anderson, C. J. (1944). The development of the pump-priming theory. *Journal of Political Economy, 52*, 144–159.

Anteby, M., Lifshitz, H., & Tushman, M. (2014). Using qualitative research for 'how' questions. *Strategic Management Journal, 3*. https://www.strategicmanagement.net/pdfs/qualitative-research-in-strategic-management.pdf. Accessed May 25, 2018.

Dosi, G. (1982). Technological paradigms and technological trajectories: A suggested interpretation of the determinants and directions of technical change. *Research Policy, 11*, 147–162.

Engau, C., & Hoffmann, V. H. (2011). Corporate response strategies to regulatory uncertainty: Evidence from uncertainty about post-Kyoto regulation. *Policy Sciences, 44*, 53–80.

Flyvbjerg, B., & Sunstein, C. R. (2016). The principle of the malevolent hiding hand: Or, the planning fallacy writ large. *Social Research, 83*, 979–1004.

Freeman, C. (1987). *Technology policy and economic performance: Lessons from Japan*. Pinter Publishers.

Gerring, J. (2006). *Case study research: Principles and practices*. Cambridge University Press.

Guan, J., & Chen, K. (2012). Modeling the relative efficiency of national innovation systems. *Research Policy, 41*, 102–115.

Hall, B., & Van Reenen, J. (2000). How effective are fiscal incentives for R&D? A review of the evidence. *Research Policy, 29*, 449–469.

Hirschman, A. O. (1967). *Development projects observed*. The Brookings Institution.

Intarakumnerd, P., & Goto, A. (2018). Role of public research institutes in national innovation systems in industrialized countries: The cases of Fraunhofer, NIST, CSIRO, AIST, and ITRI. *Research Policy, 47*(7), 1309–1320.

Jaffe, A. B., Newell, R. G., & Stavins, R. N. (2005). A tale of two market failures: Technology and environmental policy. *Ecological Economics, 54*, 164–174.

Jalonen, H. (2012). The uncertainty of innovation: A systematic review of the literature. *Journal of Management Research, 4*, 1–47.

Kellogg, R. (2014). The effect of uncertainty on investment: Evidence from Texas oil drilling. *American Economic Review, 104*, 1698–1734.

Kemp, R., Schot, J., & Hoogma, R. (1998). Regime shifts to sustainability through processes of niche formation: The approach of strategic niche management. *Technology Analysis and Strategic Management, 10*, 175–196.

Knight, F. H. (1921). *Risk, uncertainty and profit*. Houghton Mifflin.

Leahy, J., & Whited, T. (1996). The effect of uncertainty on investment: Some stylized facts. *Journal of Money, Credit and Banking, 28*, 64–83.

Levy, J. S. (2008). Case studies: Types, designs, and logics of inference. *Conflict Management and Peace Science, 25*, 1–18.

Leyden, D., & Link, A. (1991). Why are governmental R&D and private R&D complements? *Applied Economics, 23*, 1673–1681.

Lundvall, B. (1992). *National systems of innovation: Towards a theory of innovation and interactive learning*. Pinter Publishers.

Marcus, A. A. (1981). Policy uncertainty and technological innovation. *The Academy of Management Review, 6*, 443–448.

McGrath, R. G., & MacMillan, I. C. (2000). *The entrepreneurial mindset: Strategies for continuously creating opportunities in an age of uncertainty*. Harvard Business Press.

McWilliams, A., & Siegel, D. (2001). Corporate social responsibility: A theory of the firm perspective. *Academy of Management Review, 26*, 117–127.

McWilliams, A., Siegel, D., & Wright, P. M. (2006). Corporate social responsibility: Strategic implications. *Journal of Management Studies, 43*, 1–18.

Ministry of Land, Infrastructure, Transport and Tourism. (2019). *Nihon no mizu shigen no genkyō (Current status of water resources in Japan)*.

Mowery, D. C. (1998). The changing structure of the US national innovation system: Implications for international conflict and cooperation in R&D policy. *Research Policy, 27*, 639–654.

Nelson, R. R. (1992). National innovation systems: A retrospective on a study. *Industrial and Corporate Change, 1*, 347–374.

Nelson, R. R. (Ed.). (1993). *National innovation systems: A comparative analysis*. Oxford University Press.

Nidumolu, R., Prahalad, C. K., & Rangaswami, M. R. (2009). Why sustainability is now the key driver of innovation. *Harvard Business Review, 87*, 56–64.

Numagami, T. (1999). *Ekishō display no gijutsu kakushinshi: Kōi rensa system to shite no gijutsu (History of liquid crystal display technology: Technology as an action chain system)*. Hakuto Shobo.

Numagami, T. (2000). *Kōi no keieigaku: Keieigaku ni okeru ito sezaru kekka no tankyū (Management of actions: Exploring unintended consequences in business administration)*. Hakuto Shobo.

Porter, M. E., & Kramer, M. R. (2011). Creating shared value. *Harvard Business Review, 89*, 62–77.

Porter, M. E., & van der Linde, C. (1995). Toward a new conception of the environment-competitiveness relationship. *The Journal of Economic Perspectives, 9*, 97–118.

Raphael, K. (1987). Recall bias: A proposal for assessment and control. *International Journal of Epidemiology, 16*, 167–170.

Schein, E. H. (1990). Organizational culture. *American Psychologist, 43*, 109–119.

Schot, J., & Geels, F. (2008). Strategic niche management and sustainable innovation journeys: Theory, findings, research agenda, and policy. *Technology Analysis & Strategic Management, 20*, 537–554.

Shimamoto, M. (2014). *Keikaku no sōhatsu: Sunshine keikaku to taiyōkō hatsuden (Emergence of planning: Sunshine plan and solar power generation)*. Yuhikaku.

Suarez, F. F. (2004). Battles for technological dominance: An integrative framework. *Research Policy, 33*, 271–286.

Takeishi, A., Aoshima, Y., & Karube, M. (2012). *Reasons for innovation: Creative legitimacy for resource mobilization*. Yuhikaku.

Tatsumoto, H. (2008). Handōtai sangyō ni okeru kyōdō kenkyū kaihatsu no rekishi (History of joint R&D in the semiconductor industry). *Akamon Management Review, 7*, 263–274.

Utterback, J. M., & Abernathy, W. J. (1975). A dynamic model of process and product innovation. *Omega, 3*, 639–656.

Waddock, S. A., & Graves, S. B. (1997). The corporate social performance-financial performance link. *Strategic Management Journal, 18*, 303–319.

Weber, M., Hoogma, R., Lane, B., & Schot, J. W. (1999). *Experimenting with sustainable transport innovations: A workbook for strategic niche management*. Universiteit Twente.

Yin, R. K. (1984). *Case study research: Design and methods*. SAGE Publications.

Part I
Overview

Chapter 2
The Increasing Demand for Water Treatment and Reverse Osmosis

Abstract In this chapter, we explain that demand for water treatment has been growing worldwide, and that the RO method is gaining in popularity as a key technology for desalination. We then provide an overview of the water treatment industry.

2.1 The Growing Demand for Water Treatment

The demand for water continues to grow around the world. However, only approximately 0.01% of the Earth's water is readily available for human use. At the World Water Forum held in South Korea in 2015, it was reported that more than 800 million people do not have access to safe drinking water. In 2018, a critical water shortage occurred in Cape Town, South Africa, due to a drought that had lasted for several years.

In response to this serious situation, various measures are being taken around the world (Oki, 2016). The desalination of seawater is one of them. Desalination plants are being built one after another in Saudi Arabia, UAE, Kuwait and Qatar in the Middle East, Algeria in North Africa, and Trinidad and Tobago in the Caribbean. In Singapore, which lacks rivers and suffers from chronic water shortages, 25% of the country's total water demand is met by two desalination plants, and a new plant is still under construction. In China, a total of 131 desalination projects were completed by the end of 2016, and the country's total desalination capacity reached 1,183,100 m^3/day (State Oceanic Administration, 2016). In Japan, desalination plants are also in operation, albeit on a medium scale, in Okinawa (in operation since 1995, with a maximum desalination capacity of 40,000 m^3/day) and Fukuoka (in operation since 2005, with a capacity of 50,000 m^3/day). Figure 2.1 shows the seawater desalination plant in Fukuoka.

The demand for water treatment is also increasing outside of drinking applications. For example, in order to obtain sufficient water for agriculture, it is necessary to desalinate slightly salty brine. In industrial applications, the development of small electronic components has led to an increase in the need for pure or ultrapure water

M. Fujiwara and Y. Aoshima, *Mechanisms for Long-Term Innovation*,
Advances in Japanese Business and Economics 31,
https://doi.org/10.1007/978-981-19-4896-1_2

Fig. 2.1 Desalination plant in Fukuoka. *Source* Taken by the authors (August 28, 2013)

with impurities removed. There is also an increasing need to remove toxic substances from factory effluent to prevent environmental pollution. Thus, water treatment has a wide range of applications, including water and sewage treatment, seawater desalination, brine desalination, purified industrial water production, and industrial wastewater treatment. When these diverse water demands are added together, it is estimated that the global water-related business will expand to the scale of 100 trillion yen by 2025.[1]

The capacity of desalination plants has been consistently increasing to meet the water demand for industrial and drinking water by desalinating brine and seawater. Particularly remarkable is the growth since the beginning of the 2000s: 1.64 million m^3/day of new capacity was actualized in 2000, while nearly four times that amount at 6.21 million m^3/day of new capacity was actualized in 2010.[2] If we assume that each person uses about 200 L of water per day, this is enough to supply more than 30 million people. Not all of this water will be used for domestic purposes, but it is clear that desalination facilities are becoming increasingly important in meeting the world's freshwater needs.

[1] RobecoSAM Study (2015).

[2] IDA (2014).

2.2 Widespread Use of the Reverse Osmosis Method

2.2.1 *From the Evaporation Method to the Reverse Osmosis Method*

Among the various desalination methods, the reverse osmosis method (RO method) using reverse osmosis membranes (RO membranes) is currently the mainstream. RO membranes are semipermeable membranes that separate water from impurities. Semi-permeable membranes are classified into several types according to pore size: microfiltration membranes [MF membranes] (pore size: 10 nm and larger), ultrafiltration membranes [UF membranes] (pore size: 1–10 nm), nanofiltration membranes [NF membranes] (pore size: about less than 1 nm), and RO membranes (pore size: approximately 0.1 nm).[3] NF membranes are sometimes called loose RO membranes because they are similar to RO membranes, but their inhibition of salinity is lower. Among the semipermeable membranes with various pore sizes, RO membranes have the finest pores.

In the past, the evaporation method was the mainstream desalination process. There are two main types of evaporation methods: multi-stage flash (MSF) and multi effect desalination (MED). In both methods, the basic idea of obtaining fresh water by evaporating raw water is the same. Middle Eastern countries, which have long suffered from drought, have long employed this evaporation method to take advantage of cheap crude oil.

The problem with the evaporation method is that it consumes a large amount of energy. According to an estimate made around 1980, for example, the energy consumption for desalination was 64.2 kWh/m^3 for the evaporation method compared to 3.69 kWh/m^3 for the RO method.[4] When the price of crude oil began to rise in the wake of the 1973 oil crisis, this energy difference became an issue that could not be overlooked. Although the RO method was initially more expensive, the economic efficiency of the two methods gradually became close to each other as the technology evolved.[5]

Figure 2.2 shows the number of desalination plants and water production by method until 1980. The number of plants with the RO method rapidly increased after

[3] The size of the hole diameter presented here is meant as a general guide.

[4] Sōgō Hōsō Publication (1980), pp. 154–155.

[5] Based on the "Report on the Evaluation of the Total System for Seawater Desalination" by the Society of Sea Water Science, Japan in 1977, Okazaki and Kimura (1983) reported that the desalination cost at that time when using seawater as raw water was 200–323 yen/m^3 for the evaporation method and 324–375 yen/m^3 for the RO method. Although there is still a difference in costs, the highest price of the evaporation method and the lowest price of the RO method are close to each other. In addition to the evaporation method and the RO method, the refrigeration method and the electrodialysis method using ion exchange resins can be mentioned as desalination methods. However, since these desalination methods did not compete with evaporation and RO methods for a long time in desalination applications, they are not included in this book.

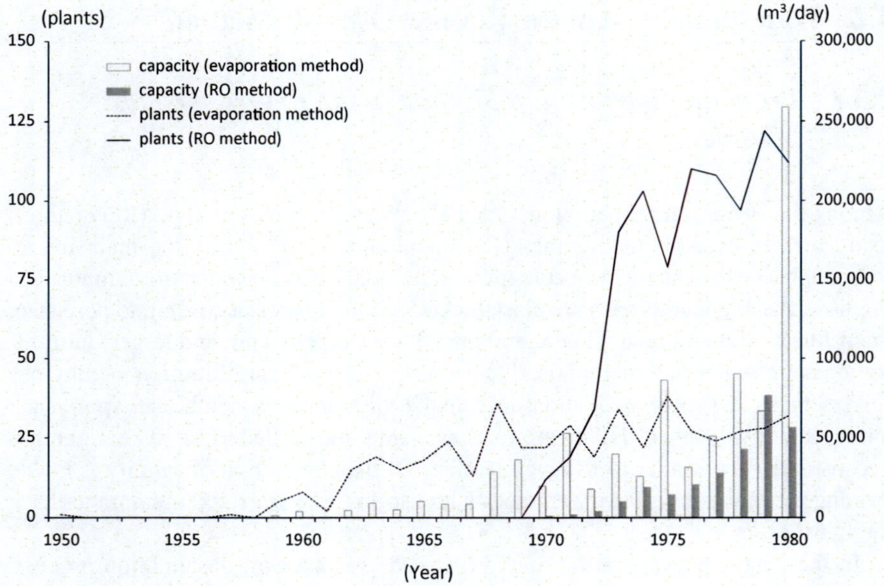

Fig. 2.2 The number and capacity of plants by method until 1980. *Source* Prepared by the authors based on National Water Supply Improvement Association (1981). *Note* As of June 1980. The total of 2,202 plants listed in the report includes plants scheduled to be in operation after 1981. Thus, after excluding missing values, we first extracted 2,134 plants up to 1980. Then, 481 plants by the evaporation method and 886 by the RO method are shown in the figure. The horizontal axis is the plant operation year or contract year

the oil crisis and surpassed the number of plants with the evaporation method though the latter's water production was still larger due to newly built large-scale plants.

Although the energy consumption of the RO method is lower than that of the evaporation method, the reduction in energy cost is still a major issue. This is because high pressure must be applied to the desalination process using a pump, and the electricity cost is a major cost factor. In the case of seawater desalination, the electricity cost accounts for 26.9% of the desalination cost according to the Kaisui Tansuika Gijutsu Hyōka Iinkai (Seawater Desalination Technology Evaluation Committee) (1998), and 41.5% according to Nagai (2010).[6] In the case of the desalination center

[6] This is an estimation of desalination costs by the RO method in Japan. It assumes that the unit construction cost is 200,000 yen/(m^3/day) and the electricity cost is 12 yen/kWh. The total desalination cost is calculated to be 144 yen/m^3. The total desalination cost varies considerably from country to country because it is greatly affected by the cost of electricity. According to Fusaoka (2004) from Toray, the electricity cost in Trinidad and Tobago is only $0.02/kWh, and the total desalination cost of the desalination plant (136,000 m^3/day) to which Toray supplied the RO membrane module is 70.73 cents/m^3 (approximately 85 yen). The electricity cost was 7.61 cents/m^3, which accounted for only 10.7% of the total desalination cost. Desalination costs for other plants include 74 cents/m^3 for Cyprus (54,000 m^3/day for operation in 2001), 55 cents/m^3 for Tampa (95,000 m^3/day for operation in 2003), 54 cents/m^3 for Ashkelon (330,000 m^3/day for operation in 2004), and 46 cents/m^3 for Chuas (136,000 m^3/day for operation in 2004) (Yamamura, 2010).

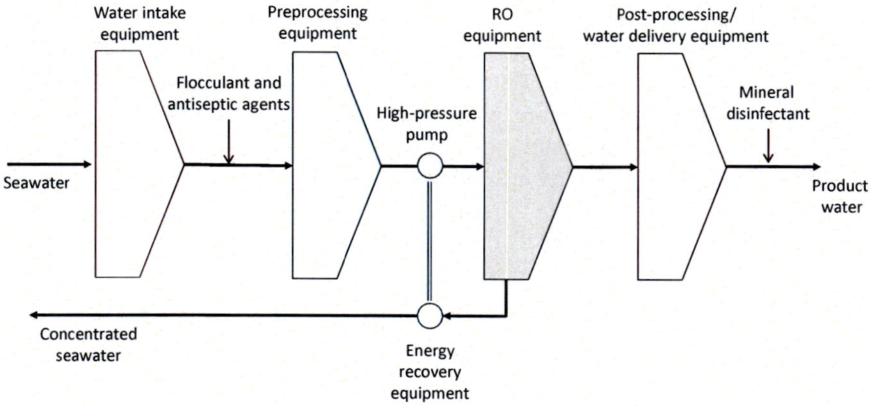

Fig. 2.3 Basic process in a seawater desalination plant. *Source* Prepared by the authors based on the Kaisui Tansuika Gijutsu Hyōka Iinkai (Seawater Desalination Technology Evaluation Committee) (1998)

in Okinawa, the electricity cost accounted for 25.9% of the total cost (FY2004 actual) and 45.3% assuming full operation (Kuniyoshi, 2007). Since the desalination process consists of a heavy variable electricity cost, an increase in water production does not necessarily lead to a decrease in cost.

The cost reduction of RO desalination has been realized mainly because of the improved performance of the RO membrane itself, desalination system innovations, the increased efficiency of the high-pressure pump, and the improvement of the energy recovery device. An energy recovery device is a technology to reuse energy by rotating a turbine with the impact of water flow to generate electricity (the Pelton turbine method) or by recovering pressure from concentrated seawater after desalination (the pressure conversion method). For example, Morita (2011) reported that about 20% of the electricity used for high-pressure pumps was reduced by the introduction of the energy recovery system at the seawater desalination center in Fukuoka, and the Ministry of the Environment recorded that about 15,100 thousand kWh of electricity was reduced annually, saving about 80 million yen in electricity costs.[7] Figure 2.3 shows the basic process of the desalination of seawater and the positioning within the process of the RO equipment (the shaded area) and energy recovery equipment.

According to Wilf and Bartels (2005), who investigated the desalination cost of the RO method in a large desalination plant, the desalination cost was approximately $2/m^3 in 1988, but was halved in about 10 years, and further halved to approximately

[7] http://www.env.go.jp/earth/ondanka/gel/ghg-guideline/water/measures/view/012.pdf (Accessed December 26, 2017). The Ministry of the Environment has published a reference to the Guidelines for Greenhouse Gas Emission Control for the Water Supply and Industrial Water Supply Sectors. The Ministry of the Environment refers to energy efficiency and conservation at a seawater desalination center in Fukuoka, drawing on the Japan Water Works Association ed. (2009) as reference information.

$0.5/m^3 in 2000.[8] As a result, it was reported in 2008 that the cost of desalination by the RO method became lower than that by the evaporation method regardless of the size of the desalination plant (Wittholz et al., 2008). In fact, the adoption of RO greatly exceeded that of evaporation in the mid-2000s.[9] This suggests that the lower cost of the RO method is promoting its popularization. In China, for example, 112 of the 131 desalination plants mentioned above have adopted the RO method, and based on the 13th Five-Year Plan for seawater use announced by the country at the end of 2016, the adoption of the RO method is expected to increase in the future.

2.2.2 *Development of Various Applications*

The RO method has been developed with desalination applications in mind from the beginning. However, it took a long time before it achieved sufficient performance for seawater desalination and was accepted by the market. In the meantime, various applications other than seawater desalination have been developed, such as food applications including fruit juice concentration, artificial kidney applications, industrial water production, and wastewater treatment.[10] Because of these applications, companies were able to continue development and achieve the level of technology necessary for seawater desalination.

According to a survey conducted by the Fuji Keizai company, the largest application of RO membranes and NF membranes in the global market by demand segment (estimated for 2016) is industrial demineralized water and pure water production, accounting for 41% of the total market at 43.7 billion yen. The next largest application is seawater and brine desalination at 25 billion yen (24%).[11] However, seawater desalination applications are mostly for large plants and tend to fluctuate up or down depending on the status of plant orders. The ultrapure water production market, which is mainly intended for use in semiconductors, accounts for 5% of the total market at 5.3 billion yen.

Singapore is one of the most proactive countries in terms of wastewater reuse among a wide range of demand sectors. In Singapore, the NEWater project, which uses RO membranes to treat and purify wastewater for drinking water, was launched in earnest in the early 2000s, and as of 2018, five plants were supplying 40% of the

[8] The desalination cost of $0.5/m^3 is consistent with the estimation results presented by Wittholz et al. (2008). In this paper, the desalination cost by the RO method for a seawater desalination plant with a capacity of about 200,000 m^3/day was reported to be about $0.5/m^3. However, the fact that cost comparisons between the two methods are still being made in the mid-2000s suggests that, depending on conditions, the evaporation method may have a cost advantage. In Saudi Arabia, for example, desalination plants using the evaporation method are still being built. This is because seawater can be desalinated at low cost by using steam, a by-product of thermal power plants.

[9] IDA (2016).

[10] In many cases, NF and UF membranes are used for artificial kidneys instead of RO membranes, so we will not discuss the application in depth in this book.

[11] Fuji Keizai (2016), p. 74.

country's water demand. When combined with the supply from the two desalination plants mentioned at the beginning of this chapter, 65% of the domestic water demand is met by the RO method. It is no exaggeration to say that Singapore's water policy is supported by the RO method. Thus, RO continues to spread to meet all water treatment demands.

2.3 Industry Structure and Competition

2.3.1 *Basic Structure of the Water Treatment Industry by the Membrane Method*

RO membranes are usually integrated with other components to form a product called a membrane element. A membrane module is a product in which several membrane elements are connected in series and stored in a pressure vessel. In general, RO membrane manufacturers design, manufacture, and sell RO membranes, membrane elements, and membrane modules.

Figure 2.4 shows the value chain of the membrane water treatment industry, including RO membranes. RO membrane manufacturers, which handle RO membrane components, are upstream. Plant engineering companies, which build water treatment systems by installing RO membrane modules in equipment, are midstream. In the downstream domain, there are companies that own plants and operate water treatment and water production.

Dow, Toray, Nitto Denko, and Toyobo are the leading companies in the upstream RO membrane industry. The RO membrane industry, to which these companies belong, is the subject of this book. The market shares of these companies and their trends will be discussed in detail in the next section.

The leading companies in the midstream water treatment equipment, plant and engineering industries differ depending on the demand sector. In the seawater desalination market, Doosan Heavy Industries & Construction of South Korea, Veolia and Suez Environment (Suez) of France, Hyflux of Singapore, and Befesa and Cadagua

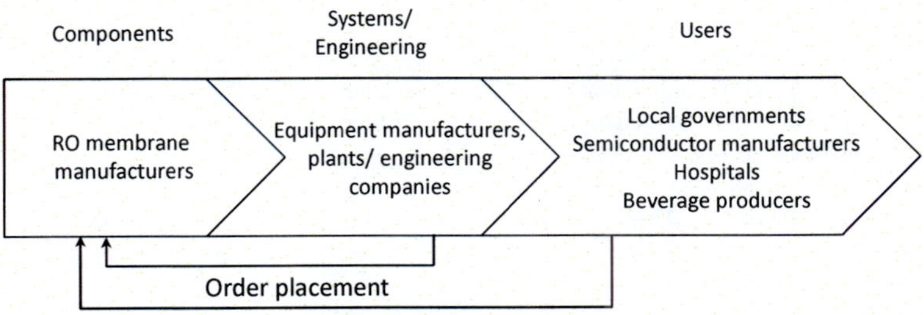

Fig. 2.4 A structure of the water treatment industry with the RO method. *Source* Prepared by the authors

of Spain have dominant positions. Among Japanese companies, Mitsubishi Heavy Industries, Sasakura, Hitachi Zosen, and Swing Corporation have a presence, of which Mitsubishi Heavy Industries stands out in the construction of plants using the RO method. These midstream projects are called EPC (engineering, procurement, and construction).

The three leading Japanese companies in the pure water and ultrapure water production equipment market are Kurita Water Industries, Organo, and Nomura Micro Science. Looking at the financial information for each company for the March 2018 accounting statement, Organo's sales were 79.2 billion yen (made up of 71.5 billion yen in the water treatment engineering business and 17 billion yen in the functional products business) and Nomura Micro Science's sales were 21.6 billion yen (made up of 20.6 billion yen in the water treatment equipment business, of which 7.9 billion yen was in maintenance, etc., and 0.9 billion yen in other businesses).[12] Among Japanese companies, Kurita Water Industries is in the lead, followed by Organo and Nomura Micro Science.

The markets for deionized water production equipment categorized by application are well-distributed: 21.6% for electric power, 16.9% for chemicals, 15.4% for food and beverages, 14.9% for electronics, and 12.9% for pharmaceuticals and medical care. In contrast, sales of ultrapure water production equipment to the electronics industry are overwhelmingly large, accounting for 93.0% of total sales. This is followed by 4.2% for electric power and only 2.8% for pharmaceuticals and medical care.

Downstream players using RO membrane modules also differ depending on the application. In the case of ultrapure water production for semiconductors, the users are semiconductor companies.[13] In the case of seawater desalination, municipalities are the main users in Japan. In countries and municipalities where water and sewage systems are privatized, the companies that manage and operate the water and sewage systems are downstream users.

---

[12] Each company's businesses are as follows.

Kurita Water Industries Ltd.

Water treatment facilities: Manufacturing and sales of water treatment facilities and equipment, and provision of maintenance services.

Water treatment chemicals: Manufacturing and sales of chemicals for water treatment and provision of maintenance services.

Organo

Water treatment engineering: Plants (manufacturing and sales of large-scale water treatment facilities), solutions (facility maintenance, operation management, modification work, etc.).

Functional products: Manufacturing and sales of standard water treatment equipment, water treatment chemicals, and food additives.

Nomura Micro Science

Water treatment facilities: Manufacturing and sales of various facilities, equipment, and chemicals related to water treatment from ultrapure water to wastewater, and maintenance of water treatment facilities and equipment.

Other: Sales of high-purity chemicals, piping materials, etc.

[13] Manufacturers of ultrapure water production equipment are not only delivering equipment, but also developing solution businesses that include post-operation maintenance, which means that they are expanding their business into some downstream areas.

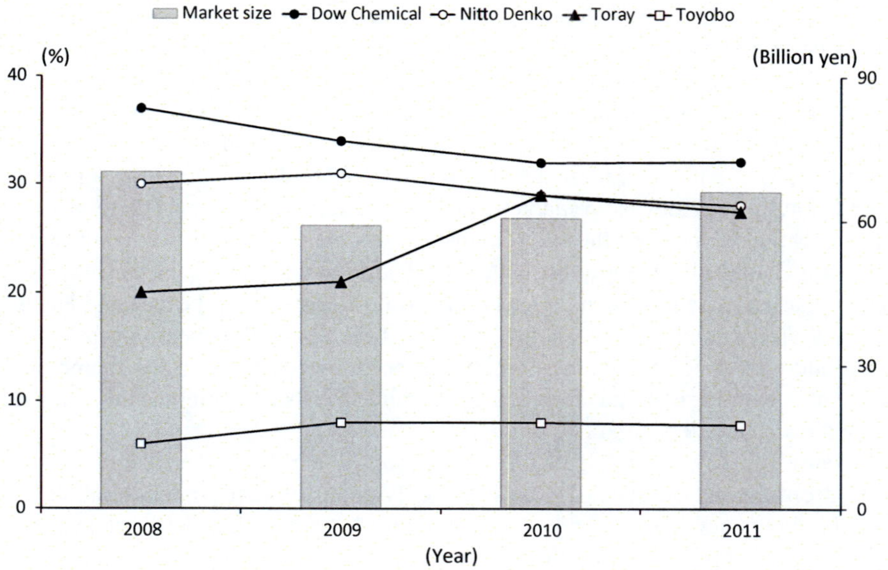

Fig. 2.5 Global market of RO membranes: market size and share. *Source* Prepared by the authors based on Nikkei Sangyō Shimbun ed. (2010, 2011, 2012). *Note* The left axis shows the line graphs and the right axis shows the bar graph

There are also vertically integrated companies that operate broadly across the value chain. For example, Veolia and Hyflux as water utilities are also engaged in plant engineering for water treatment facilities. General Electric (GE) has acquired an RO membrane manufacturer Osmonics and has integrated its membrane business with its plant engineering business. The company's water treatment business was strong for industrial water used in oilfields, food, power, and other plants, with sales of €2 billion in FY2016. However, in 2017, GE sold its water treatment business to Suez. This allowed Suez to establish a vertically integrated business structure across the entire value chain, from RO membrane production to the water supply business.

2.3.2 Market Share Trends

Figure 2.5 shows the market size and market share of RO membranes from 2008 to 2011. It shows that the market is an oligopoly with the combined market share of the top three companies Dow, Nitto Denko, and Toray reaching nearly 90%. Toray's market share increased significantly in 2010 because it received orders for large-scale seawater desalination plants in Algeria, Spain, and Bahrain.[14]

[14] Nikkei Sangyō Shimbun (2011), p. 47.

Although the available market share data only goes up to 2011, there seems no major change in the market shares after that.[15] According to the Nikkei newspapers and authors' estimates based on interviews with industry insiders, Dow is in the lead with just over 30%, while Toray and Nitto Denko are considered to be in close competition at around 25–30%.[16] However, if one considers the characteristics of the industry, in which the presence or absence of orders for large plants can cause significant ups and downs in market share, it would not be appropriate to place a clear pecking order among the three companies. Compared to these three companies, Toyobo's market share is low, estimated to be approximately 10%. However, it has a high market share in certain regions of the Middle East, as will be discussed later.

Dow, Toray, and Nitto Denko have been developing RO membranes called polyamide flat sheet membranes, which are made by rolling up a flat membrane made of polyamide material. Today, this polyamide flat sheet membrane has become the mainstream in the market.

Meanwhile, Toyobo has established a unique position in the market, distinguishing itself from the mainstream. Toyobo is the only company that produces cellulose-based hollow fiber RO membranes, which are made from cellulose triacetate and a thin straw-like hollow fiber. Although the company's overall market share is low, it is highly competitive in the market for large-scale seawater desalination plants in Saudi Arabia, where it can maximize its technological superiority. If we look only at the Middle East region, Toyobo is thought to have a market share of over 50%.[17]

The lifetime of both types of RO membranes is about seven years. Therefore, commercial opportunities for membranes arise not only when new water treatment plants are built, but also when membranes need to be replaced. The replacement membranes are sometimes ordered by user companies in the water treatment business, and sometimes by plant engineering companies that build water treatment equipment and systems.

Since the construction of large-scale desalination plants is not a regular process, the new demand fluctuates considerably from year to year. On the other hand, the replacement demand is not negligible in the long term because it is based on the accumulated number of plants already built, and the replacement time can be predicted in advance.

A comparison of the values of new demand and replacement demand at three points in time (FY2009, FY2013, and 2016 forecast) shows that new demand was 25 billion yen, 54 billion yen, and 31.8 billion yen respectively, while replacement demand was 29.9 billion yen, 31 billion yen, and 74.2 billion yen respectively.[18]

---

[15] Toray acquired Woongjin Chemical in 2014, which might contribute to Toray's increased market share.

[16] For example, the Nikkei Bulletin News Archive (May 24, 2012; January 21, 2015; May 2, 2017), Nikkei Veritas (December 2012, p. 49), Nihon Keizai Shimbun (September 27, 2013, p. 9; February 19, 2014, p. 3; July 10, 2014, p. 13; January 21, 2015, p. 12), and Nikkei Sangyō Shimbun (February 20, 2014, p. 15; May 26, 2014, p. 19; August 31, 2015, p. 19).

[17] Nihon Keizai Shimbun (March 14, 2017, p. 15).

[18] Fuji Keizai (2010), p. 110; Fuji Keizai (2014), p. 70; Fuji Keizai (2016), p. 75.

While new demand has been fluctuating greatly, replacement demand has been steadily increasing which captured 70% of the market in 2016.

2.3.3 Price Competition

One may conjecture that the membrane manufacturer can monopolize subsequent replacement demand once it captures the new demand. In reality, however, there is competition for every replacement. Replacement demand is not always guaranteed.

Since the sizes of elements and modules for RO membranes are different between the polyamide flat sheet membrane and cellulose acetate hollow fiber membrane groups, switching costs are incurred when the customer switches to the membrane of the other group. Thus, there is not much competition between the two groups, and customers tend to be fixed in each group.

However, it is a different story when it comes to competition within a group. Competition is unlikely to occur for the cellulose acetate hollow fiber membranes, which are supplied only by Toyobo. On the other hand, the polyamide flat sheet membrane market, where there are several membrane manufacturers, is likely to be exposed to competition. This is because the de facto standard for elements and modules has been formed and there is no significant difference in membrane performance between companies. Therefore, it is easy for users to switch to other company products for polyamide flat sheet elements.

Users prefer to maintain the status quo unless they face a major problem, as they fear the risk of losing water treatment efficiency by switching to another company's membrane. As a result, there has not been a dramatic reversal of market share with every membrane change so far. However, the mere presence of a potential conversion risk tends to increase tensions between the companies and inspire fierce price competition within the group.

It is difficult for incumbent membrane manufacturers to maintain the first mover advantages. As a result, the market has induced many new entries and active M&A activities. For example, GE acquired Osmonics in 2003 to achieve vertical integration, but sold the business to Suez in 2017. In South Korea, Woongjin Chemical acquired Saehan to acquire its membrane technology.[19] In China, Motimo and Vontron entered the RO membrane market. Flat sheet RO membranes with an acceptable level of performance can be produced by purchasing automated equipment that handles the membrane production process. Therefore, emerging companies are steadily following leading companies. Although the market share of new companies is small at present, their existence itself is a factor that intensifies market competition.

The market competition and the threat of new entrants have caused severe price competition in each application market. According to Fuji Keizai, the price of RO membrane modules for seawater desalination (on a daily water volume basis) in 2010

---

[19] As mentioned earlier, Woongjin Chemical was acquired in February 2014 through a subsidiary of Toray.

ranged from 8,000 to 20,000 yen per m^3 per day. On the other hand, the prices of flat sheet membranes and hollow fiber membranes in 2016 were 3,000–4,000 yen per m^3 per day and 1,000–8,000 yen per m^3 per day, respectively.[20] Compared with the highest price in 2010, the price of flat sheet membranes has been reduced to one-fifth in seven years, and the price of hollow fiber membranes has been reduced to less than half.

For deionized water production and brine desalination, the price of RO membrane modules has fallen from 3,000–10,000 yen per m^3 per day (for general deionized water production) to about 2,000 yen per m^3 per day (for flat sheet low-pressure membranes, deionized water, and brine desalination) in the same period. The price of ultrapure water production, mainly for the semiconductor industry, has also fallen from 7,000–25,000 yen per m^3 per day to about 5,000 yen per m^3 per day (flat sheet low-pressure membranes).

While there is an overall downward trend in prices, the downward pressure on prices is somewhat different by application depending on the per-unit commercial value and the lot size. For example, in the case of food enrichment applications, the per-unit commercial value is large whereas the production scale is small. Therefore, it is difficult for customers to pressure the price down and the price decline is likely to be relatively slow. In contrast, in desalination applications, the per-unit commercial value is low, but the lot size is large due to the large scale of desalination plants, so customers tend to pressure prices down.

2.4 Summary

In this chapter, we presented an overview of the RO membrane market that has been developed in response to the increasing demand for water treatment. The development of efficient desalination technology has been expected as a key technology to solve social issues as drought areas spread all over the world due to the progress of industrialization, population growth, and changes in the natural environment. In the Middle East, where the supply and demand of water tends to be tight due to the large number of arid regions, the procurement of water has long been a central issue for society, and fresh water was procured by the evaporation method using crude oil, which is available at low cost. However, the price of crude oil has risen and the need to conserve oil as a strategic commodity has increased while the cost of desalination by the RO method has declined. Reflecting such situations, the construction of seawater desalination plants with the RO method has progressed especially since the late 1990s.

This book aims to uncover the historical process that the RO membrane was developed to create a new water treatment industry. As preparatory information, we will summarize the technical outline of the RO method in the next chapter.

[20] Fuji Keizai (2010), p. 111 and Fuji Keizai (2016), p. 77.

References

Fuji Keizai. (2010). *Kōkinō **bunrimaku/filter kanren gijutsu, Shijō** no zembō to shōraiyosoku 2010 **(Market overview and future forecast of high-performance separation membrane/filter technology** 2010).*

Fuji Keizai. (2014). *Kōkinō **bunrimaku/filter kanren gijutsu, Shijō** no zembō to shōraiyosoku 2014 **(Market overview and future forecast of high-performance separation membrane/filter technology** 2014).*

Fuji Keizai. (2016). *Kōkinō **bunrimaku/ filter kanren gijutsu, Shijō** no zembō to shoraiyosoku 2016 **(Market overview and future forecast of high-performance separation membrane/ filter technology** 2016).*

Fusaoka, Y. (2004). Trinidad and Tobago no rei (Example of SWRO plant in Trinidad and Tobago). *Nihon Kaisui Gakkaishi (Bulletin of the Society of Sea Water Science, Japan), 58*, 264–267.

http://www.env.go.jp/earth/ondanka/gel/ghg-guideline/water/measures/view/012.pdf. Accessed December 26, 2017.

IDA. (2014). *Desalination yearbook 2014–2015.*

IDA. (2016). *Desalination yearbook 2016–2017.*

Japan Water Works Association (Ed.). (2009). *Suidō shisetsu ni okeru energy taisaku no jissai (Actual energy measures in water supply facilities).* Japan Water Works Association.

Kaisui Tansuika Gijutsu Hyōka Iinkai (Seawater Desalination Technology Evaluation Committee). (1998). *Kaisui tansuika gijutsu kaihatsu* (Gyakushintō hō kaisui tansuika chōsa) Chūkan hōkokusho (*Interim report on seawater desalination technology development (Reverse osmosis seawater desalination technology development study)).*

Kuniyoshi, N. (2007). Operation status of seawater desalination facility of Okinawa prefectural enterprise bureau. *Journal of Water Re-Use Technology, 31*, 35–39.

Morita, Y. (2011). Operating conditions of the sea water desalination plant in Fukuoka district. *EICA, 15*, 48–51.

Nagai, M. (2010). Dai 2-hen gijutsu-hen (Part 2 technology): Chapter 4 Maku o mochiiru kaisui tansuika system no genjō to kongo no tenkai (Current status and future development of seawater desalination systems using membrane). In S. Nakao & Y. Watanabe (Eds.), *Water treatment technology using membranes: Popular edition* (pp. 119–138). CMC Publishing.

National Water Supply Improvement Association. (1981). *Desalination plants inventory report* (No. 7).

Nihon Keizai Shimbun. (September 27, 2013; February 19, 2014; July 10, 2014; January 21, 2015; March 14, 2017).

Nikkei Bulletin News Archive. (May 24, 2012; January 21, 2015; May 2, 2017).

Nikkei Sangyō Shimbun (Nikkei Business Daily) (Ed.). (2010). *Nikkei market share 2011.* Nikkei Business Publications.

Nikkei Sangyō Shimbun (Nikkei Business Daily) (Ed.). (2011). *Nikkei market share 2012.* Nikkei Business Publications.

Nikkei Sangyō Shimbun (Nikkei Business Daily) (Ed.). (2012). *Nikkei market share 2013.* Nikkei Business Publications.

Nikkei Sangyō Shimbun (Nikkei Business Daily). (February 20, 2014; May 26, 2014; August 31, 2015).

Nikkei Veritas. (December 2012).

Okazaki, M., & Kimura, S. (1983). Recent advances of reverse osmosis technology for seawater desalination. *Nihon Kaisui Gakkaishi (Bulletin of the Society of Sea Water Science, Japan), 37*, 149–158.

Oki, D. (2016). *Mizu no Mirai: Global risk to Nihon (The future of water: Global risks and Japan).* Iwanami Shoten.

RobecoSAM Study. (2015). *Water: The market of the future.* RobecoSAM. http://www.robecosam.com/images/Water_Study_en.pdf. Accessed April 9, 2018.

Sōgō Hōsō Publication. (1980). *Kōbunshi takōshitsu maku no kaihatsu dōkō to shijō jittai: Gyakushintō maku, gengai roka maku, seimitsu roka maku, ion kōkan maku (Development trend and market status of polymer porous membranes: Reverse osmosis, ultrafiltration, microfiltration, and ion exchange membranes)*. Sōgō Hōsō Publication.

State Oceanic Administration. (2016). *The national seawater utilization report of 2016*.

Wilf, M., & Bartels, C. (2005). Optimization of seawater RO systems design. *Desalination, 173*, 1–12.

Wittholz, M., O'Neill, B., Colby, C., & Lewis, D. (2008). Estimating the cost of desalination plants using a cost database. *Desalination, 229*, 10–20.

Yamamura, H. (2010). Dai 3-hen ōyō-hen (Part 3 applications): Dai 3-shō kaisui tansuika shisetsu 1. Sekai no kaisui tansuika shisetu (Chapter 3: Seawater desalination facilities 1. Seawater desalination facilities in the world). In S. Nakao & Y. Watanabe (Eds.), *Maku o mociita mizushori gijutsu: Fukyū-ban (Water treatment technology using membranes: Popular edition)* (pp. 256–259). CMC Publishing.

Chapter 3
Technical Overview of the RO Membrane

Abstract In this chapter, we provide a technical overview of the RO method and its implication to competition in the industry. After briefly reviewing the RO phenomenon, we classify various types of RO membranes by material, structure, and shape. We then illustrate how companies have historically adopted different types of membranes. Finally, we explain that the element shape affects switching cost for customers.

3.1 Reverse Osmosis Phenomenon

In this chapter, we explain the basic principles and definitions of RO membranes as a basis for future discussion and summarize the types of RO membranes. Then, we specify several technical issues involved in RO membrane development.

To understand the working principle of RO membranes, it is first necessary to recall the term osmotic pressure. In general, when water with high salinity and water with low salinity are separated by a semipermeable membrane, water permeates from the side with low salinity to the side with high salinity, and the liquid level of the side with high salinity rises, eventually reaching an equilibrium. At this time, pressure is generated by the difference in water head due to the water that has moved. This is the osmotic pressure. On the other hand, if a pressure higher than the osmotic pressure is applied to the side with the higher concentration, water will permeate from the side with the higher concentration to the side with the lower concentration. This can be thought of as suppressing the force of water flowing from the low concentration side to the high concentration side, and instead, "pushing out" water strongly from the high concentration side to the low concentration side. It is a phenomenon known as reverse osmosis. The reverse osmosis method uses this principle to remove impurities (see Fig. 3.1). Normally, in reverse osmosis, this process is carried out by applying a pressure approximately twice the osmotic pressure.

RO membranes are semipermeable membranes with a pore size of approximately 0.1 nm that can remove salts and ionic components. RO membranes can be further

M. Fujiwara and Y. Aoshima, *Mechanisms for Long-Term Innovation*,
Advances in Japanese Business and Economics 31,
https://doi.org/10.1007/978-981-19-4896-1_3

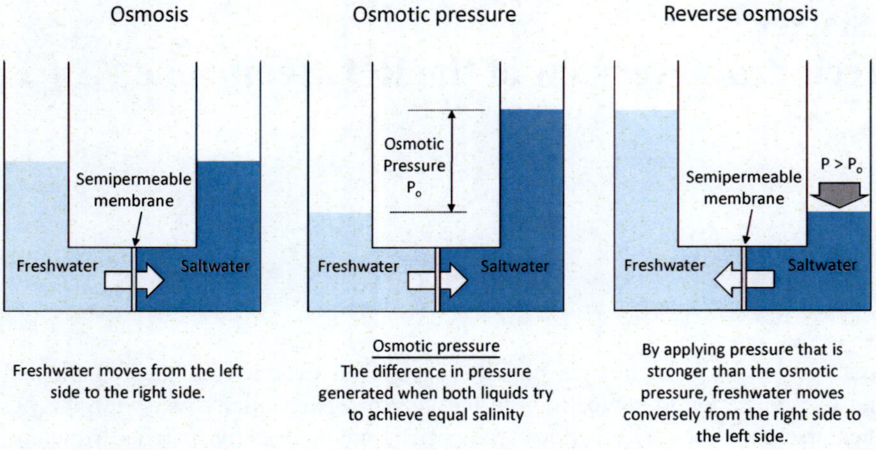

Fig. 3.1 Osmosis and reverse osmosis phenomena. *Source:* Prepared by the authors

classified into several types based on the following three factors: the material used for the membrane, the membrane structure, and the membrane shape.

3.2 Material, Structure, and Shape

3.2.1 Material

The first axis for classifying RO membranes is the type of material used for the membrane. In 1960, Srinivasa Sourirajan and Sidney Loeb at UCLA succeeded in developing cellulose acetate asymmetric membranes with 10 times higher water permeability than conventional membranes.[1] This membrane was called the Loeb-Sourirajan membrane (hereafter referred to as the L-S membrane) after their names. With this remarkable development, many researchers and companies attempted to commercialize cellulose acetate asymmetric membranes.

Although the L-S membrane was revolutionary at the time, it did not have high enough permeability to be widely used in practical applications. The search for a membrane material that could simultaneously achieve high levels of desalination and water permeability continued, and polyamide-based materials eventually came to be the focus of attention. Polyamide-based materials became widespread, partly because they were compatible with the production of flat membranes with a composite structure, as described below. Meanwhile, the number of companies

[1] Permeability mainly refers to the amount of water permeated through the membrane (the amount of water that passes through the membrane in a certain period of time) or flux. The term "permeability" will be used in this sense broadly throughout this book, except in cases where different specific terminology should be specified.

developing cellulose acetate-based membranes has gradually decreased, and only Toyobo is commercializing them today.

The advantage of the cellulose acetate material, which was developed in the early stage, is that it has excellent chlorine resistance. Normally, when water treatment such as desalination is continued, contamination gradually adheres to the membrane. A particular problem is that microorganisms thrive on the membrane, and the substances derived from the bacteria clog the membrane. This is called biofouling.

If the membrane is not chlorine-resistant, the chlorine contained in the treatment chemical will damage the membrane and cause its performance to deteriorate. Although it is possible to prevent deterioration using special chemicals, that leads to an increase in cost. In the 1980s, the allowable chlorine concentration in polyamide RO membranes was 0.1 ppm, and sufficient dechlorination was necessary to prevent damage to the membrane (Ohya, 1985).

On the other hand, a chlorine-resistant membrane, such as cellulose acetate, can withstand repeated sterilization. In addition, it has high flexibility in the chemicals used for disinfection, and the cost can be kept low because general-purpose chemicals can be used. Since the frequency of disinfection treatment increases for heavily polluted waters, chlorine resistance is a particularly important point of appeal.

Compared to chlorine-resistant cellulose acetate, the superiority of polyamide materials lies in their high heat resistance, wide applicable pH range, and high permeability performance under low-pressure operation.

First, polyamide-based materials have high heat resistance. Ohya (1985) reported that the heat resistance of cellulose acetate membranes at that time was 35 °C (or 55 °C in some manufacturer data), whereas the heat resistance of polyamide membranes was 40 °C (or 100 °C in some manufacturer data). Second, the applicable pH range of polyamide-based materials is wide, from 4 to 11. In the case of cellulose acetate materials, which hydrolyze under alkaline conditions, the applicable pH range is limited to around 4–7. And third, polyamide membranes have high water permeability under low-pressure operation and can pass a large amount of water without high pressure. If high water permeability can be obtained under low-pressure operation, the electricity cost for desalination can be kept low. Since the ratio of the electricity cost to the desalination cost is high, as described in Chap. 2, companies have been actively working on the development of membranes with high performance under low-pressure operation.

However, in the case of seawater desalination, polyamide-based materials are not necessarily advantageous to cellulose acetate materials since operating pressure corresponding to high osmotic pressure is required in any way due to the high salinity. Seawater desalination requires a minimum pressure of about 50 kg/cm^2 while brine desalination requires only about 28 kg/cm^2 of pressure (Wada, 2004). Where the required osmotic pressure is low, the economic effect of lowering the pressure is greater, and the superiority of polyamide materials is apparent.

3.2.2 *Structure: The Choice of Asymmetric or Composite Membranes*

The second axis for classifying RO membranes is the structure of the membrane. There are two major types of membrane structures: asymmetric and composite. Figure 3.2 shows a comparison of the two types of membrane structures.

The usefulness of asymmetric membranes was demonstrated at UCLA, and cellulose acetate-based asymmetric membranes became widely known as L-S membranes. In the early days of RO membrane development, asymmetric membranes were the mainstream.

However, asymmetric membranes have the fundamental problem of "consolidation" in which the soft sponge-like support layer is crushed by high pressure during operation. This problem is especially troublesome in the case of seawater desalination which requires high-pressure operation.

Composite membrane was developed to solve this problem. In contrast to asymmetric membranes that use a single material, composite membranes have a separation layer and a support layer made of different materials. The separation layer, which functions as a separator, is made as thin as possible to improve desalination and water permeability, while the support layer is made of a different material and overlapped with the separation layer to prevent the membrane from being crushed by pressure during operation. The initial concept of composite membranes that increase permeability while maintaining strength by overlaying a very thin separation layer on a support layer, was conceived in the 1960s, and its practical application was developed in the United States. The developers of General Atomic (ROGA) developed composite membranes (PA-300 and RC-100) in the mid-1970s, and the thickness of

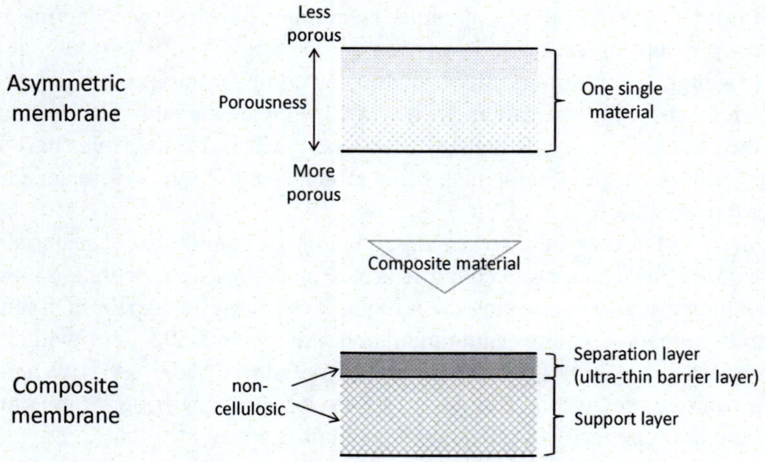

Fig. 3.2 Transition from asymmetric membrane to composite membrane and non-cellulosic. *Source* Prepared by the authors

Fig. 3.3 Differences in membrane shapes. *Source* Prepared by the authors. *Note* The flat sheet membrane is represented here as a composite membrane

the separation layer was generally 10–30 nm (Zōsui gijutsu Henshū kikaku Iinkai Hen (Desalination Technology Editorial Planning Committee ed.), 1983).

In the 1970s, the development of RO membrane materials diversified, and the search for materials other than cellulose acetate was promoted. The first step was to develop a non-cellulosic support layer, and polysulfone became the mainstream support layer. Next, the separation layer became non-cellulosic. After exploring various materials for the separation layer, cross-linked aromatic polyamide came to be considered a promising material, and subsequently became the mainstream material in the market today. The thickness of the separation layer in Toray RO membranes using the same material is about 200 nm, and the thickness of the support layer using polysulfone is about 45 μm.[2] The reason why the separation layer of today's membranes is thicker than before is that the membranes now have a pleated structure. The pleated structure increases the surface area of the membrane and thus improves the water permeability.[3] In general, the polyamide material is superior to the cellulose acetate material in terms of water permeability.

3.2.3 *Shape: Tubular, Hollow Fiber, and Flat Sheet Membranes*

The third axis of classification is the shape of the membrane. There are three main shapes: tubular, hollow fiber, and flat sheet. A visual representation of each type is shown in Fig. 3.3.

Both tubular and hollow fiber RO membranes are thin, straw-like with a hollow interior. The development of RO membranes started with the tubular type. The L-S membrane described above was also manufactured as a tubular RO membrane and used for demonstration. The hollow fiber RO membrane, which was introduced later, has a smaller inner diameter than the tubular membrane. The outer diameter of the hollow fiber membrane is about 40–200 μm and the wall thickness is 10–50 μm (Desalination Technology Editorial Planning Committee ed., 1983).

[2] Although it is a rough calculation, the separation layer is only about 1/225th as thin as the support layer.

[3] For more on this importance, see Part 3.

RO membrane elements are made by arranging several layers of these straw-like RO membranes in a bundle.[4] These elements are then connected in series and stored in a pressure vessel to form an RO membrane module. The hollow fiber membrane is thinner than the tubular type, so it is more efficient to fill the element with the membrane. However, in general, the higher the packing density is, the more difficult it is to remove contaminants from the element.

The flat sheet type literally refers to a flat sheet of RO membrane. Among several elementalization methods, the most common one today is to wrap the flat sheet membrane around a pipe, which is called "spiral wound element". Thus, the term "flat sheet type membrane element" is used in the same sense as "spiral wound element" in this book. In the case of flat sheet membranes, it is necessary to insert a spacer between the membranes to promote the flow of raw water, so the filling efficiency of the membranes must be reduced.

In the case of the flat sheet type as well, the membrane module is constructed by connecting several elements in series and storing them in a pressure vessel. The differences between the hollow fiber membrane and the flat sheet membrane inside the RO membrane element are shown in Fig. 3.4a and b.

In terms of the permeability of the membrane itself, the polyamide material is higher than the cellulose acetate material. However, the polyamide flat sheet membrane element requires a spacer that reduces the filling efficiency while cellulose acetate hollow fiber membranes can be filled more into the same size of element because of no need for a spacer. In this way, the polyamide flat sheet membrane element with high water permeability and low filling efficiency, and the cellulose acetate hollow fiber membrane element with inferior water permeability and high filling efficiency, are almost the same in terms of water permeability per element.

3.2.4 Convergence of Technological Focuses by Firms

Table 3.1 summarizes RO membrane technologies along the three axes of material, structure, and shape.

Today's industry giants, Dow, Toray, and Nitto Denko, all originally adopted cellulose acetate materials for their early development. The shape of the material differed depending on the company: Nitto Denko adopted the tubular membrane, Toray adopted the flat sheet membrane, and Dow and Toyobo adopted the hollow fiber membrane. They were located in one of the three upper-left cells in Table 3.1. Meanwhile, only DuPont chose the polyamide hollow fiber membrane located in the lower left of the Table. DuPont developed its own standard RO membrane module and was in a strong position to lead the market in the early days. However, DuPont withdrew from the market in 2000 for technical and commercial reasons.

[4] For example, according to the Fukuoka Seawater Desalination Center, where Toyobo membrane elements are housed, about 1.5 million hollow fibers are woven into each membrane element (http://www.f-suiki.or.jp/facility/kaitan-center/kaitan-facility/about-maku/ accessed May 27, 2018).

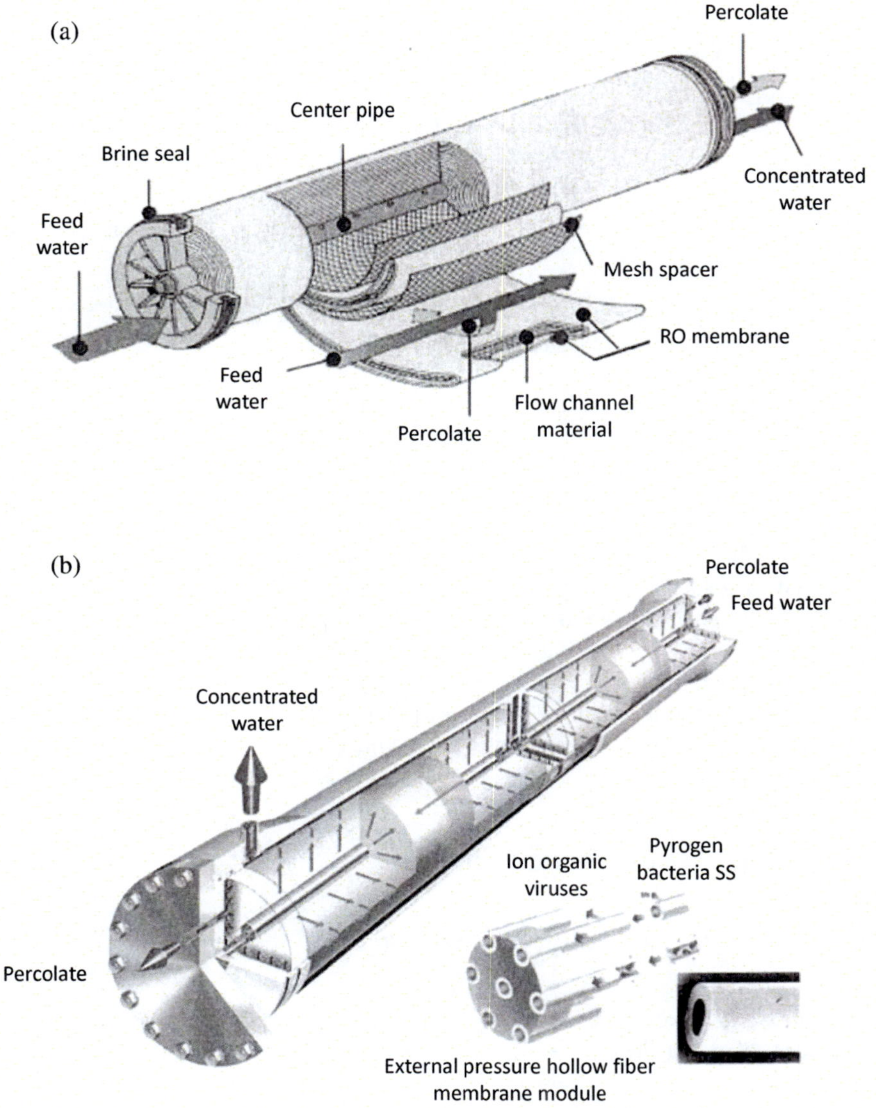

Fig. 3.4 **a** Flat sheet type (spiral wound element). **b** Hollow fiber type. *Source* Makubunri Gijutsu Shinkō Kyōkai/ Maku Jōsui Iinkai (Kanshū)/Jōsui Maku (Dai 2-han) Henshū Iinkai (Hen) (Membrane Separation Technology Promotion Association, Membrane Water Purification Committee (Supervisor), Water Purification Membrane (2nd Edition) Editorial Committee (ed.) (2008) p. 86 and p. 88

Table 3.1 Major technological categories

		Membrane structure	
		Asymmetric membranes	Composite membranes
Membrane material	Cellulose acetate	Tubular	Tubular
		Hollow fiber	Hollow fiber
		Flat sheet	Flat sheet
	Polyamide	Tubular	Tubular
		Hollow fiber	Hollow fiber
		Flat sheet	Flat sheet

Source Prepared by the authors

In the beginning, many companies actively developed asymmetric membranes using cellulose acetate materials (upper left), but gradually shifted to composite membranes using polyamide materials (lower right). As mentioned above, various types of membrane shapes were developed in the early stage, such as tubular membranes, hollow fiber membranes, and flat sheet membranes, but eventually converged to flat sheet membranes. In this way, many companies shifted to the polyamide flat sheet membrane shown in the lower right of the Table, and that convergence is the state of the industry today.

Toyobo is a company that has taken a very special position. The company has consistently supplied hollow fiber asymmetric membranes made of cellulose triacetate, which is one of the cellulose acetate materials.[5]

3.3 Differences in Element Shape

3.3.1 Elements, Modules, and Trains

The difference between today's polyamide flat sheet type and cellulose acetate hollow fiber one is not only in the membrane technologies, but also in the shape of the element and module.

For the polyamide flat sheet membrane, there is a de facto standard for the shape of the membrane element, with an outer diameter of 4″ or 8″ and a length of 40″.[6]

[5] In the early days of the industry, many companies used cellulose acetate with a degree of substitution of about 2.5 as a material for membrane production. Although this material is easy to handle, it has major drawbacks in the case of seawater desalination, such as poor separation performance. Although cellulose triacetate is difficult to dissolve and difficult to handle, it is a material that can actualize separation performance for seawater desalination.

[6] In this book, the concept of dominant design is used for the materials, shapes, and structures of RO membranes, which are the mainstream of the industry. The concept of de facto standard is used for membrane elements and modules, which are also the mainstream of the industry. This

Recently introduced 16-inch products also have the same shape among companies. Several of these membrane elements are connected in a series and placed in a pressure vessel to form a membrane module, which is then stored in a metal frame called a train to form a single unit. Then, the desalination plant is built up by installing a number of units according to the desired desalination volume.

Figure 3.5a shows a photograph of the plant of the Okinawa Desalination Center that uses polyamide flat sheet membrane modules. The cylindrical objects neatly arranged in this photograph are pressure vessels and membrane modules. The metal framework that holds them in line is a train. In the case of this plant, there are 63 membrane modules installed in one train. Since each membrane module contains six membrane elements in a series, the number of membrane elements per train is 378.

In contrast, the cellulose acetate hollow fiber membrane element supplied by Toyobo has an outer diameter of 10 inches, which is larger than that of the polyamide flat sheet membrane. Toyobo makes RO membrane modules by connecting two of these membrane elements in a series and storing them in a pressure vessel. Since the size of the membrane module increases with the size of the membrane element, the train for storing the membrane module is different from that for the polyamide flat sheet membranes. Figure 3.5b is a photograph of the desalination process of a seawater desalination center in Fukuoka, Japan, where Toyobo RO membranes are installed.

3.3.2 Switching Costs

Differences in membrane elements and module geometries make it difficult for customers to switch to a different group when it is time to replace their membranes. Switching to another group requires more than just replacing the membrane element. Pressure vessels and trains must also be replaced. Since the train is very robust in both groups to withstand heavyweight, it cannot be easily replaced.

Taking the desalination plant in Okinawa as an example, there are 378 membrane elements installed in one train. Each membrane element weighs about 16 kg, so the total load per train is more than six tons for the membrane elements alone. Besides the weight of 63 pressure vessels added to this, all the membrane modules are filled with water during operation. Thus, a considerable amount of weight is placed on the train.

At the desalination center in Fukuoka, 200 membrane modules, each containing two membrane elements, are stored in one train, and a total of five trains are deployed.[7] These modules weigh about 470 kg when full of water, which means

is because the former is mainly concerned with the design concept of the product itself, while the latter is concerned with standards for connectivity with other components, which are of considerable importance.

[7] The membrane module product name is HB10255FI. For more information on the business development of Toyobo, refer to Chapter 7.

(a)

(b)

Fig. 3.5 **a** Okinawa seawater desalination center. *Source* Taken by the authors (June 22, 2012). **b** Fukuoka Seawater Desalination Center. *Source* Provided by Toyobo

that each train can withstand a weight of at least 94 tons during operation, according to simple calculations. Switching to membranes from another group would mean replacing the entire train. Unless there are serious water quality or desalination problems, there will be no incentive to switch.

On the other hand, it is easy to switch membrane manufacturers within the same polyamide-based flat sheet type, because the module shape is standardized and there is no need to replace the train. Therefore, there is always a risk of switching when the membrane is replaced.

3.4 Challenges in Technological Development and Their Recent Changes

3.4.1 Fundamental Challenges in RO Membrane Development

The primary performance indicators of RO membranes are desalination rate and water permeability. The developers have been working on, for a long time, to simultaneously achieve these two performances at a high level.

You need to remove salt to supply drinking water from seawater. The salinity of seawater is about 3.5%, which must be reduced to less than 0.05% for daily drinking use. On the other hand, it is necessary to secure high permeability and to increase productivity to supply drinking water at a low cost. A membrane that blocks inorganic ions but allows water molecules to pass through is necessary.

However, the desalination rate and the water permeability often contradict each other. For example, to improve the desalination performance, the membrane should be made finer, but this would make it difficult for water to pass through. On the other hand, if the membrane is made coarse to improve the water permeability, the desalination rate will decrease. Overcoming this fundamental trade-off and developing a practical membrane has always been the goal of developers.[8]

Various other kinds of performance were also required for the practical application of RO membranes in addition to the desalination rate and permeability. There were many trade-offs to be solved. For example, the membrane must have adequate resistance to pressure applied by a high-pressure pump. In the case of high salinity seawater that requires high pressure for desalination, pressure resistance is particularly important. One way to increase the pressure resistance is to make the membrane thicker. However, this will reduce the water permeability. Therefore, developers need to find a way to increase the pressure resistance without reducing the water permeability.

---

[8] Shintani (2011) used permeability and desalination rate as the axes when depicting the technological progress of RO membranes over 40 years. This confirms that desalination performance and permeability performance were two important development targets for a long period of time.

Table 3.2 Key performance indicators required for RO membranes

Separation performance	Separation-related performance	Operation-related performance
Desalination	Permeability	Stain resistance
TOC removal	Pressure resistance	Chlorine resistance
Boron removal	Heat resistance	Chemical resistance

Source Prepared by the authors
Note Other kinds of separation performance include removal of bacterial-origin endotoxins

Furthermore, countermeasures against membrane contamination have always been an important issue. A wide variety of microorganisms and substances are dissolved in seawater. The problem of these materials contaminating the membrane and causing clogging has been a difficult hurdle for developers. A particularly troublesome problem is clogging caused by microbial growth. One way to deal with biofouling is chlorine disinfection. If chlorine disinfection is not possible, the plant must be shut down, and the membranes must be cleaned, which causes significant downtime and increases the cost of desalination.

For chlorine disinfection, the membrane must have a high chlorine resistance to prevent damage from chlorine. Polyamide composite membranes, which have low chlorine resistance, have a weak point here. This is one of the reasons why Toyobo, which supplies cellulose acetate hollow fiber membranes with high chlorine resistance, has been able to maintain a certain level of competitiveness. Membrane design with strong resistance to contamination as well as that with chemical resistance in cleaning (compatibility with a wide pH range) is also important. However, both are difficult to make compatible with other performance factors.

Table 3.2 summarizes key performance factors required for RO membranes. They are classified into three categories: separation performance, separation-related performance, and operation-related performance. Separation performance includes total organic carbon (TOC) removal and boron removal as well as desalination performance. Separation-related performance includes water permeability and pressure resistance as well as heat resistance that is critical in food applications and seawater desalination. The operation-related performance includes stain resistance, chlorine resistance, and chemical resistance. The history of RO membrane development has been to overcome the trade-offs between these different performance factors and to search for membranes that can simultaneously achieve them at a higher level.

3.4.2 Changes in Technical Issues

Although the central issue in the development of RO membranes has always been to achieve both desalination and permeability at a high level, there are other performance factors that have become important with time, one of which is boron removal.

In the 1990s, the World Health Organization (WHO), which was concerned about the adverse effects of boron on the human body, set strict interim guidelines for boron in drinking water and demanded the development of membranes with excellent boron removal functions. Polyamide membranes were able to meet this demand. Cellulose acetate-based membranes could not sufficiently meet this requirement. Therefore, in a desalination plant where cellulose acetate RO membranes are adopted in the desalination process as in Fukuoka, polyamide RO membranes responsible for boron removal are applied to the post-process.

Since the WHO loosened its restrictions on boron removal afterward, the performance race in this respect has subsided. However, the polyamide-based group still lists the boron removal rate as one of the important performance indicators.

Low-pressure membranes are also being developed to reduce the cost of electricity, which accounts for most of desalination cost. In recent years, the development of desalination by Forward Osmosis membranes (FO membranes) instead of RO membranes has been achieving results, and FO membranes are gradually being used. In Japan, the membrane engineering group of Kobe University is conducting research activities on FO membranes.[9] Toyobo is also actively promoting joint development of FO membranes with the Saline Water Conversion Corporation (SWCC) of Saudi Arabia.

As the seawater desalination market started up in earnest, one of the most important issues was the prevention of membrane contamination. As an increasing number of large-scale desalination plants have been installed around the world, the issue of membrane fouling has become a serious problem that has attracted more attention than ever before. Chlorine disinfection is an efficient countermeasure, but polyamide RO membranes are vulnerable to chlorine. Therefore, the development of chlorine-resistant membranes and fouling-resistant membranes has been enthusiastically promoted in the polyamide group, and fouling prevention is still a major theme of membrane development today.[10]

Improving membrane elements has also been an issue in the industry. The polyamide group has been trying to increase the efficiency of membrane filling into the element by making the spacer thinner, and the thickness of the spacer has come to be listed in the technical specification.

3.4.3 Changes in Desalination Systems

RO membrane development was first targeted to the compatibility between desalination performance and permeability, then to the boron removal and low-pressure

[9] For the membrane engineering group at Kobe University, see, for example, Chung et al. (2011) and Takahashi and Matsuyama (2016). The joint development between Toyobo and SWCC was reported in Nihon Keizai Shimbun (March 14, 2017).

[10] In the RO membrane session held at the conference of the Society of Chemical Engineers, Japan (September 16–18, 2013), which the authors attended, most of the presentations were related to fouling countermeasures.

membranes, and then to the anti-fouling measures. As membrane performance has improved through these efforts, the desalination system has changed.

When desalination performance was low, the mainstream approach was a two-stage system in which the desalination process is carried out twice. However, improvement of desalination performance and permeability has made it possible to desalinate water sufficiently even in a single stage, and to save the additional energy consumption required for two-stage desalination.

On the other hand, a two-stage desalination system with a different purpose from the conventional system has appeared. In a sea area where the raw water environment is unstable, forcing single-stage desalination may increase energy consumption. Therefore, a two-stage desalination system has been introduced to obtain freshwater while reducing energy consumption by using a low-pressure membrane in the second stage. Two-stage desalination has also been used to increase the freshwater recovery rate by desalinating the concentrated water from the first stage again with a high-pressure membrane. In some cases, another RO membrane is used for the boron removal process after desalination. Desalination plants with a three-stage system have also been designed. In such ways, RO membranes are used in different ways in the same desalination process.

References

Chung, T.-S., Zhang, S., Wang, K. Y., Su, J., & Ling, M. M. (2011). Forward osmosis processes: Yesterday, today and tomorrow. *Desalination, 287*, 78–81.

Makubunri Gijutsu Shinkō Kyōkai/ Maku Jōsui Iinkai (Kanshū)/Jōsui Maku (Dai 2-han) Henshū Iinkai (Hen) (Membrane Separation Technology Promotion Association, Membrane Water Purification Committee (Supervisor), Water Purification Membrane (2nd Edition) Editorial Committee (ed.). (2008). ***Jōsui maku (Dai 2-han) (Membrane water purification*** (2nd edn.). Tokyo: Gihodo Publishing.

Nihon Keizai Shimbun. September 27, 2013; February 19, 2014; July 10, 2014; January 21, 2015; March 14, 2017.

Ohya, H. (1985). A review on recent progress in reverse osmosis membranes. *Membrane, 10*, 101–106.

Shintani, T. (2011). Large scale seawater desalination and municipal reclamation treatment by RO membrane process. *Membrane, 36*, 227–232.

Takahashi, T., & Matsuyama, H. (2016). Forward osmosis technology using seawater for filtration concentration of municipal wastewater. In *Nihon Kaisui Gakkaishi.* (*Bulletin of the Society of Sea Water Science, Japan, 70*, 340–346).

Wada, Y. (2004). *Zōsui no Gijutsu (zōho-ban) (Technology of creating water (enlarged edition)).* Tokyo: Chijin Shokan.

Zōsui gijutsu Henshū kikaku Iinkai Hen (Desalination Technology Editorial Planning Committee ed.). (1983). *Zōsui Gijutsu Journal of water re-use technology.* Tokyo: Water Reuse Promotion Center. http://www.f-suiki.or.jp/facility/kaitan-center/kaitan-facility/about-maku/. Accessed 27 May 2018

Part II
The Development of RO Membranes in the United States

Chapter 4
The Beginnings of Research in the Public Sector

Abstract In this chapter, we describe the history of RO membrane development by academic and public institutions in the United States. We first clarify that development was launched at UCLA and other US universities, and then focus on the US government's massive policy support that played a critical role in the early history. We also point out that research and development at North Star Research Institute had a particular contribution to the advancement of the RO membrane method.

4.1 Introduction

The purpose of this chapter and Chap. 5, which constitute Part 2, is to clarify the history of the development and commercialization of RO membranes in the United States. The history can be roughly divided into two periods: one in which public institutions such as universities and the government played active roles, and the other in which private companies were active. The first half of the history is discussed in this chapter, and the second half is discussed in the next chapter.

The development of RO membranes was widespread throughout the United States in the early days. Among the key players were the universities such as UCLA and the University of Florida, the Office of Saline Water in the Department of the Interior as the government agency, and the North Star Research and Development Institute (referred to below as North Star) as the non-profit contract research organization. In addition to the two universities, various other universities throughout the United States participated in the early development. Behind this progress in membrane research in many places was the massive public support of the U.S. government. Concerned about future water shortages, the U.S. government enacted the Saline Water Conversion Act in 1952 and began to provide public support. The support extended not only to universities, but also to non-profit research institutions.

Below, we first clarify a series of RO membrane research and development processes promoted mainly by public organizations in the 1950s and 1960s, the dawn of industrial development.

M. Fujiwara and Y. Aoshima, *Mechanisms for Long-Term Innovation*,
Advances in Japanese Business and Economics 31,
https://doi.org/10.1007/978-981-19-4896-1_4

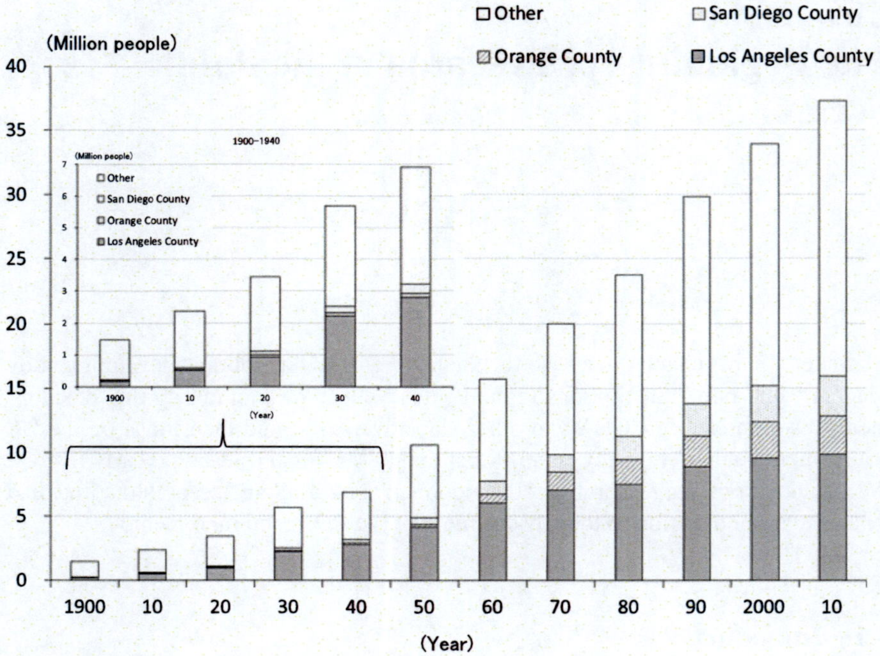

Fig. 4.1 Population trends in California (1900–2010). *Source* Prepared by the authors based on the Department of Finance, California

4.2 Development in Universities

4.2.1 The Beginning of Research at UCLA

The state of California, located on the West Coast of the United States, is famous for its long history of water shortages. In particular, Los Angeles County in the southern part of the state is a semi-arid region, with an annual precipitation of only about 300 mm. That is about one-fifth that of Tokyo. In the summer, the weather is clear and sunny, and rainfall is limited to one day a month.

Los Angeles County is the most populous county in California. Figure 4.1 shows that population growth in Los Angeles County stands out among the three major counties in the state. The county's population, which was only 170,000 in 1900, grew rapidly in the early twentieth century, reaching 2.78 million in 1940.[1]

The concentration of population in a drought-stricken region naturally led to the challenge of securing water resources. The county has managed to do so by drawing water from Northern California and the Colorado River, hundreds of miles to the east,

[1] Today, Los Angeles County has around 10 million people.

but this has caused friction with surrounding communities. The search for alternative water sources in the immediate area was a key issue for policymakers.[2]

UCLA's early involvement in desalination research was understandable, given the region's urban environment plagued by water shortages. Desalination research at UCLA began in 1949 when Gerald Hassler, a lecturer in Engineering, wrote a report entitled "The Sea as a Source of Fresh Water". In 1950, Hassler mentioned the idea of desalination by membrane treatment. And by 1954, he had completed an experimental apparatus based on the membrane treatment method to verify his idea (Glater, 1998).[3]

4.2.2 Development of Cellulose Acetate Membranes at the University of Florida

The University of Florida, located in the southeastern part of the United States, had also begun to develop membranes. In 1953, Reid proposed desalination by the RO method, and in 1955, the U.S. Department of the Interior launched the Office of Saline Water (OSW). With the support of the Office, Reid and his graduate student Ernest Breton began research on membranes under the theme of "Osmotic Membranes for Demineralization of Saline Water". The possibility of the RO method had already been recognized at that time. They had high hopes for polymeric membranes and repeated desalination experiments using various commercially available semipermeable membranes (Reid & Breton, 1959).

After repeated experiments, they found that cellulose acetate semipermeable membranes had the ability to desalinate water. Among the many membranes they tried, DuPont's cellulose acetate membrane achieved a high desalination rate of 97.4%, which was revolutionary at the time. The results of the experiments using various membranes were published in a paper in 1959,[4] and this marked a major step forward in the research on the practical application of RO membranes. Many of the RO membranes commercialized in the early stage of industrial development were made of cellulose acetate materials. Even now, Toyobo has secured a certain market position with its cellulose acetate membranes. The discovery by Reid and Breton that cellulose acetate materials are suitable was a major contribution to lay the foundation of the history of RO membrane development.

However, the cellulose acetate membrane discovered by Reid and Breton had a very low flux at a pressure of about 10 MPa as well as poor durability. it was far

[2] The water problem in California was so serious that a book, *The Water Problem of Southern California*, was published in 1934 by the distant University of Illinois (Bogart, 1934).

[3] Glater (1998) notes that the term reverse osmosis may have been first used by Hassler in the UCLA Engineering Report in August 1956.

[4] Reid and Breton (1959).

from a practical level.[5] After that, there were no more groundbreaking results from the University of Florida, and the research came to an end.

4.2.3 Development of the L-S Membrane at UCLA

It was not only the University of Florida that faced difficulties in realizing the membrane technology; a group led by Hassler was also struggling with the practical application at UCLA. From around 1955, when the initial idea was verified, Hassler's group had been seeking to develop a narrow gap membrane. However, this method failed to achieve satisfactory performance, and his project was disbanded in 1960 (Glater, 1998).

The breakthrough at UCLA came from the group of Professor Samuel Yuster, who had been working separately from Hassler's group in the School of Engineering. In Yuster's laboratory, which was supported by the state of California, graduate student Srinivasa Sourirajan discovered the semi-permeability of cellulose acetate membranes in 1958. However, performance was unstable. That same year, Sidney Loeb, who had entered graduate school at UCLA at the age of 41 after a long career in business, began working with Sourirajan on RO membranes. After Sourirajan and Loeb tried various membranes without success, they decided to focus their research on cellulose acetate.

They found that the desalinating performance of cellulose acetate membranes depended on the way they were heat-treated. What bothered them was that the experimental results fluctuated. As they repeated the experiment, they began to realize that it was not happening randomly. As shown in Fig. 3.2 in the previous chapter, the semipermeable membrane of cellulose acetate is asymmetric in the thickness direction with fine and coarse surfaces. To obtain stable performance, desalination had to be carried out in consideration of the direction.[6] In 1959, they succeeded in producing an asymmetric cellulose acetate membrane that had both desalination and water permeability at a high level. In the following year, 1960, they announced their achievement and applied for a patent.[7] In 1962, the results were published in a paper.[8]

This asymmetric membrane, called the L-S membrane after the initial letters of their names, was a revolutionary membrane at that time. It showed a desalination rate

[5] Zōsui gijutsu Henshū kikaku Iinkai Hen (Desalination Technology Editorial Planning Committee ed. (1983), p. 79.

[6] Loeb (1981), p.5.

[7] US Patent 3,133,132: High flow porous membranes for separating water from saline solutions (application date: November 29, 1960). The appearance of L-S membranes is often marked as 1962 (e.g. Baker, 2004). However, Sourirajan himself states in a footnote that he announced the L-S membrane in 1960 (Sourirajan, 1981), referring to the year of publication in the journal. Therefore, the year 1960, when the results were published, is considered as the year of development in this book.

[8] Loeb and Sourirajan (1962).

comparable to that of Reid at the University of Florida, but with a permeability nearly 50 times higher.[9] Based on these results, Loeb was involved in the demonstration of water desalination using tubular L-S membranes[10] when the world's first desalination plant using the RO method (desalination capacity: approximately 19 m^3/day) went into operation in 1965 in Coalinga, located between San Francisco and Los Angeles.

With the advent of the L-S membrane, the practical application of RO membranes finally became a reality, and many companies began to actively develop the technology for commercialization. Especially in San Diego, an industrial cluster was being formed.

4.3 Massive Government Support

4.3.1 Enactment of the Saline Water Conversion Act and Establishment of the OSW

The response to water shortages in the United States was a national effort, not limited to the West Coast. Concerned about future water shortages, the Harry S. Truman administration enacted the Saline Water Conversion Act (Public Law 82–448) on July 3, 1952, which provided $2 million over five years for research and development funding for desalination programs under the Department of the Interior (MacGowan, 1963; U.S. Congress, Office of Technology Assessment, 1988).

Initially, the government had an optimistic outlook and planned to end public support by bringing desalination technology to practical use quickly, but this did not proceed as expected. Therefore, Dwight D. Eisenhower, who took power from Truman, enacted Public Law 84–111 on June 29, 1955. This extended the support period until 1963 and increased the research funds to $10 million (MacGowan, 1963).

At the same time, the OSW was established in the Department of the Interior to provide systematic support for desalination research.[11] The OSW gave subsidies to research institutes in various regions and started to provide full-scale government support for desalination research. The desalination program was truly a major national project, and the research of Hassler at UCLA and Reid at the University of Florida were also bolstered by this public support.

---

[9] Boehiert (Annual editions), *Legislative History: Saline Water Conversion Act, Vol. 6*, Pt. 1, p. 190.

[10] Stevens and Loeb (1967); Loeb and Selover (1967). Sourirajan left the U.S. to take a position with the National Research Council Canada shortly after the announcement of the L-S membrane. Loeb was involved in the demonstration of the L-S membrane at the Coalinga desalination plant before leaving the U.S. for Israel in 1967.

[11] Although there are some papers and books that place the establishment of the OSW in 1952 or 1953, the year the Saline Water Conversion Act was enacted, we follow the U.S. National Archives and Records Administration's description of the OSW (https://www.archives.gov/research/guide-fed-records/groups/380.html) and give the year as 1955 in this book. However, it is assumed that the research activities started in 1952 when the Saline Water Conversion Act was enacted.

Support for commercialization began in the late 1950s, with the enactment of Public Law 85–883 (Demonstration Plant Act) in 1958, which appropriated $10 million for research and development, plus an additional $10 million to construct demonstration plants in five areas across the United States. Accordingly, five desalination demonstration plants were constructed in Freeport, Texas; Webster, South Dakota; Roswell, New Mexico; San Diego, California; and Wrightsville Beach, North Carolina. The methods used in these demonstration plants were evaporation, electrodialysis, and refrigeration, but not RO. L-S membranes had not yet been introduced, and RO technology was not even in its infancy.

4.3.2 Expansion of Support

In the 1960s, government support was further expanded. In particular, Public Law 87–295 (the Anderson-Aspinall Act), enacted in September 1961, was an influential law that not only extended Public Law 85–883 through 1970, but also provided for the investment of as much as $75 million in desalinization research from 1962 to 1967. Figure 4.2 shows the amount of support for desalination research by the OSW. In 1965, *Desalination,* an academic journal, was published as a medium to present scientific findings related to desalination. The 1960s was a time when the U.S. government provided ample support for desalination research.[12]

John F. Kennedy, who assumed the presidency in January 1961, was the driving force behind this massive aid. He was famous for his goal of landing a man on the moon, but he was also a strong proponent of seawater desalination. The Democratic administration led by Kennedy placed desalination on the same level of importance as landing a man on the moon and enacted Public Law 87–295. By enacting Public Law 87–295, the Democratic administration provided massive support for desalination research programs.

The Lyndon B. Johnson administration, which succeeded Kennedy after his unfortunate assassination in November 1963, continued the generous public support for desalination research. Stewart Udall, who had served as Secretary of the Interior under both Kennedy and Johnson, recalled the generous support.

> Both President Kennedy and President Johnson had the same attitude. They like the program. This was a glamour program to them. This was a way we could do something for the world. And therefore, I got very good support. We got very big budget increases, right along. We got the program up six, eight times, I believe, to what they had in 1960 in the Eisenhower administration.[13]

---

[12] Not all of the amount shown in the figure was allocated to the RO method, because the OSW supported not only the RO method but also a wide range of research and development on desalination.

[13] Udall (1970) p.153.

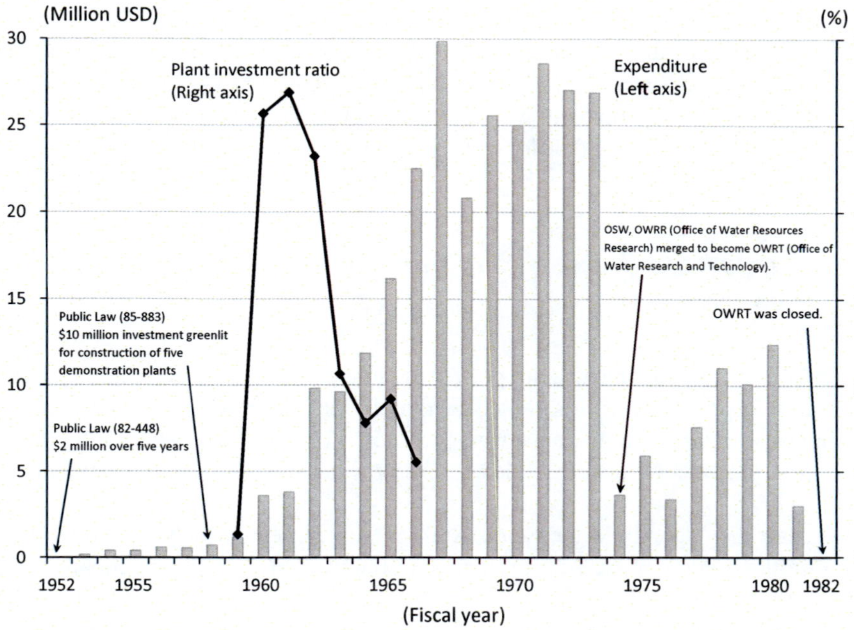

Fig. 4.2 Government budget for desalination programs. *Source* Prepared by the authors based on *Legislative History, Vol. 6*, Pt. 1, p. 67; MacGowan (1963), p. 26; General Accounting Office (1979), p. 3; U.S. Congress, Office of Technology Assessment (1988)

As of 1964, just after the Johnson Administration took over, the OSW had a total of 124 contract research projects.[14] Of these, 57 (totaling $3,761,081) were commissioned to universities and public research institutions, and 67 (totaling $6,592,576) were commissioned to private companies. Looking at the names of the research themes, most of the research entrusted to public research institutes were in the upstream area, such as research on core components that play a role in the desalination function and membrane forming processes. Conversely, most of the research entrusted to private companies were in the downstream area, such as construction of desalination plants and development of operation technologies, which require large amounts of money.

Amidst a loose division of labor in which the public sector is responsible for upstream basic research and the private sector is responsible for downstream practical application research, it was the General Atomic (GA) Division of General Dynamics

[14] Legislative History (Vol. 5, Pt. 2). In some cases, multiple projects were commissioned to the same organization, so this does not mean the Office commissioned projects to 124 different organizations. Consignments to other divisions of the same organization are considered as separate projects, and if complete institutional duplication is excluded, the total number of consigned institutions is 93. Since we were unable to obtain the amount of money for each individual case of commissioned research before that time, this is the oldest data available to the authors. This chronological progression is described in detail in Chapter 10.

that was responsible for large projects in the upstream area despite its private status. The division received about $100,000 from the OSW to conduct research and development of RO membrane modules. The organizational unit in the GA Division, which was involved in the development, was called ROGA. It played an important role in the history of RO membrane development as described in the next chapter.

4.3.3 Support for Overseas Plants and Downsizing of Public Support

The OSW not only financially supported desalination research activities in the U.S., but also promoted efforts to secure outlets for research results by cooperating in desalination plant demonstration projects.[15] These efforts were not limited to domestic projects but also extended to overseas desalination plant construction projects.

Overseas, they supported the Saudi Arabian government, which was suffering from a drought, and conducted a preliminary study in 1963 on the practicality of a dual-purpose plant with both power generation and desalination capabilities in Jeddah, a city along the Red Sea. A dual-purpose plant here refers to a facility where a desalination plant is attached to a thermal power plant. The steam produced in the thermal power generation process is used for the desalination process.[16] The combination of power generation and desalination was necessary to reduce the cost of desalination for the spread and penetration of the desalination technology. In 1965, two years after the completion of the feasibility study, the U.S. and Saudi governments agreed to build a dual-purpose plant, with the OSW acting as an intermediary. The project was completed in 1971.

At the same time, the U.S. government also conducted a feasibility study with Israel. Discussions between President Johnson and Prime Minister Levi Eshkol in 1964 led to agreement in 1965 on conducting a feasibility study of a dual-purpose plant for electricity generation and desalination. In 1965, another agreement was signed with Mexico to study the feasibility of a dual-purpose nuclear power and desalination plant.

In October 1965, the First International Symposium on Water Desalination was held in Washington, D.C.. In May 1967, the "International Conference on Water for Peace" was held in Washington, D.C. In response to a request from the U.S. government, 12 members of the Japan Water Agency and other public organizations participated in the conference (Cho, 1967).

However, due to the nature of domestic policy, overseas assistance at the expense of the OSW was gradually discouraged. When the U.S. fell into a trade deficit in 1971, the Republican Richard M. Nixon administration began to cut spending, and

[15] Office of Saline Water (1972).

[16] Such dual-purpose plants, combining power generation and desalination, are still being built today.

public support for the desalination program was also reduced. In 1974, the OSW was merged with the Office of Water Resource Research and reorganized into the Office of Water Research and Technology (OWRT), and its budget was quickly reduced. The OWRT was finally closed in 1982.

4.4 North Star's Contributions

4.4.1 Development of Composite Membranes

The next development challenge posed by the L-S membrane was to improve its water permeability. Although the permeability of the L-S membrane was much higher than that of the previous membranes, it was not high enough to produce a practical amount of water.

The L-S membrane is an asymmetric membrane in which the separation layer and the support layer are made of the same material. The disadvantage of the L-S membrane is that the high pressure applied during operation causes the spongy support layer to collapse and consolidate, and the permeability does not last long. As long as the asymmetric structure was adopted, it was thought that there was a limit to maintaining the high performance over a long time. In order to improve the desalination performance, it is desirable to make the separation function of the membrane as thin as possible, but this would impair the pressure resistance of the membrane and cause it to lose pressure during operation. The idea of a composite membrane was devised to solve this problem, in which a separation layer and a support layer made of different materials are layered on each other.

One of the first people to publish results leading to the realization of composite membranes was Peter Francis of North Star, a non-profit contract research organization established in Minneapolis, Minnesota in 1963. Francis, who was one of the first recruits to the institute and who was tasked with establishing a chemistry department there, developed cellulose acetate-based composite membranes while improving the L-S membrane deposition process. A year later, in 1964, he succeeded in developing a composite membrane with enhanced water permeability. John E. Cadotte, a fellow researcher, wrote that "this achievement led to an initial research contract with the Office of Saline Water".[17]

[17] Cadotte and Petersen (1981) p. 306. Unfortunately, there was no record specifying the reason why North Star, which had just been established, paid attention to the RO research. However, since the government was strongly supporting desalination research at that time, they might have thought that RO research would provide opportunities for commissioned research.

4.4.2 Development of Non-Cellulosic Composite Membranes

At North Star, Cadotte was also developing composite membranes while searching for a material to use for the support layer. In 1967, Cadotte and his colleagues announced the development of a membrane with a support layer made of polysulfone instead of cellulose acetate.[18] As a result, the water permeability of the membrane was increased about five times compared with that of the cellulose acetate asymmetric membranes. The results of their development were so revolutionary that polysulfone is still used as the support layer for composite membranes today, more than 50 years later.

The next challenge was to make the separation layer non-cellulosic. Cellulose acetate-based materials have problems with low water permeability and degradation of the membrane by bacteria. Cadotte therefore searched for and developed new materials, and succeeded in making the support and separation layers non-cellulosic. The results were published in 1970 as NS-100, an acronym for North Star. Cadotte further developed NS-200 in 1972 and NS-300 in 1977.[19]

However, there were still some problems to be solved in mass production, such as the variation in performance when scaled up. Another problem was the difficulty of performing chlorine disinfection during plant operation due to reduced chlorine resistance. NS-200 also lacked long-term stability of performance (Cadotte, 1985).

Cadotte took on these challenges, and in 1979 succeeded in developing a revolutionary cross-linked aromatic polyamide composite membrane produced by interfacial polycondensation. This was the real breakthrough, as its basic composition is still mainstream in the industry today, more than 40 years later (Baker, 2004). The new composite membrane was marketed as FT-30 by FilmTec, which Cadotte had founded with his colleagues. It was a revolutionary membrane with high levels of both desalination and water permeability, as well as excellent chlorine resistance and durability.

4.5 Summary

This chapter describes the early history of the research and development of the RO method, which was born in the United States. The development of RO membranes began in Los Angeles, a city that had historically been plagued by drought. In 1952, the U.S. government enacted the Saline Water Conversion Act, which provided substantial financial support. The development of RO membranes paved the way for their commercialization with the advent of the L-S membrane by Loeb and Sourirajan

[18] Cadotte et al., (1980); Rozelle et al. (1968).

[19] Tomaschke (2000) notes the year of development of NS-300 as 1975. It is true that Cadotte (1985) also reported the basic technical results of NS-300 in 1975. However, since that was a presentation of the basic thrust of the concept, we set 1977 as the year of development of NS-300 according to Cadotte et al. (1980).

at UCLA. The idea of a composite membrane was developed by Francis and Cadotte at North Star, and the support and separation layers were made non-cellulosic. The resulting cross-linked aromatic polyamide composite membrane, FT-30, was a groundbreaking innovation that set the course for the subsequent development of RO membranes.

Public support also stimulated research activity in the private sector, which was a major driving force in advancing early RO technology. In San Diego in the West Coast, the development unit of GA, a division of the military contractor General Dynamics, led the way in commercializing the technology. In the Midwest, near Minneapolis, FilmTec, founded by retired North Star researchers, produced a series of innovative results. Michigan-headquartered Dow developed technology in the north and Delaware-based DuPont did so in the east. A timeline of RO technologies developed by universities and public institutions and by private companies in the United States is shown in the chronology at the end of this book.

In the next chapter, we will clarify the process by which the results of the development carried out by these public institutions were transferred to private companies in the United States, and the process of progression for practical applications and commercialization.

References

Baker, R. W. (2004). *Membrane technology and applications* (2nd ed.). John Wiley & Sons Ltd.

Bogart, E. L. (1934). *"The water problem of Southern California" in illinois studies in the social sciences, 19.* The University of Illinois for Claremont Colleges Library.

Cadotte, J. E. (1985). Development of composite reverse osmosis membranes in retrospect. *Membrane, 10*, 117–118.

Cadotte, J. E., Petersen, R. J., Larson, R. E., & Erickson, E. E. (1980). New thin-film composite seawater reverse osmosis membrane. *Desalination, 32*, 25–31.

Cadotte, J. E., & Petersen, R. J. (1981). Thin-film composite reverse-osmosis membranes: Origin, development, and recent advances. In Turbak, A.F. (Ed.), *Synthetic Membranes: Based on the 20th Anniversary Symposium Honoring Drs. Loeb and Sourirajan Vol. 1 Desalination, American Chemical Society* (pp. 305–326).

Cho, T. (1967). Heiwa no tame no mizu riyō (Report on the International Conference on 'Water Use for Peace') in Chūtekkan (Cast Iron Pipe), 4, 12–13.

General Accounting Office. (1979). *Desalting Water Probably will not Solve the Nation's Water Problems, but can Help. Report to the Congress/ by the Comptroller General of the United States.* Washington: General Accounting Office.

Glater, J. (1998). The early history of reverse osmosis membrane development. *Desalination, 117*, 297–309.

Loeb, S., & Selover, E. (1967). Sixteen months of field experience on the Coalinga pilot plant. *Desalination, 2*, 75–80.

Loeb, S., & Sourirajan, S. (1962). Sea water demineralization by means of an osmotic membrane. *Advances in Chemistry Series, 38*, 117–132.

Loeb, S. (1981). The Loeb-Sourirajan membrane: How it came about. IN *ACS symposium series (USA)* (pp. 1–9).

MacGowan, C. (1963). History, function and program of the office of saline water. In *Proceedings of the Eighth Annual New Mexico Water Conference* (pp. 24–32).

Office of Saline Water. 1972. *Saline Water Conversion Summary Report.*

Reid, C. E., & Breton, E. J. (1959). Water and Ion flow across cellulosic membranes. *Journal of Applied Polymer Science, 1*, 133–143.

Rozelle, L. T., Cadotte, J. E., Corneliussen, R. D., & Erickson, E. E. (1968). Development of new reverse osmosis membranes for desalination. In *Office of Saline Water Research and Development Progress Report, 359.*

Sourirajan, S. (1981). Reverse osmosis: A new field of applied chemistry and chemical engineering. In Turbak, A.F. (Ed.), *Synthetic Membranes: Based on the 20th Anniversary Symposium Honoring Drs. Loeb and Sourirajan Vol. 1 Desalination, American Chemical Society* (pp. 11–62).

Stevens, D., & Loeb, S. (1967). Reverse osmosis desalination costs derived from the Coalinga pilot plant operation. *Desalination, 2*, 56–74.

Tomaschke, J. E. (2000). Interfacial composite membranes. In *Membrane Preparation* (pp. 3319–3331).

Udall, S. L. (1970). *Recorded Interview by W.W. Moss,* John F. Kennedy Library Oral History Program.

U.S. Congress, Office of Technology Assessment. (1988). *Using Desalination Technologies for Water Treatment, OTA-BP-O-46.* Washington D. C.: U. S. Government Printing Office.

Zōsui gijutsu Henshū kikaku Iinkai Hen (Desalination Technology Editorial Planning Committee ed.). (1983). *Zōsui Gijutsu Journal of water re-use technology.* Tokyo: Water Reuse Promotion Center.

Chapter 5
Development for Commercialization by Private Companies

Abstract In this chapter, we describe the history of RO membrane development by US private companies. We shed light on the specific role taken by ROGA, a part of General Atomic, in the progress of the RO membrane technology in the San Diego area as well as critical roles taken by FilmTec and Dow Chemical in the Minneapolis area. We also elucidate that DuPont began development in the East Coast area and expanded its business into the Middle Eastern market.

5.1 Introduction

In Los Angeles, Florida, and Minneapolis, the initial development of RO membranes was carried out by universities and research institutes under the auspices of the government, and was later continued through the process of commercialization by private companies. Due to the large amount of government support, many private companies embarked on desalination projects after the announcement of the L-S membrane at UCLA. The scope of the project was wide ranging, from material development to plant development. This chapter clarifies the series of commercialization and development processes promoted by these private companies.

The development and commercialization of RO membranes took place in various parts of the United States. In San Diego on the west coast, more than 30 companies conducted desalination-related businesses at one time, and an industrial cluster formed in the area.[1] Among them, General Atomic (GA), which was established as a division of perennial defense company General Dynamics, was the most enthusiastic in development. Behind this was the national policy of the peaceful use of nuclear energy. President Eisenhower, who felt threatened by the rapid nuclear arms race caused by the Cold War, made a speech entitled "Atoms for Peace" at the United Nations General Assembly in New York in 1953, and set a clear policy to promote the peaceful use of nuclear energy as a national policy. GA was looking

[1] https://www.sandiego.gov/sites/default/files/legacy/water/pdf/purewater/060501.pdf (Accessed February 10, 2017).

M. Fujiwara and Y. Aoshima, *Mechanisms for Long-Term Innovation*,
Advances in Japanese Business and Economics 31,
https://doi.org/10.1007/978-981-19-4896-1_5

for nuclear energy applications in accordance with this policy. Desalination that consumes massive amounts of energy was a promising and favorable field for GA.

In Minneapolis, Minnesota, development was also advanced by FilmTec, a venture company spun out of North Star, where the composite membrane had been developed. In neighboring Michigan, Dow, a major chemical manufacturer headquartered in Midland, started its development. Dow later acquired FilmTec to obtain key patents that it held.

The first practical application of RO membrane was the L-S membrane installed in the Coalinga demonstration plant. While the initial technological development was focused on the West Coast, it was DuPont, headquartered in Delaware on the East Coast, that led the commercialization effort. DuPont developed and commercialized distinctive membranes by applying its own materials technology. DuPont launched Permasep, an RO membrane that became a powerful alternative to L-S membranes and a driving force in the development of the industry. On the East Coast, the Koch Group, another energy-related company, also entered the RO membrane business.

In this chapter, we describe this initial development and commercialization process by the private sector in the United States.

5.2 Development in the San Diego Area

5.2.1 The Beginning of Development at ROGA

GA was established in 1955 by General Dynamics as a division to conduct research and development on the peaceful use of nuclear energy. General Dynamics has a long history as a company founded in 1899 as Electric Boat, which was renamed in 1952.

It was in 1961 that GA began research on RO membranes in the John Jay Hopkins laboratory within the division.[2] The vice president of GA, who was a member of the California State Water Board, heard that Loeb and his colleagues at UCLA had developed a revolutionary membrane. On the vice president's orders, Robert Riley and Harold Lonsdale began investigating membrane development at GA. They visited Loeb to understand the performance and development status of the L-S membrane, and then decided to elucidate its desalination mechanism in detail and develop further improvements. To receive support for their development, they applied to the OSW for a membrane development project that applies the L-S membrane.

The following year, in April 1962, the application was approved, and a contract development agreement was signed with the OSW. GA had not been able to obtain sufficient resources for the development of RO membranes because of its focus on nuclear power development projects. Therefore, public support was a great help. All

[2] Legislative History vol. 5, p. 304.

the GA's early contract development work on RO membranes was based on contracts signed with the OSW.

The development team for RO membranes officially existed under the umbrella of the Basic Development Department of GA. It was led by Ulrich Merten and dubbed ROGA, an acronym for Reverse Osmosis General Atomic.

Once it was realized that the asymmetric membranes had limited water permeability, the development of new composite membranes with high desalination and water permeability performance and the development of technology for efficiently converting RO membranes into elements became targets. Of these, the elementalization technology was the first to be developed.

There are two major types of RO membranes: the hollow fiber type and the flat membrane type. The flat membrane elements of the time, which were manufactured by layering membranes, were significantly less efficient to manufacture. In response to this, ROGA developer Julius Westmoreland devised a method known as the spiral method. In this method, flat membranes are wrapped around a pipe to form an element, rather than laying the membranes flat on top of each other. He applied for a patent in 1964. The germinal idea for the spiral membrane was originally devised by Alan Michaels, an associate professor at MIT at the time, and a patent application was filed in 1962.[3] However, this element was different in shape from Westmoreland's and was not intended for use in RO membranes. It can thus be said that it was Westmoreland who invented the spiral element for RO membranes.[4]

Later, Donald Bray, who was also a developer at ROGA, applied for a patent based on this idea in 1965 for an element made by wrapping multiple flat membranes around a pipe.[5] The element shape was improved while maintaining the basic idea, and is still widely used today. Riley recalls a series of events related to the shape of the membrane element as follows.

> The commercial potential of the membrane was quickly recognized. Thus, we had to scale-up the flat sheet membrane manufacturing process to large scale production to produce a forty-inch wide membrane. In addition, a membrane packaging configuration was required to make the process commercially economical that exhibited a high membrane packing density. That is, a configuration that incorporated a large membrane area in a small volume. Julius Westmoreland, a chemical engineer in the group at General Atomic, conceived, invented and patented the concept of the spiral-wound membrane element. The original patent by Westmoreland incorporated a single membrane leaf. Shortly thereafter, a second patent by his colleague, Donald Bray, expanded the patent to include multiple membrane leaves. At

---

[3] US3173867A (Membrane separation device: filing date September 28, 1962). Strictly speaking, there is a difference between the flat membrane type in the old sense and the spiral type newly developed here, in that the flat membrane is piled up and the spiral type is wrapped around. However, since the spiral type is now the mainstream and there is no need to distinguish between the two, the spiral type will also be referred to as the flat membrane type in this book. For further information, see https://www.nap.edu/read/12473/chapter/36#208 (Accessed June 7, 2018).

[4] The patent filed by Westmoreland is US3367504 (Spirally wrapped reverse osmosis membrane cell: filing date December 21, 1964).

[5] US3417870 (Reverse osmosis purification apparatus: filing date 22 March 1965); Bray joined GA midway through 1958 as a senior staff engineer and was an engineer responsible for the design of small reactors for military use (Capone, 2008).

that time, the research and development conducted by the General Atomic reverse osmosis membrane group (ROGA) was funded by the U.S. Department of the Interior, Office of Saline Water (OSW). As a result, the patents were granted to and owned by the U.S. Department of the Interior. General Atomic was granted free-use of the patents for their own commercialization.[6]

One of the reasons why the ROGA developers chose the flat membrane type instead of the hollow fiber type was that the hollow fiber type was said to present difficulty in removing contamination and could not overcome the fouling problem. Another reason was that they had been developing a composite membrane in parallel. It is difficult to realize hollow fiber composite membranes by overlapping the separation layer and the support layer. In terms of manufacturability, the flat membrane type was much more compatible with the composite membrane.

5.2.2 *Development and Commercialization of Composite Membranes by ROGA*

The performance of the membranes at this time was not yet sufficient for desalination applications, so ROGA developers looked for other applications. The first major commercial achievement was the delivery of an RO system to the semiconductor giant Texas Instruments (TI) in 1969. Randy Truby, who joined GA in 1969 as a research assistant and was involved in the development of the RO membrane element and its delivery to TI, recalls:

> Seawater was not our primary focus. Mostly it was brackish water. And very quickly, as a result of the pilot studies we did, it became obvious that ultra-pure water for Semiconductors is going to be a big market. That was the first big commercial market. We sold 100,000-gallon per day system to Texas Instruments at Dallas in 1969. That was the biggest RO system in the world at that time. About a year later, we sold, through Ajax and Kurita, 800,000-gallon per day system for North Power Station in Japan. That plant was built by Ajax and shipped to Kurita. It was the biggest plant in the world at that time.[7]

Buoyed by these commercial successes in ultrapure water production for semiconductors and the desalination of brackish water, Riley and his colleagues worked to further improve membrane performance. They continued to develop composite membranes by experimenting with a variety of film deposition methods, including dip coating, while also exploring non-cellulosic membrane materials. As described in the previous chapter, Cadotte at North Star succeeded in making the separation layer and support layer non-cellulosic and developed NS-100, opening up the possibility of composite membranes. In 1975, Riley and his colleagues succeeded in developing PA-300, a composite membrane in which polyetheramide is formed by interfacial polymerization with high production efficiency (Mattson, 1979; Riley et al., 1976). PA-300 was the first composite membrane commercially available for

[6] From the authors' interview with Robert (Bob) Riley at Morena Boulevard on March 5, 2015.

[7] From the authors' interview with Randy Truby at Carlsbad on March 4, 2015.

seawater desalination applications (Wada, 2004). It was used for brine desalination at a demonstration plant in Roswell, New Mexico, and for seawater desalination at a demonstration plant in Wrightsville, North Carolina. Both of these performed well. At this time, Riley and his colleagues also succeeded in developing RC-100, which is made of polyether urea.

PA-300 and RC-100 were installed in a seawater desalination plant in Saudi Arabia. Saudi Arabia was suffering from a drought in the 1970s, and established the Saline Water Conversion Corporation (SWCC) in 1974. There was growing momentum for energy conservation under the oil crisis, and plans were underway to build a seawater desalination plant using the RO method. In 1978, the world's largest seawater desalination plant (12,000 m^3/day) was constructed in Jeddah. PA-300 and RC-100 were selected for this plant, which began operation in January 1979.[8]

While ROGA was developing and commercializing membranes, GA was frequently acquired by other companies and changed its name multiple times. In 1967, it was acquired by Gulf Oil and became Gulf General Atomic, and in 1970, it changed its name to Gulf Environmental Systems. In 1974, the company was acquired by Universal Oil Products (UOP (founded in 1914)) and merged with UOP's Fluid Sciences division in San Diego to form Fluid Systems (Fluid).[9] New RO elements developed by Riley and his team, such as the PA-300, were launched by Fluid, a subsidiary of UOP. UOP, which got ROGA and established Fluid, was acquired by Signal in 1975. In 1985, it became part of Allied Signal, a new company of Allied that acquired Signal.[10] Fluid was sold to Anglian Water in the UK, and in 1998, it was sold to Koch Membrane Systems (Koch), which engaged in the membrane business as a subsidiary of the energy-related company Koch Industries. Although the site itself remained the same, the ownership of the site has changed over the years.

Thus, ROGA was continuously at the mercy of corporate acquisitions while it played a leading role in the history of RO membrane development, and lost talented members on each occasion that an acquisition happened. For example, Bray, the inventor of the multi-leaf spiral element, left the company just before it was acquired by Gulf Oil and became an entrepreneur, founding Desalination Systems Incorporated (DSI) in 1966 and Nimbus Water Systems in 1968. After moving to Alza Corporation, Lonsdale and Richard Baker founded Bend Research, a contract research

---

[8] This plant uses a two-stage desalination system. The size of the installed RO membrane element (TFC-1501PA) is 4 inches in diameter by 40 inches in length. According to Taniguchi (1981), the number of membrane elements installed per unit of RO membrane system in the plant is 336. Since there are 12 RO units in the plant, including one spare unit, a simple calculation shows that 4,032 membrane elements were used. For the specifications of major seawater desalination plants at that time, including the Jeddah plant, see Senoo and Kimura (1983).

[9] The process was very complex and difficult to understand in detail. Spiegler and Liard eds. (1980), which compiled Principles of Desalination, also wrote, "Names and affiliations of both government and industrial organizations engaged in desalination work change with bewildering frequency, and we have not tried to keep up."

[10] Based on https://www.uop.com/about-us/uop-history/a-friendly-acquisition/ (Accessed June 6, 2018).

company that aimed to conduct research on behalf of the U.S. government.[11] Riley also became independent in 1985 and began research and development activities with government support. Truby moved from Koch to Hydranautics (Nitto Denko) in 2002, and then to Toray Industries, Inc.

5.2.3 *Entering of Diverse Companies*

In addition to ROGA, a number of other companies have established RO-related businesses in San Diego and the Los Angeles area.

Founded in 1963 in Santa Barbara, California, Bardex Hydranautics was engaged in the heavy-duty equipment business for drilling applications and the RO membrane business. Of these, the RO membrane business was separated as Hydranautics Water Systems and proceeded membrane development in San Diego. In 1970, it entered the RO water treatment business and later changed its name to Hydranautics.

The company first came to prominence in 1977. In September of the same year, the Bureau of Reclamation, which was working on a desalination project in Yuma, Colorado, selected Fluid (with a production capacity of 280,000 m^3/day and a contract amount of $20,610,924) and Hydranautics (with a production capacity of 83,000 m^3/day and a contract amount of $7,276,630) as membrane manufacturers for the desalination plant.[12] Since Dow and DuPont had also bid for the Yuma project, a great deal of attention was paid to this Hydranautics as an emerging company that defeated the two giants. Dow and DuPont challenged the decision, but the outcome was unchanged.[13] After expanding its business, Hydranautics was acquired by Rohm and Haas, and then by Nitto Denko in 1987.[14] The fact that Hydranautics was involved not only in the membrane business but also in the downstream system business was later of great significance to Nitto Denko.

The business of desalination systems also arose in San Diego, where Glen Havens, who founded Havens Industries, developed porous glass fibers with asymmetric membranes in 1961 for practical use in tubular form. Then, in 1964, he demonstrated that potable water could be obtained from seawater by a two-stage desalination process (Merton, 1970). This achievement attracted the attention of a major local power company San Diego Gas & Electric (SDG&E), which announced that it would

[11] From the authors' interview with Richard Baker on March 3, 2015 at Membrane Technology and Research, Inc. in Newark. Baker appears in Nitto Denko's development history, as described in Chap. 9.

[12] The contract amounts are based on Mattson (1979), p. 216 and Kamisawa (1980). Throughout the demonstration, neither Dow nor DuPont, which had chosen the hollow fiber type, achieved positive results. Only Fluid (ROGA) and Hydranautics, which had chosen the flat membrane type, achieved positive results.

[13] Bickell (1999), pp. 11–12.

[14] After leaving Koch in 2002, Truby moved to Hydranautics, where he served as vice president of operations until 2006. Many people from Fluid and Koch moved to Hydranautics.

cooperate in the construction of a pilot plant.[15] Later, Havens Industries was acquired by UOP.[16] Another company, King Lee, was founded in San Diego in 1977 to handle chemical cleaning agents for RO membranes.

In Los Angeles as well, military company Aerojet General (Aerojet) turned to the desalination business as it sought to diversify into non-military businesses. Aerojet planned to engage in a wide range of businesses, from membranes to plants, and in July 1962 began contract research with the OSW. The company received substantial public support: about $160,000 for basic research on RO membranes, about $160,000 for further basic research, about $60,000 for the development of new polymers for RO membranes, and about $180,000 for the design, construction, and operation of a pilot plant using the RO method. The fact that military companies such as GA and Aerojet received government support for the desalination program also reflected another policy intent: to find a peaceful use for nuclear energy. We will discuss this point in Chap. 10.

The total amount of contract research received by Aerojet exceeded the amount of support received by GA.[17] However, while ROGA left a significant mark on the history of RO membrane development, Aerojet and its subsidiary, Envirogenics, did not achieve much and are rarely mentioned in history.

5.3 Development in the Minneapolis Area

5.3.1 Establishment of FilmTec and Emergence of Breakthrough Technology

At the same time that Riley was developing composite membranes at ROGA, Cadotte was developing composite membranes at North Star in Minneapolis.[18] In 1975, North Star was integrated into and became a part of the Midwest Research Institute (MRI), based in Kansas.[19] The following year, North Star signed a research contract with the OSW. Under this contract, Cadotte conducted research on RO membranes. His work included NS-100, followed by NS-200 developed in 1972, which showed high desalination performance.[20] However, both membranes showed high fluctuation in performance and lacked chlorine resistance. NS-300, which was developed in 1977

[15] Legislative History, vol. 6, p. 205 (638).

[16] Senoo and Kimura (1983) p. 79.

[17] Legislative History, Vol. 5, pp. 740–741.

[18] Lonsdale (1982), p. 110.

[19] http://id.loc.gov/authorities/names/n50057680 (Accessed February 23, 2015).

[20] The year of development of NS-200 was adopted in this book as 1972 based on Cadotte and Petersen (1981, p. 309), but Cadotte et al. (1980) wrote it as 1973.

to cope with these problems, also achieved high performance at the laboratory level, but presented difficulty in scaling up.[21]

Eventually, Cadotte, along with colleagues Eugene Erickson, Roy Larson, and Robert Petersen, began to think about starting their own business. One of the reasons was that the MRI, which had integrated North Star, asked the researchers to relocate to Kansas, more than 600 km away from Minneapolis.[22] They began discussions in the summer of 1977 and established FilmTec that same year. Ericsson served as president. At the time of the establishment of FilmTec, Cadotte was still working at North Star. Cadotte left the MRI on December 31, 1977 and joined FilmTec in 1978.

The birth of a spin-out company from a research institute is not a rare occurrence in the United States. However, the six-month period from the summer when Cadotte began to think about starting a company to the end of the year when he actually left the MRI had a decisive meaning in the patent disputes that occurred later. This point will be described in Chap. 9.

At that time, Cadotte was working on the formation of composite membranes by interfacial polymerization of a separation layer of cross-linked aromatic polyamide on a support layer of polysulfone. Having successfully developed a new composite membrane, Cadotte filed a patent application on February 22, 1979 for a patent that would prove to be pivotal in the direction of the RO industry, which is US patent 4,277,344.[23] He named this new polyamide RO membrane FT-30 after his company name, and presented it at the International Congress on Desalination and Water Reuse held in France in October of the same year, attracting a great deal of attention. The patent he applied for was published on July 7, 1981, and became commonly known as the '344 patent after the last three digits of the patent number 4,277,344.

The '344 patent attracted attention because this RO membrane not only had high desalination performance and water permeability, but also had high chlorine resistance, could tolerate a wide pH range, and could be easily produced by the method called interfacial polymerization as described above. Motohiro Okazaki and Shoji Kimura of the Institute of Industrial Science at the University of Tokyo, who obtained and evaluated FT-30, wrote, "Compared with cellulose acetate membranes and conventional polyamide membranes, FT-30 has greater pH resistance and chlorine resistance, and can be used in pH ranges from 3 to 11" (Okazaki & Kimura, 1983). FT-30 was used in the one-stage desalination process of the demonstration plant in Wrightsville.

Since interfacial polymerization was more suitable for flat membrane production than hollow fiber production, the '344 patent marked a shift in the mainstream from hollow fiber to flat membrane production. FilmTec's business was off to a good start. However, as will be discussed below, FilmTec was acquired by Dow in 1985. It is said

[21] Petersen (1986), p. 130, Cadotte (1985a), Cadotte et al. (1980).

[22] From the authors' interview with Peter Eriksson (GE Water Technologies & Solutions) on March 6, 2015 in Vista. He was an engineer who joined FilmTec in 1982, and the founding members told him why they started the company.

[23] US Patent 4,277,344 (Interfacially synthesized reverse osmosis membrane: filing date Feb. 22, 1979). According to Cadotte (1985a, b), FT-30 itself was developed in 1978.

that Dow's objective was to acquire the '344 patent. This shows just how immense the impact of Cadotte's development results was.[24]

Near Minneapolis, in addition to FilmTec, Osmonics was founded in 1969. Osmonics expanded its vertically integrated business to include not only RO membranes but also downstream plants, and in 1996 acquired DSI to expand its business, but was acquired by GE in 2003.[25]

5.3.2 Development by Dow Chemical Co.

It is not clear what year Dow formally began membrane development. However, since Reid at the University of Florida was using Dow's semipermeable membranes in his experiments, it is thought that Dow had begun some kind of development before the L-S membrane was introduced. Since the patent application for RO membranes by Dow researcher Henry Mahon was filed on September 19, 1960, just after the announcement of the L-S membrane on August 23, 1960, it is clear that Dow had been developing the membrane from early on.

According to the company's patent information, Mahon and Earl Mclain were the two people behind the early research. 14 patents related to RO membranes were filed by Dow in the United States in the 1960s, five were filed with Mclain's name and four were filed with Mahon's name. Two of these patents were co-invented by Mclain and Mahon.[26]

In 1960, Mahon filed a patent application for a hollow fiber RO membrane based on cellulose acetate.[27] Mahon and Mclain developed a hollow fiber RO membrane with an outer diameter of 100 μm and a thickness of 2–10 μm using cellulose triacetate as the material.[28] Later, Mclain became involved in the development of artificial kidneys.

After falling behind Hydranautics and Fluid in the Yuma project, Dow decided to make a comeback by developing membranes that can work at the lower pressure. The company developed Dowex, a low-pressure RO membrane that could desalinate water at about half the pressure of conventional RO membranes, and successfully installed it in a brine desalination plant in Venice, Florida, in 1981. However, when Cadotte's '344 patent was announced in the same year, the polyamide-based flat membranes showed rapid progress in lowering the operating pressure of membranes.

---

[24] Cadotte, who made a great contribution to the history of RO membranes, retired from Dow in 1994 and passed away in 2005.

[25] http://www.fivecitieswater.com/Well_Water_Treatment/0803Executive_Insight.pdf (Accessed June 3, 2018). Note that GE's water treatment division was also sold to Suez in 2017.

[26] Found using Google Patents with the search term reverse osmosis and Dow Chemical as the assignee (search date February 12, 2017).

[27] US3228876 (Permeability separatory apparatus, permeability separatory membrane element, method of making the same and process utilizing the same: Application date September 19, 1960).

[28] Lonsdale (1982), p. 107, Nagasawa and Takizawa (1975), p. 18.

Dow, which had been improving cellulose acetate hollow fiber membranes, found itself in a difficult situation.

Even though Dow saw the limitations of developing cellulose acetate hollow fiber membranes, it was already too late to develop polyamide flat membranes on its own with FilmTec's '344 patent standing in the way. So, in 1985, Dow acquired the '344 patent by purchasing FilmTec for $75 million, and decided to move into polyamide flat membrane. Since polyamide flat membrane was a new area for Dow, it respected and preserved FilmTec's existing R&D activities. As a result, unlike ROGA, which experienced an exodus of personnel with each acquisition, FilmTec did not experience a series of spin-offs.

In 1990, Dow filed a patent dispute against Nitto Denko, which had acquired Hydranautics in 1987. It was no exaggeration to say that Nitto Denko's business as well as the businesses of other companies were in danger of extinction, because if Dow's claim was accepted, they would not be able to manufacture and sell similar polyamide composite membranes.

The patent dispute between Dow and Nitto Denko will be discussed in Chap. 9, but the bottom line is that Dow did not win this dispute because the '344 patent was recognized as the property of the U.S. government. Dow could not prohibit competitors from using it. As a result, each company used the '344 patent to compete homogeneously, and the competition in the polyamide composite membrane group remains fierce to this day.

In 1994, FilmTec under the Dow umbrella made the decision to end its primary R&D activities, leaving only production-related R&D support. In June of that year, Cadotte left the company to become a part-time contract consultant.[29] Dow continued to lead the global market for RO membranes in the face of fierce competition. It is believed that this is due to the company's ability to use durability and reliability as its basic weapons to gain a competitive advantage not only in the seawater desalination market, but also in a wide range of other applications including urban water for residential use and industrial water for power plants.

5.4 Market Development by DuPont

5.4.1 The Dawn of Development

The early RO market was led by DuPont, which is headquartered in Wilmington, Delaware, on the East Coast, far from San Diego and Minneapolis.

The exact date that DuPont began research and development of membranes is not known, but Herbert Hoehn, who was responsible for research and development at DuPont, recorded development as beginning in 1962 (Hoehn, 1985).

[29] Water Desalination Report (1994), Vol. 30, Issue 19. Eriksson, who had moved from FilmTec to Dow as a result of the acquisition, moved in 1995 to DSI, which was founded by Bray after he left ROGA.

However, DuPont's cellulose acetate membranes were used in desalination experiments conducted by Reid and Breton at the University of Florida in the late 1950s, and Breton joined DuPont after writing up his doctoral dissertation.[30] These facts, whether they were official or not, suggest that DuPont had been developing semipermeable membranes from the early days, showing interest in RO membranes. It is considered that DuPont, like Dow, had been developing membranes of some kind before the advent of L-S membranes, and that development officially began in 1962, partly due to the influence of L-S membranes.

The main application market for RO membranes that DuPont envisioned in the early stages of development was gas separation, and Hoehn notes that the R&D and commercialization of a gas separation membrane (B-1) was approved internally in 1964. At that time, the company already had the capability to procure a variety of polymers and melt-spin them into hollow fiber membranes with an outer diameter of less than 100 μm and a thickness of less than 25 μm. Hoehn wrote of DuPont's development of membranes at the time, "Enthusiasm was sufficiently high that hollow fiber membranes for water desalination were also included in the venture," indicating that the company at least envisioned desalination as a future project.[31]

DuPont was the first company in the world to successfully synthesize nylon in 1935 and has since developed a wide range of synthetic fiber businesses. Irving Moch, who joined DuPont in 1954 and was in charge of membrane development, wrote,

> In the 1960s DuPont, utilizing its nylon chemistry knowledge, discovered that polyamides possessed unique properties for the separation of inorganic salts from water. The Corporation was also deeply involved in textile fiber technology and so it was natural that its pioneering research efforts were directed towards the utilization of both of these technical innovations.[32]

Initially, DuPont's development team used nylon 6 (an aliphatic polyamide) as the material for membrane development. However, as they proceeded to develop membranes using nylon 6 under the codenames B-1, B-2, B-3, and B-4, they were confronted with the insurmountable disadvantage of short life span and sought to switch materials. As a result, the development team focused on aramid (aromatic polyamide) as a new material and developed the codename B-5.

---

[30] Breton (1957b), p. 10. Breton completed his doctoral dissertation in 1957. Later, at the time Reid and Breton (1959) was published, Breton's address was written as E. I. du Pont de Nemours & Co., Wilmington, Delaware.

[31] Hoehn (1985), p. 83.

[32] Moch (1989), p. 172.

5.4.2 Development of Polyamide Hollow Fiber Membranes

In May 1967, DuPont presented Permasep, an RO membrane based on B-5, at the International Conference on Water for Peace sponsored by the U.S. government.[33] The Permasep was big news and surprised the conference participants who were visiting from Japan. In July, two months after the conference, the Japanese Science and Technology Agency's Natural Resources Research Council reported as follows in its report on the development of desalination technology.

> The most notable one is Du Pont's, which is a systematic device developed by itself, unlike the products of other companies . . . DuPont's method uses completely synthetic material for its semi-permeable material, so there is immense room for improvement, and future development will be an extremely noteworthy arena.

The impact of DuPont's announcement on the industry was so great that it prompted Toray to start developing RO membranes, as described in Chap. 7. At the time of the announcement, DuPont was already developing a new RO membrane, B-9, using a polyamide material. This was originally developed for use in gas separation applications.[34] However, the situation changed when William Richter, a colleague of Hoehn's, discovered that it was also effective for brine desalination and seawater desalination. Development using aromatic polyamide was promoted intensively[35] and a patent application was filed on October 11, 1967.[36]

In 1969, DuPont introduced a hollow fiber RO membrane (B-9) made of aromatic polyamide as a Permasep for brine (Moch, 1989). In field tests in Plains, Texas, B-9 not only performed better than B-5, but also reduced the cost of desalination per 1,000 gallons (about 3.78 m^3) of brine from \$1.1 to \$0.65.[37]

After getting a good response from B-9, DuPont aggressively expanded its business into desalination systems with industrial applications. In December 1973, DuPont delivered B-9 to a Siemens semiconductor plant near Regensburg, Germany, for use in the production of ultrapure water. In December of the same year, the company delivered 8-inch RO membrane elements to Tewa works in Poland. And

---

[33] Aoki and Takeyama (1970) from Industrial Materials Laboratory, Toray Fibers & Textiles Research Laboratories cited DuPont's "A Research Development in the Field Test Stage" No. A-54545 (1967) and A-54557 (1967), and mention Permasep. As described in Chap. 7, Permasep became known to Toray in 1967. In the chronology of the history of polymer science published on the web by the Society of Polymer Science in Japan, DuPont's development of the hollow fiber RO membrane in 1967 is also listed (http://main.spsj.or.jp/nenpyo/1966-1967.htm: Accessed February 18, 2015). At the conference where DuPont presented B-5, companies such as GA, Aerojet, Westinghouse, and Havens Industries also exhibited their desalination equipment.

[34] Petersen (1986), pp. 131–132.

[35] For the work of Hoehn and Richter, see also the following patent filed in 1969: US3567632: Permselective, aromatic, nitrogen-containing polymeric membranes.

[36] A continuation application for this patent was filed in 1969 and patented in 1970 as US3551331A. The inventors were Lawrence Anthony Cescon and Harvey Herbert Hoehn.

[37] Glover (1972), p. 364.The cost of desalination includes operating costs (power, chemicals, labor, and maintenance) as well as capital depreciation.

the following year, in September 1974, it delivered 8-inch RO membrane elements to the IBM Mainz plant in Germany for use in cleaning small electronic components.[38]

While B-9 was being developed for a variety of applications, DuPont announced B-10 RO membrane for seawater desalination at the International Seawater Desalination Symposium in September 1973. B-10 was newly developed by improving B-9. The performance of B-10 was high, with a desalination rate of 98% at a maximum operating pressure of 800 psi, compared with 90–95% at a maximum pressure of 400 psi for B-9. The outer diameter of the hollow fiber membrane was only 98 μm, so the filling efficiency of the membrane element was also increased, and the desalination capacity of 1500 GPD (5.68 m^3/day) was achieved. This performance was superior to that of the cellulose acetate RO membrane HR4160 series sold by ROGA at that time.[39]

B-10, which has both high desalination performance and water permeability, was such an excellent RO membrane that it could reduce the desalination process from two stages to one. The introduction of B-10 made it possible to achieve a single-stage desalination process, which led to the prospect of energy saving and cost reduction in seawater desalination, and significantly increased the practical potential of RO method for seawater desalination.[40] With the introduction of B-9 and B-10, polyamide materials were added to a market dominated by cellulose acetate materials as a promising alternative.

In October 1973, just after the release of B-10, the Fourth Arab–Israeli War broke out, causing an oil crisis. Attention began to focus on RO method instead of evaporation that consumes large amounts of crude oil. The market for RO membranes for seawater desalination finally entered its introductory phase, creating an excellent business opportunity for DuPont. DuPont launched the 4-inch and 8-inch Permasep for seawater desalination in 1974 and 1977, respectively, and quickly succeeded in developing the market.[41]

Figure 5.1 shows the cumulative installed capacity of seawater desalination plants and the market share of each company from the mid-1970s, when the market was

---

[38] A series of implementation results are based on Shields (1979).

[39] Field test results are presented in Environmental Protection Agency, Office of Research and Development, Industrial Environmental Research Laboratory (1978), p. 84.

[40] The introduction of B-10 certainly advanced the use of RO membranes in seawater desalination, but it still took some time before single-stage desalination systems became widely used in desalination. In the first demonstration tests conducted by the Water Reuse Promotion Center in Japan in the 1970s, single-stage desalination systems using membranes from UOP, DuPont, and Dow were tested. As a result, Kunisada and Hirai (1978) reported that the performance of UOP and DuPont membranes was relatively positive, but that of Dow membranes was not so positive, and that, "It may be said that there are still some concerns that the single-stage method cannot be used with confidence at the present technological level." Senoo and Kimura (1983) involved in the demonstration tests wrote in the first tests "We had no choice but to conduct the experiment using foreign modules that claimed to be capable of single-stage desalination of seawater." However, regarding the second tests, they wrote "What was conspicuous was the rapid improvement in the performance of Toyobo's hollow fibers. In 1978, a membrane capable of single-stage desalination was completed."

[41] Moch (1989), p. 171, Hoehn (1985), p. 82.

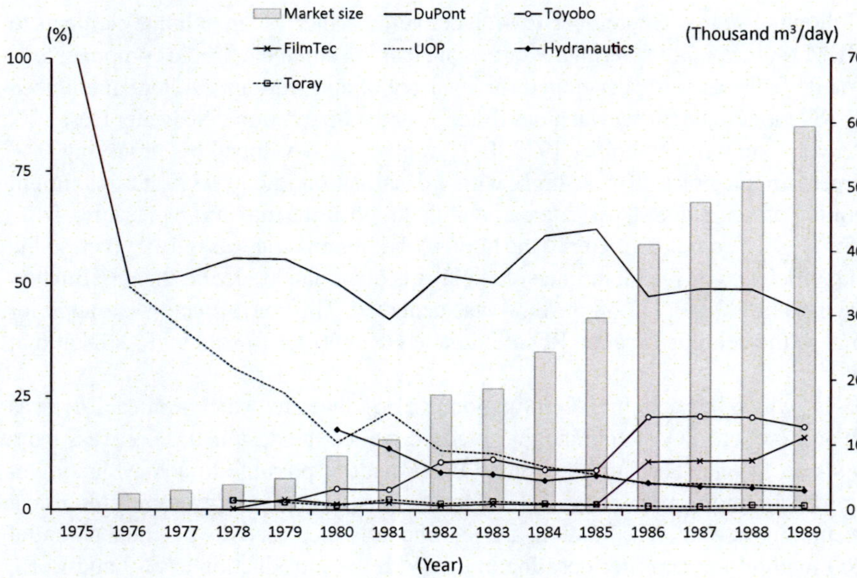

Fig. 5.1 Market size and share of RO membrane systems for seawater desalination. *Source* Prepared by the authors based on Kumano (1992). *Note* Cumulative facility capacity data. The left axis shows the line graphs and the right axis shows the bar graph

introduced, to the 1980s. DuPont, which entered the market early, maintained a share of nearly 50% in the expanding market. This indicates that the company has been a driving force behind the expansion of the seawater desalination market. On the other hand, UOP, which owns Fluid, gradually lost its market share.

Moch, who was deeply involved in the development of B-10 and who contributed to the expansion of the business, became the head of technical development for RO membranes in 1978. He held this position for 18 years until his retirement in 1996.

As described above, DuPont's development of RO membranes has evolved from gas separation to desalination applications. In the desalination application, DuPont has progressively developed products for various applications, including water purification, brine desalination, industrial applications, and seawater desalination. It is interesting to note that DuPont, which did not necessarily aim to desalinate seawater, opened up the market for RO membranes for seawater desalination ahead of its competitors through gradual technological innovation.

5.4.3 Expansion to the Middle East

DuPont immediately set about embarking on a foray into the Middle Eastern market. In 1974, Saudi Arabia established the SWCC and began planning the construction of seawater desalination plants. The following year, in 1975, the Ministry of Agriculture,

Forestry and Fisheries of Saudi Arabia launched a project to desalinate groundwater in a 1,400-m underground aquifer by the RO method (Rovel & Daniel, 1987). Due to the oil crisis, interest in the RO method as a new desalination method was increasing in Saudi Arabia. DuPont was the first company to take advantage of this trend.

According to Moch, by 1989, Saudi Arabia had the largest number of B-9 installations, with 17 plants. Following them were Algeria (5 plants), the United States (4 plants), and Iraq (3 plants).[42] The Moch's list also includes an unnamed "Gulf region" plants with a capacity of 151,400m^3/day. It implies that DuPont's target was the Middle East rather than the United States. In fact, the DuPont Permasep was used in five of the eight large-scale brine desalination plants with a capacity of more than 10,000 m^3/day using the RO method, which were constructed in Saudi Arabia in the 1970s and 1980s (Al-Mutaz, 1996).

However, DuPont's RO business did not expand. In 1991, DuPont sold its flat membrane RO business to Trisep, headquartered outside Los Angeles, and focused on the hollow fiber membranes business. The following year, DuPont launched B10-Twin, and it looked as if the company was going to focus more on expanding its hollow fiber business. However, the mainstream of the market had shifted to the flat membrane type, as Dow acquired FilmTec and switched to the polyamide flat membrane type. This led to fierce price competition in the polyamide flat membrane group. And that also worsened the business environment for DuPont, which had been competing in the high price range in a different group. In addition, the company faced technological problems, such as leaks in RO membrane modules. Finally, in 1997, DuPont decided to withdraw from the market and dissolved the Permasep group in 1999. The company then decided to close its manufacturing sites in Waynesboro and Glasgow by 2004.

5.5 Summary

This chapter clarifies the process by which various U.S. companies fully developed and commercialized RO membranes after the emergence of L-S membranes. DuPont, a major chemical manufacturer, was the driving force behind the initial market expansion. The company developed polyamide-based hollow fiber RO membranes based on its own technology and deployed them in the brine desalination and seawater desalination markets, primarily in the Middle East.

In the early days of the market, hollow fiber RO membranes were the mainstream, but Cadotte's groundbreaking invention changed the situation. As a result, the market shifted from the hollow fiber type to flat membrane type. Dow, which had been operating with cellulose acetate hollow fiber membranes, acquired FilmTec and obtained the '344 patent, which enabled it to switch to flat membrane composite membranes.

Although ROGA, which had been developing the technology in San Diego, made a significant contribution to the development of key technologies for practical use such

[42] Moch (1989), p. 178.

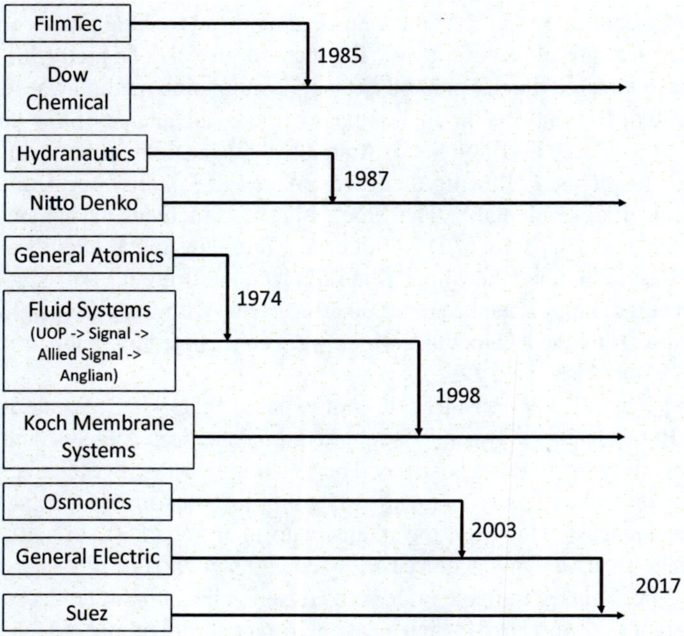

Fig. 5.2 U.S. Companies in transition. *Source* Prepared by the authors

as composite membranes and spiral elements, it did not become a major player in the market because of the outflow of talented personnel fueled by repeated acquisitions. Thus, the players that make up the RO membrane industry in the U.S. have been changing through many mergers and acquisitions as summarized in Fig. 5.2.

As described throughout Part 2 up to this point, U.S. universities, research institutes, and private companies were the pioneers in the development and commercialization of RO membranes. However, among the U.S. companies, Dow is the only one that got a high market share. Japanese companies such as Toray, Nitto Denko, and Toyobo replaced many American companies and maintained stable operations over the long term.

In the following Part 3, we will clarify the development activities and business development process of RO membranes by these Japanese companies.

References

Al-Mutaz, I. S. (1996). A comparative study of RO and MSF desalination plants. *Desalination, 106*, 99–106.

Aoki, M., & Takeyama, T. (1970). "(2) Sangyō shizai" ((2) Industrial Materials). *Sen-i Kōgaku (Fiber Engineering), 23*, 21–26.

Bickell, L. (1999). *Colorado river basin salinity control project*.

Breton, E. J. (1957b). Water and ion flow through imperfect osmotic membranes. *Research and Development Progress Report*, No. 16. Retrieved from http://babel.hathitrust.org/cgi/pt?id=mdp.39015078499574;view=1up;seq=3.

Cadotte, J. E. (1985a). Development of composite reverse osmosis membranes in retrospect. *Membrane, 10*, 117–118.

Cadotte, J. E. (1985b). Evolution of composite reverse osmosis membranes. In D. R. Lloyd (Ed.), *Materials science of synthetic membranes* (pp. 273–294). American Chemical Society.

Cadotte, J. E., & Petersen, R. J. (1981). Thin-film composite reverse-osmosis membranes: origin, development, and recent advances. In A. F. Turbak (Ed.), *Synthetic membranes: Based on the 20th anniversary symposium honoring Drs. Loeb and Sourirajan vol. 1 desalination, American chemical society* (pp. 305–326).

Cadotte, J. E., Petersen, R. J., Larson, R. E., & Erickson, E. E. (1980). New thin-film composite seawater reverse osmosis membrane. *Desalination, 32*, 25–31.

Capone P. (2008). Living in interesting times: Donald T. Bray. ***Water Conditioning & Purification,*** *50*. Retrieved June 3, 2018, from http://www.fivecitieswater.com/Well_Water_Treatment/0803Executive_Insight.pdf.

Environmental Protection Agency, Office of Research and Development, Industrial Environmental Research Laboratory. (1978). *Assessment of best available technology economically achievable for synthetic rubber manufacturing wastewater* (vol. 1). Cincinnati.

Glover, R. M. (1972). Hollow fiber ni yoru gyakushintō hō" (Reverse osmosis with hollow fiber). *Nihon Kaisui Gakkai shi* (*Bulletin of the Society of Sea Water Science, Japan*), *25*, 363–365.

Hoehn, H. H. (1985). Aromatic polyamide membranes. *Lloyd, D. R. ed. Materials Science of Synthetic Membranes, ACS Symposium Series, 269*, 81–98.

Kamisawa, C. (1980). Advanced membrane desalination technology and newly developed membrane for desalination. *Membrane, 5*, 348–356.

Kumano, A. (1992). Recent trends in water desalination technology by reverse osmosis. *Journal of the Society of Fiber Science and Technology Japan, 48*, 28–34.

Kunisada, Y., & Hirai, M. (1978). Shō energy gata kaisui tansuika gijutsu kaihatsu sono 1 (Development of energy-saving seawater desalination technology, Part 1). *Journal of Water Re-use Technology, 4*, 39–49.

Lonsdale, H. K. (1982). The growth of membrane technology. *Journal of Membrane Science, 10*, 81–181.

Mattson, M. E. (1979). Significant development in membrane desalination: 1979. *Desalination, 28*, 207–223.

Merton, U. (1970). Makuhō ni yoru datsuen (Desalination by the membrane method). *Bulletin of the Society of Sea Water Science Japan, 23*, 206–209.

Moch, I., Jr. (1989). A twenty year case history: B-9 hollow fiber permeator. *Desalination, 74*, 171–181.

Nagasawa, M., & Takizawa, A. (1975). *Mizu shori no kōbunshi kagaku to gijutsu (jō): Kōbunshi maku (Polymer science and technology of water treatment (1): Polymer membrane)*. Tokyo: Chijin Shokan.

Okazaki, M., & Kimura, S. (1983). Recent advances of reverse osmosis technology for seawater desalination. *Nihon Kaisui Gakkaishi (Bulletin of the Society of Sea Water Science, Japan), 37*, 149–158.

Petersen, R. J. (1986). Membranes for desalination. In M. B. Chenoweth (Ed.), *Synthetic membranes: MMI press symposium series, 5, MMI Press* (pp. 129–154).

Reid, C. E., & Breton, E. J. (1959). Water and ion flow across cellulosic membranes. *Journal of Applied Polymer Science, 1*, 133–143.

Riley, R. L., Fox, R. L., Lyons, C. R., Milstead, C. E., Seroy, M. W., & Tagami, M. (1976). Spiral-wound poly (ether/amide) thin film composite membrane systems. *Desalination, 19*, 113–126.

Rovel, J. M., & Daniel, L. (1987). Start-up of R.O. modules after a very long storage time: A case history. *Desalination, 65*, 373–379.

Senoo, M., & Kimura, S. (1983). *Shinkinō zairyō "Maku": Seitai maku no kinō o mezashite (New functional material "membrane": Toward the function of biomembranes)*. Tokyo: Kogyo Chosakai.

Shields, C. P. (1979). Five year's experience with reverse osmosis systems using Du Pont 'Permasep' permeators. *Desalination, 28*, 157–179.

Spiegler, K. S., & Liard, A. D. D. (Eds.). (1980). *Principles of desalination (Part A and Part B)* (2nd ed.). Academic Press.

Taniguchi, Y. (1981). Desalting technique applications in water treatment processes. *Sen'i Gakkaishi (Journal of Fiber Science and Technology), 37*, 23–30.

Wada, Y. (2004). *Zōsui no Gijutsu (zōho-ban) (Technology of creating water (enlarged edition))*. Tokyo: Chijin Shokan.

Water Desalination Report. (1994). *30*, Issue 19.

Part III
The Development of RO Membranes in Japan

Chapter 6
The Rise of Japanese Companies

Abstract In this chapter, we provide an overview of RO membrane development in Japan. We first describe the activities of Tokyo Kōgyō Shikenjo (the Government Chemical Industrial Research Institute, Tokyo) and Zōsui Sokushin Center (the Water Reuse Promotion Center) in the public sector. For the private sector, we summarize the development efforts of three major companies: Toray, Toyobo, and Nitto Denko. We also clarify that a strong demand for ultrapure water production in the growing semiconductor industry enabled these companies to continue their development activities.

6.1 Introduction

The development of RO membranes in the United States, which began in response to water shortages on the West Coast, received national support when the OSW was established in 1955. Development toward practical application subsequently accelerated in the 1960s when the L-S membrane was introduced. This trend also reached Japan. At the private level, the Nagasaki Research Institute of Mitsubishi Heavy Industries began investigating the RO method in 1965, and in 1969 planned to construct a seawater desalination plant using the RO method with L-S membranes (Suzuki, 1972).

Japan has an abundance of rainwater, so it was not necessarily expected that there would be a large demand for seawater desalination. However, many Japanese companies started to develop RO membranes to respond to the increasing global demand for water, to follow the lead of American companies, and to diversify their own businesses. Toray, Nitto Denko, and Toyobo, which accumulated technological capabilities throughout the 1970s and 1980s and commercialized membranes mainly for industrial applications, grew their membrane businesses in line with the expansion of the seawater desalination market from the late 1990s onward.

As seen in Chap. 2, Dow, Toray, and Nitto Denko have a high share of the global RO membrane market. Although Toyobo possesses only about 5% of the total market, it has a high share in the Middle East, especially in Saudi Arabia. How did Japanese

M. Fujiwara and Y. Aoshima, *Mechanisms for Long-Term Innovation*,
Advances in Japanese Business and Economics 31,
https://doi.org/10.1007/978-981-19-4896-1_6

companies that were latecomers to the market, manage to catch up with and overtake the leading U.S. companies? Until the latter half of the 1990s when the desalination market began to grow substantially, the RO membrane business was not contributing to company's overall sales and profits. How was such a business recognized and nurtured, and then blossomed in the companies?

In the next and subsequent chapters, histories of development and commercialization in Toray, Nitto Denko, and Toyobo will be described in detail, respectively. This chapter provides a brief overview of those histories of Japanese companies.

6.2 Market Entry by Japanese Companies

In the 1970s, the development of RO membranes in the United States moved from the research stage to the commercialization stage. DuPont's polyamide hollow fiber membranes went on the market in 1969, and cellulose acetate flat membranes developed by ROGA began full-scale sales under Gulf Oil around the same time. In Japan, these overseas-made membranes were initially imported, systematized, and sold. However, from the mid-1970s onward, the performance of Japanese-made membrane modules improved and gradually came to be produced domestically.

As a result of early development efforts, five chemical and textile companies, namely Daicel Chemical Industries (now Daicel), Nitto Electric Industrial Co. (now Nitto Denko Corporation), Toray Industries, Toyobo, and Teijin, began to sell their own RO membranes in the mid-1970s, as shown in Table 6.1. In addition to these companies, other chemical and textile companies such as Asahi Kasei, Mitsubishi Rayon, and Kurashiki Boseki also entered the market with MF and UF membranes, which are coarser than RO membranes. Asahi Kasei has maintained a high share of the market for UF membranes used in the production of ultrapure water for semiconductors, a market that went on to expand later.[1]

Table 6.1 Japanese company participation in the RO membrane business

Company name	Type	Material	Commenced selling
Daicel Chemical Industries	Tubular	Cellulose acetate	1975
Nitto Electric Industrial Co. (Nitto Denko)	Tubular	Cellulose acetate	1979
Toray Industries	Flat sheet	Cellulose acetate	1976
Teijin	Tubular	Polybenzimidazolone polymer (PBIL)	1979
Toyobo	Hollow fiber	Cellulose acetate	1976

Source Prepared by the authors based on Sōgō Hōsō Publication (1980), p. 36
Note The company names are the names used at the target time

[1] Nihon Keizai Shimbun (June 23, 1984, p. 6).

The first company to commercialize RO membranes in Japan was Daicel. Daicel succeeded in developing a cellulose acetate tubular type RO membrane in 1974 and began to sell it under the "MOLSEP" brand in 1975 by collaborating with Mitsui Engineering & Shipbuilding. In the early 1980s, Daicel developed hollow fiber membranes with the seawater desalination market in mind,[2] but later developed food applications, including efforts such as establishing a technology for concentrating tomato juice using RO membranes in collaboration with Kagome in 1984.[3] In 1985, Daicel commercialized an RO winemaking system and delivered it to winemakers in Japan and South Korea.[4] In 1994, Daicel established Daicen Membrane Systems as a three-way joint venture with Central Filter Industries and Central Maintenance, and established a system to provide integrated services from the development of membrane materials to systems and maintenance.[5] However, rather than focusing on the seawater desalination market as the main battleground for RO membranes, Daicel has been developing its UF membranes and other products for applications such as pure water and ultra-pure water production, recovery of valuable materials, and wastewater treatment.

In 1977, Teijin developed a unique RO membrane using a polybenzimidazolone polymer (PBIL), and proceeded with the initial business development by applying it to metal surface treatment, small marine desalination equipment, and fruit juice concentration equipment.[6] Although PBIL showed high separation performance, it did not become widespread in semiconductor manufacturing applications and seawater desalination, which are the mainstream markets for RO membranes, mainly due to mass production cost problems.

Figure 6.1 shows the number of domestic patent applications related to RO membranes filed by Japanese nationals and the number of patents applications filed by Daicel and Teijin.[7] The former increased in the 1970s, reaching its first peak in 1977, indicating that technological development aimed at domestic production of overseas technologies accelerated during this period. Although Daicel and Teijin led the development of RO membranes at the dawn of RO membrane history in Japan, their presence gradually weakened.

In contrast to these companies, Toray, Nitto Denko, and Toyobo have been developing products for a long time. These companies have also been competing with foreign companies in desalination applications that require the most stringent performance in many aspects, including desalination rate and water permeability.

---

[2] Nikkei Sangyō Shimbun (July 10, 1982, p. 1).

[3] Nikkei Sangyō Shimbun (March 10, 1984, p. 9).

[4] Nikkei Sangyō Shimbun (August 26, 1986, p. 18; August 10, 1985, p. 9).

[5] Chemical Daily (August 18, 1995, p. 1).

[6] Nikkei Sangyō Shimbun (April 5, 1980, p. 8; November 25, 1977, p. 8), Nihon Keizai Shimbun (May 9, 1978, p. 8).

[7] The number of patent applications filed by Japanese nationals under the F-term theme code 4D006G/A03 (separation/reverse osmosis using semipermeable membranes). An F-term is a classification code assigned by the JPO to various characteristics of each patent.

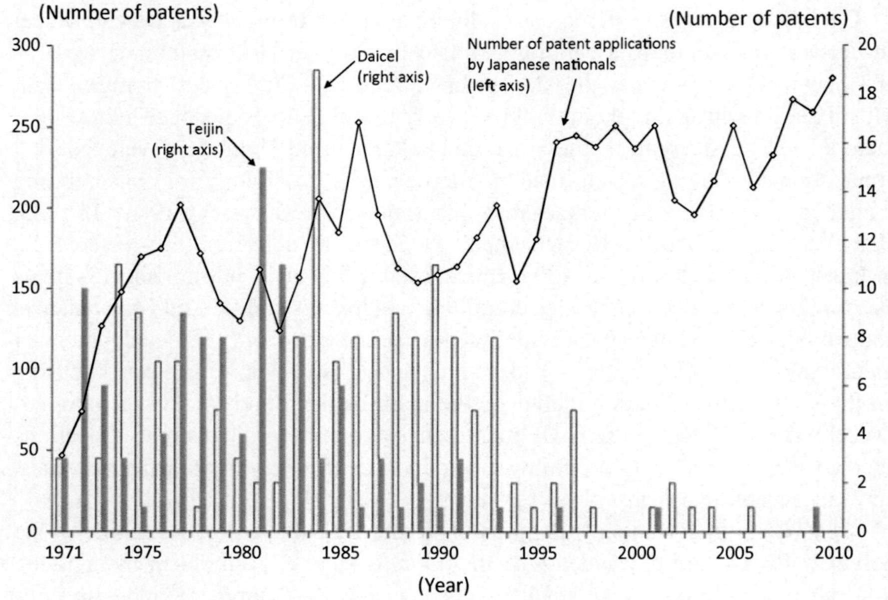

Fig. 6.1 Number of RO membrane patent applications by year. *Source* Prepared by the authors based on data from UltraPatent

6.3 Early RO Membrane Development by Three Major Companies

6.3.1 Toray Industries, Inc.

Toray was the earliest Japanese company to begin full-scale development of RO membranes.

In May 1967, DuPont presented its RO membrane Permasep at the "International Conference on Water for Peace" sponsored by the U.S. government. This triggered Toray to start development. The membranes announced by DuPont were hollow fiber synthetic membranes that showed commercial feasibility to textile companies whose core technology is synthetic chemistry. The fact that the membranes were already at a practical level was a major shock to the attendees, including those from Japan. In the same year, DuPont applied for a patent for aromatic polyamide RO membranes that would later become the company's core technology. And in 1969, DuPont launched Permasep B-9 for brine applications.

Toray, which had benchmarked DuPont as a role model, immediately reacted to these developments. Especially at that time, when it was becoming clear that the textile industry was in a structural recession, textile companies, including Toray, were aggressively developing new projects to diversify their businesses. For Toray, the development of RO membranes was not only a way to learn from DuPont, it was also a way to develop new businesses necessary for structural reform.

Toray RO membrane development officially began in 1968. Initially, the developers were divided into several divisions to explore the possibilities of various technologies, ranging from asymmetric cellulose acetate membranes to polyamide composite membranes. As for the membrane shape, a flat membrane type was chosen instead of a hollow fiber type to differentiate it from DuPont. They thought they could not beat DuPont by using the same method. Since DuPont's RO membranes had been developed to desalinate seawater from brine, Toray's development efforts from the beginning were aimed at realizing seawater desalination. Since the technical hurdles for seawater desalination applications were high, Toray adopted a strategy of commercializing cellulose acetate-based materials first, which were relatively easy to develop, to accumulate peripheral technologies there, and waited for the development of more difficult composite membranes. In line with this strategy, at the end of 1976, Toray succeeded in receiving an order for a cellulose acetate-based RO membrane module for wastewater treatment at IBM's Yasu Plant. Subsequently, the market for ultra-pure water production for semiconductors opened up significantly.

On the other hand, the development team searching for composite membranes as the main target developed PEC-1000, a polyether-based membrane with both desalination performance and water permeability at a high level. However, PEC-1000 remained in the laboratory stage and could not be commercialized because of its low resistance to chlorine and durability against dissolved oxygen.

6.3.2 Toyobo

Toyobo began developing RO membranes in 1971, about three years later than Toray. As it became apparent that the mainstay textile business was in a structural recession, Toyobo needed to quickly develop new businesses that could replace their textile business to ensure the long-term survival of the company. The development of RO membranes began in the midst of this company-wide trend. Initially, when the level of technology was low, the target was the concentration and recovery of valuable materials, such as the recovery of wool grease produced in the wool washing process for use in cosmetics. However, the final goal for the developers was to realize seawater desalination.

In promoting business development, Toyobo tried to effectively leverage technologies of the textile business. It thus focused on a hollow-fiber RO membrane using cellulose acetate materials from the beginning of development. However, in order to achieve both high desalination performance and water permeability in a hollow fiber membrane, it was necessary to create a homogeneous ultra-fine hollow structure with a diameter of 0.2 mm and an inner diameter of 0.1 mm. For this purpose, a series of tacit knowhow inherited by their skilled technicians contributed to the establishment of the advanced production technology.

While adopting a general-purpose material based on cellulose acetate, Toyobo devised an original method for bundling hollow fibers. In contrast to the method used by Dow and DuPont, in which hollow fibers are arranged in parallel and bundled, Toyobo adopted a method called "cross-winding", in which hollow fibers are woven in such a way that they cross each other. The hollow fiber membranes rubbed against each other to remove dirt, making the membrane elements less dirty and easier to clean. The hollow fiber RO membrane module developed in this way was named HOLLOSEP.

In 1977, HOLLOSEP was applied to the desalination demonstration test at the Water Reuse Promotion Center (WRPC) in Chigasaki, led by the Ministry of International Trade and Industry (MITI). In the second phase of the demonstration test that began in 1978, Toyobo adopted cellulose triacetate membranes that were still in the basic research phase. The cellulose triacetate membrane had the advantage of being able to withstand the high-pressure operation of seawater desalination. It was also expected to obtain a distinctive market position due to special knowhow required in controlling the spinning process. Toyobo has continued to use these cellulose triacetate-based materials up to the present.

In parallel with these experiments, in November 1978, Toyobo established the RO Business Development Department and stepped up its efforts to commercialize RO membranes. In October 1979, it announced the construction of a new RO membrane plant in Iwakuni with an investment of approximately 300 million yen, and production began in May 1980. Initially, the company developed its business in small seawater desalination systems for use on ships. And in 1983 it established the Functional Membrane Division. Although the product was well received in shipboard applications, it was still a small-scale business.

6.3.3 Nitto Denko Corporation

Research into RO membranes officially began at Nitto Denko in 1973. Originally, Nitto Denko supplied parts for the pressure vessels that contain RO membranes to Kobe Steel. However, when Kobe Steel withdrew from water treatment, Nitto Denko took over the production of RO membranes.

Saburo Hijikata, who became president in 1974, set a company-wide policy to establish three new business pillars: electronics, medical care, and membranes, and aimed to establish new businesses. The RO membrane business was launched under this company-wide policy and was supported by the vision of top management.

Nitto Denko initially paid a license fee to Kobe Steel to produce cellulose acetate tubular RO membrane modules. The first product was delivered to Hokkaido in 1976 for the treatment of potato starch effluent. In the same year, Nitto Denko participated in a seawater desalination demonstration test conducted by the MITI. However, Nitto Denko's tubular RO membranes were clearly inferior in volumetric efficiency to those of other companies. Thus, after a year they were no longer included in the experiment. This experience prompted Nitto Denko to embark on the development of a new membrane module to replace the tubular type. Since Nitto Denko had film technology, it proceeded with the development of a flat membrane. A variety of materials were considered, including those from overseas, without being bound by the company's own framework.

Unlike Toray and Toyobo, seawater desalination was not necessarily the main target for the development team at Nitto Denko. For Nitto Denko, which advocated itself as a "department store of membranes", RO membranes for seawater desalination were just one of many kinds. They conducted an extensive search by looking at a wide range of applications.

In 1979, the R&D organizations that had been dispersed throughout the company were integrated into a single location, and a company-wide membrane development project was launched. In 1980, the company established the Membrane Module Development Department within the Technical Research Laboratory. And in the same year, it succeeded in developing composite membranes NTR-7197 and NTR-7199 using polyurea-based materials.

In 1981, the Membrane Module Development Department was spun off as the Membrane Module Business Promotion Department, and the developers continued to explore a variety of applications.

6.4 National Projects for Seawater Desalination

6.4.1 Efforts of the Government Chemical Industrial Research Institute, Tokyo

As in the United States, public funds have long been invested in the development of seawater desalination technology in Japan to address anticipated future water shortages. In particular, the Tokyo Kōgyō Shikenjo (Government Chemical Industrial Research Institute, Tokyo (GCIRIT)) (established in 1900) of the Sangyō Gijutsu Sōgō Kenkyujo (Agency of Industrial Science and Technology (AIST)) under the MITI began research on seawater desalination in April 1961 to secure water to meet the needs of industrialization (GCIRIT, 1971).

Table 6.2 shows the special research themes and the allocated budget in the early years of the GCIRIT. In 1963, 13 million yen was spent on this theme, indicating high expectations for this project. In addition to these special research projects, it had a budget for research related to seawater desalination as a regular research theme.

In 1964, the U.S.-Japan Conference on Development and Utilization of Natural Resources (UJNR) was established, and desalination was taken up as one of the panel topics. Three GCIRIT members joined this panel, as well as each one from Japan Tobacco and Salt Corporation and the Government Chemical Industrial Research Institute, Shikoku. Since then, information on desalination research was exchanged between the U.S. and Japan once a year. The GCIRIT absorbed knowledge of the

Table 6.2 Early special research at the GCIRIT

Year	Research theme	Allocated budget (million yen)
April 1961–March 1966	Research on production of industrial water from saline water	5
1963	Research on production of industrial water from saline water	13
1964	Research on the desalination of saline water	4.5
1965	Research on the desalination of saline water	4.7
1966	Research on the desalination of saline water	4.7

Source AIST

RO method through these information exchanges, and in 1967 began research on this method as part of chemical process engineering.[8]

Around this time, the AIST launched the National Research and Development Program of MITI (Large-Scale Project) as a national industrial policy. A research project on "Seawater Desalination and Byproduct Utilization" was selected as one of the targets of this program, and the research started in 1969 with a total budget of about five billion yen for seven years. The purpose of the project was to address future water shortages in cities. In the first year of the project, plant tests were commissioned to private companies such as Sasakura Machinery Works, Ishikawajima-Harima Heavy Industries (IHI), Hitachi, Japan Gasoline Co. (JGC), and Mitsubishi Heavy Industries.[9] In 1970, the Waterfront Research Institute for Seawater Desalination was established in Chigasaki under AIST, and a large-scale seawater desalination plant using the evaporation method (multi-stage flash) was constructed (Yarisawa, 1970). The large-scale project on "Seawater Desalination and Byproduct Utilization" set an ambitious goal of realizing seawater desalination at a cost of only about 30 yen/m^3. The project period was extended by one year until FY1976. A total of about 6.7 billion yen was invested in the project.[10]

6.4.2 *Demonstration at the Water Reuse Promotion Center*

The oil crisis occurred in 1973 brought a growing awareness that energy-saving desalination technology was needed instead of the oil-intensive evaporation method. In response to this trend, the WRPC was established in 1974. It took over the activities of the aforementioned Waterfront Research Institute, which already existed in Chigasaki, to develop and test seawater desalination technology using the RO

[8] Tokyo Kōgyō Shikenjo (The Government Chemical Industrial Research Institute, Tokyo) (1975).

[9] Kōgyō Gijutsuin (Agency of Industrial Science and Technology) (1969) pp. 68–70.

[10] If we include the "Study on the Total System of Desalination" conducted in 1977 according to Goto (1979), the implementation period is one year longer.

method. The demonstration test in Chigasaki promoted by the WRPC was, in effect, a successor to the large-scale project. The first phase of the basic experiment was from June 1975 to March 1977. The second phase of the demonstration test began in 1978.

In the initial stage of the first phase, the performance of foreign-made RO membranes, such as UOP's PA-300 (composite flat membranes), DuPont's Permasep B-10 (hollow fiber membranes), and Dow's XFS-4167.08 (hollow fiber membranes), was evaluated using a small-scale system with a desalination capacity of about 10 m^3/day. UOP and Dow membranes were cellulose acetate membranes while DuPont's ones were polyamide membranes.[11] The outer diameter of the membrane module was 2 inches for UOP, 4 inches for DuPont, and 8 inches for Dow. In this demonstration test, a single-stage desalination system was tested. Pretreatment technology and an energy recovery system were also developed. The operation results of the single-stage system were good for the UOP and DuPont membranes, but not for Dow (Kunisada & Hirai, 1978).

In the final year of the first phase, tests using Japanese RO membranes began. Toray's SC-5000A/SC-5000B (flat membranes) and Nitto Denko's NRO98-A/NRO98-B (tubular membranes) were used first, followed by Toyobo's HR5350/HR5350S (hollow fiber membranes) a year later. All three companies used cellulose acetate as the membrane material. The modules of Toray and Nitto Denko were 4 inches in diameter, and those of Toyobo were 5.5 or 6 inches in diameter. The performance advantage of the Toray and Toyobo membranes was confirmed in this test.

In the second phase beginning in 1978, a large-scale demonstration plant with a desalination capacity of 800 m^3/day was constructed to verify the economic feasibility of the plant in practical use. Toray's SC-5200 and Toyobo's HR8350/HR8650 were adopted. Both membrane modules were enlarged to 8-inch diameters. Tests were conducted by alternately switching the modules about once a month. Toray's membranes were used in a two-stage desalination system while Toyobo's membranes were used in a single-stage desalination system. Mitsuyoshi Hirai, who joined the WRPC in 1975 and was involved in the demonstration tests, looks back on the second phase of the demonstration tests as follows.

> In 1979, we decided to build a large plant, an 800 ton (per day) demonstration plant. We have operated this plant for 10 years and conducted demonstration tests. We first employed cellulose acetate hollow fiber and spirals alternately, each of which was adopted to 800 ton plant. In the middle of the test, (Toray's) spirals changed the material to polyamide. In the end, we switched plants to the 400 ton and 400 ton combination and ran them in parallel since we could not have enough demonstration time.[12]

The desalination capacity of the demonstration plant newly built in the second phase was the third largest in the world at that time. It is not easy for a single company to own such a large-scale demonstration facility in the early phase of industrial development. It is thus believed that the state-sponsored large-scale demonstration

---

[11] The empirical results of the first period are detailed in Kunisada and Hirai (1978).

[12] From the authors' interview with Mitsuyoshi Hirai at the WRPC, March 16, 2015.

tests had certain contributions to the RO membrane industry development in Japan. Through this demonstration test, Toyobo was able to find a way to commercialize cellulose triacetate membranes that had still been in the basic research stage. Nitto Denko, through the demonstration tests, recognized that the tubular type was not competitive and turned to the development of a new membrane. Toray was able to test the operational performance of PEC-1000.

6.5 Market Expansion in Industrial Applications: Ultrapure Water Production Applications for Semiconductors

6.5.1 The Rise of the Market for Semiconductors

In the 1970s, membrane companies developed several applications, including those for wastewater treatment, food, and small ships. But all of these markets were small or did not have a significant future. Despite the monumental and socially significant development goal of alleviating water shortages, it was not easy to secure profitability as a business.

The development of the semiconductor industry transformed these difficult conditions. The global semiconductor market began to take off in the mid-1970s. And from the mid-1980s onward, it grew rapidly as the miniaturization of semiconductors progressed. The production volume of Japanese semiconductor companies also expanded substantially, riding the wave of this growth.

As the miniaturization of semiconductors progressed rapidly, the purity requirements for the cleaning water used in the cleaning process also increased. As shown in Table 6.3 for example, in the 16 Kb era, 100 to 200 particles of 0.2 μm and up to 1000 ppb of Total Organic Carbon (TOC) were allowed per 1 mL. However, in the 1 Mb era, only 10 to 20 particles of 0.1 μm and 30 to 50 ppb of TOC were allowed per 1 mL. As the demand for high purity cleaning water increased, the production of ultrapure water by the RO method came into the limelight.

Toray and Nitto Denko succeeded in capturing this application of ultrapure water production for semiconductors. Toray was the first to respond. Although it was not able to immediately market the composite membranes that it had been developing as its core product, it was able to meet the demand of the semiconductor industry with cellulose acetate membranes that it had developed earlier. Having established a stable business foundation in the semiconductor market, Toray was able to continue the development of new composite membranes following PEC-1000.

In 1984, when Toray was on the verge of developing PEC-2000, which would follow PEC-1000, it established the Membrane Business Division and set the ambitious goal of achieving sales of 10 billion yen in three years. Katsunosuke Maeda, then plant manager of the Ehime Plant and later president, saw the potential of RO membranes and lobbied to set up a mass production line at the Ehime Plant.

Table 6.3 Performance of semiconductor and required quality of water

	Specific electrical resistance	Particulates (per mL)						Microbial count	TOC	Silica	Dissolved oxygen
	(MΩcm)	0.2 µm	0.2–0.1 µm	0.1 µm	0.05 µm	0.03 µm	0.02 µm	(per L)	(ppb)	(ppb)	(ppb)
16 Kb	15–16	100–200						5,000	1,000	20–30	–
64 Kb	15–16		50–150					500–1,000	500–1,000	20–30	100–500
256 Kb	17–18			30–50				20–200	50–200	10	100
1 Mb	17.5–18			10–20				10–50	30–50	5	<50
4–16 Mb	>18			<5	<10			<10	<10	<1	<50
16–64 Mb	>18.1				<5			<1	<5	<1	<10
64–256 Mb	>18.2				<1	<10		<0.5	<2	<0.5	<5
256 Mb–1 Gb	>18.2					<5	<10	<0.1	<1	<0.1	<1

Source Prepared by the authors based on Wada (2004), p. 275 and p. 341

Nitto Denko, on the other hand, in the course of its efforts to develop a variety of applications, grasped the market for ultrapure water production in the semiconductor industry. Through repeated experiments, the developers accidentally created a membrane with inferior demineralization performance but superior ability to remove other substances and water permeability. Kurita Water Industries (Kurita), an engineering company, was impressed by the potential of this membrane for semiconductor manufacturing, and in 1983, Nitto Denko started selling it to Kurita. In 1984, it set a target of increasing sales of its membrane business, including RO membranes, to 6 billion yen by 1987, and established a Membrane Division and built the Shiga Plant in 1986.

Soon, however, things began to look bleak. Just as Nitto Denko's state-of-the-art Shiga Plant was about to start operations, the semiconductor recession hit, delaying the start of operations. At Toray, a vulnerability was found in PEC-2000 that had reached the sample shipment stage, and sales were abruptly halted. Whereas both companies were excited by the wave of semiconductor demand, it was not easy to expand their businesses continuously.

6.5.2 Search for New Solutions

Both companies needed a way out. Toray proceeded at a rapid pace with the development of new membranes in parallel with PEC-2000. It succeeded in developing UTC-70 in 1987, which was made of polyamide. This new membrane was delivered to Toshiba's 1 Mb DRAM plant, which at the time featured the most advanced technology. Since then, Toray's RO membranes for semiconductor manufacturing have switched from the cellulose acetate type to the polyamide type.

Reducing the cost of membranes is an important issue in ultrapure water production applications for semiconductors. For this, it is desirable to improve membrane performance under low-pressure operation to save the electricity cost of high-pressure pumps. Toray thus developed low-pressure membranes mainly for industrial applications, and succeeded in developing the ultra-low-pressure SUL-G20 in 1996 and the super-ultra-low-pressure SLU-H20 in 1999.

The success of UTC-70 once again opened a way for Toray to pursue seawater desalination applications. In 1996, Toray delivered the membranes to the seawater desalination center in Chatan on the main island of Okinawa.

On the other hand, Nitto Denko aimed at global expansion through M&A. Believing that business expansion was unlikely as long as the company remained in Japan, the head of the business drew up a scenario in which the company would enter the world market by acquiring foreign companies. In 1987, Nitto Denko acquired Hydranautics in the U.S. and established a foothold for global expansion. This was the company's first ever acquisition.

The acquisition of Hydranautics was a major leap forward for Nitto Denko's RO membrane business, as it opened a way into the global market by attaining access to the Hydranautic customer base. In addition, Nitto Denko discovered that Hydranautics had a new composite membrane technology that it had not expected at the time of acquisition. Based on this technology, Nitto Denko's developers commercialized a new composite membrane, NTR-759, in January 1988. NTR-759 was a membrane that maintained the same water permeability as the old membrane but achieved overwhelmingly higher desalination performance. NTR-759 helped solve the problems that had been caused by another membrane in May of the same year, and also gave Nitto Denko a competitive edge in industrial applications.

6.5.3 Toyobo's Response

Toyobo did not simply watch the expansion of ultra-pure water production applications for semiconductors. In 1991, Toyobo announced the launch of a polyamide-based hollow-fiber asymmetric membrane. However, development did not proceed as planned, and the launch was delayed until 1993. The company then proceeded to develop a hollow-fiber RO membrane with a composite structure to improve water permeability, but even this failed to raise performance to a level superior to that of competitors. Toyobo was not able to benefit from the expansion of demand for semiconductors.

The 1990s were a period of hardship for Toyobo. The textile business was struggling while the non-textile business was not expanding as expected. Junji Tsumura who took over as president in 1999 made the public declaration, "Losses are bad and profits are good," and began withdrawing from, and reforming, businesses that were suffering losses.

Fortunately for the RO membrane business, it was lumped in with the highly profitable artificial kidney membrane business as a "membrane-related business". As long as the artificial kidney membrane business continued to perform well, the sluggish RO membrane business was not accused. However, investment in expansion was not permitted, and measures were taken to survive by downsizing the workforce to the very limit.

In order to compete effectively with fewer resources, it was necessary to narrow down the market. Therefore, Toyobo stopped the development of ultra-pure water production applications for semiconductors and narrowed down the target market to seawater desalination applications. At the same time, the company stopped developing polyamide-based materials and decided to use only cellulose acetate materials. Even for seawater desalination applications, Toyobo decided to focus only on large-scale plants because expanding its business to include small and dispersed plants would require a large amount of resources to establish a maintenance system. On this basis, it chose a path to thoroughly demonstrate its competitive advantage in the area where chlorine resistance, which is the strength of cellulose acetate materials, can be utilized.

Saudi Arabia was thought to be a promising market for large seawater desalination plants, where chlorine resistance is needed. Since Saudi Arabia had been suffering from drought for a long time, many companies had been paying attention to the region. In particular, the Arabian Gulf (Persian Gulf) to the east of Saudi Arabia and the Red Sea to the west, are notably polluted waters for which the strength of Toyobo's membranes, easily sterilized with chlorine, could be realized. Seeing the potential strategic advantage there, Toyobo set its sights on a large-scale seawater desalination plant in Saudi Arabia and concentrated its resources there.

6.6 Development and Competition in Seawater Desalination Applications

The market for RO membranes had finally been launched by its application to the ultrapure water production for semiconductors. However, in order for Japanese companies to succeed in seawater desalination applications that had been their main focus from the beginning, they needed to push their technologies further.

In the early 1990s, Toray's president Katsunosuke Maeda was concerned that desalination applications were not growing as fast as expected. In 1994, under Maeda's direction, developers devised a two-stage system that increased the recovery rate of freshwater by subjecting concentrated water that had been desalinated once to a second desalination process. Since, in the second stage, it is necessary to apply higher pressure to the concentrated water to obtain fresh water, improving the pressure resistance of the membrane was critical. The fact that Maeda's name was included in the patent for this two-stage seawater desalination system shows the strong enthusiasm of the top management for this project.

The new system was called "Brine Conversion Seawater Reverse Osmosis Desalination System (BCS system)". A pilot plant was successfully opened in Ehime in 1997, and the system was delivered to Maspalomas in Spain and Curacao in the Caribbean in 1999.

Nitto Denko, on the other hand, started a project in 1992 with the goal of achieving the same amount of permeate and desalination performance at 1/3 the pressure of NTR-759. This development culminated in 1995 in the form of ES10 ultra-low-pressure membrane. The same performance at 1/3 the pressure means that the same pressure will produce 3 times the performance. This achievement of ES10 opened the way to desalination applications for Nitto Denko. In Japan, ES10 was adopted by the seawater desalination center in Okinawa along with Toray. Seawater desalination then became recognized as a real business at Nitto Denko.

Nitto Denko's acquisition of Hydranautics in 1987 contributed significantly to enter the global seawater desalination market that saw full-scale expansion in the 2000s. Nitto Denko was awarded large projects such as Carboneras in Spain (120,000 m^3/day), Fujairah in the United Arab Emirates (170,000 m^3/day), and Tampa in the United States (100,000 m^3/day).

By the late 1990s, Toyobo that had chosen to focus its business on desalination applications had established a relatively stable position in the Saudi Arabian market.

In the 1980s, Toyobo formed a caravan to demonstrate water desalination in the region. In addition, when their RO membrane elements delivered in the late 1980s experienced performance problems, Toyobo solved the problem extremely quickly. This earned the company a high level of trust from the local plant managers.

In the 2000s, the price of crude oil began to soar, and the Saudi Arabian government's finances began to grow considerably. With the aim of saving crude oil for export, the Saudi government actively promoted the use of RO and the construction of huge seawater desalination plants. Toyobo's RO membrane elements were used in great abundance in these plants. Although expensive, Toyobo's RO membranes were unrivaled in terms of chlorine resistance. If chlorine sterilization can be easily performed due to a high chlorine resistance, the number of cleaning cycles and the frequency of plant shutdowns can be reduced. This in turn increases the plant's operating rate and lowers desalination costs. In the heavily polluted waters of Saudi Arabia, this high resistance to chlorine was an immensely strategic characteristic. As the only company in the world to offer hollow fiber RO membranes, Toyobo also benefited from a monopolistic position on exchange membranes, unlike Dow, Toray, and Nitto Denko, which offer flat membrane elements that have become the de facto standard.

6.7 Summary

Japanese companies began to develop RO membranes in the late 1960s, and market introduction began in the mid-1970s. Of the five companies that entered the market, Toray, Nitto Denko, and Toyobo were the three Japanese companies that established a stable position in the seawater desalination market, the main market for RO membranes. However, the three companies did not necessarily follow the same path of development.

Toray aimed to accumulate peripheral know-how by realizing cellulose acetate membranes ahead of other companies. However, the scale of business in the 1970s was small compared to the grand goal of solving future global water shortages, and it was not expected to generate sufficient profits. The same was true for Toyobo in this respect. Nitto Denko, which had been attempting to develop a variety of applications under strong earnings pressure, was also having difficulty making a breakthrough in expanding its business.

This situation changed drastically in the late 1970s, when the production of ultra-pure water for semiconductors began to take off. Toray and Nitto Denko were the beneficiaries of this demand. Toray capitalized on this demand with cellulose acetate-based membranes and earned profits from them, which led to the development of next-generation polyamide-based membranes. Nitto Denko succeeded in penetrating the ultra-pure water production application for semiconductors because customers recognized the value of its unique membranes, which were created by chance in the course of developments.

However, the scale of these businesses was still unsatisfactory. Nitto Denko, which had a vision of making the membrane business the next pillar of its corporate growth, embarked on the acquisition of Hydranautics in 1987 as the company's first M&A deal, and became serious about global expansion. Despite repeated failures, Toray pushed forward with the development of polyamide composite membranes. It eventually succeeded in developing UTC-70 that combined high water permeability with desalination performance. After that, Toray aggressively expanded its business by building a two-stage seawater desalination system.

Toyobo was unable to successfully compete in ultra-pure water applications for semiconductors. Furthermore, as the company's overall business conditions deteriorated, pressure was placed on the RO membrane enterprise as a low-profitability business. Although the strong performance of the membrane business for artificial kidneys helped, it was necessary to reduce the number of personnel in the division.

In the face of resource constraints, Toyobo narrowed down its strategy. The company focused on the sea area where its chlorine-resistant technology would be most effective and pursued business development in Saudi Arabia. Fortunately, the Saudi Arabian economy was buoyed by high crude oil prices, and the company's RO membrane business gradually expanded in scale.

In the next and subsequent chapters, the development and commercialization processes of these companies will be clarified in more detail.

References

Gotō, F. (1979). Kaisui tansuika no gijutsu kaihatsu wa ikani okonawaretaka (How was the technological development of seawater desalination carried out?). *Denki gakkai zasshi (Journal of the Institute of Electrical Engineers of Japan), 99*, 40–44.

Kagaku Kōgyō Nippō (Chemical Daily). (1995, August 18).

Kōgyō Gijutsuin (Agency of Industrial Science and Technology). (1969). Ritchi news: kaisui tansuika to fuku sanbutsu riyō (Ōgata project seido shinki theme) no kenkyū kaihatsu keikaku oyobi kenkyū kaihatsu tantō jisshi kikan no kettei ni tsuite (Location news: Determination of R&D plan and implementing agency in charge of R&D for seawater desalination and byproduct utilization (New theme for large project system)). *Kōgyō Ritchi (Industrial Location), 8*, 68–70.

Kunisada, Y., & Hirai, M. (1978). Shō energy gata kaisui tansuika gijutsu kaihatsu sono 1 (Development of energy-saving seawater desalination technology, Part 1). *Journal of Water Re-use Technology, 4*, 39–49.

Nihon Keizai Shimbun. (1978, May 9; 2018, June 23).

Nikkei Sangyō Shimbun (Nikkei Business Daily). (1977, November 25; 1980, April 5; 1982, July 10; 1984, March 10; 1985, August 10; 1986, August 26).

Sōgō Hōsō Publication. (1980). *Kōbunshi takōshitsu maku no kaihatsu dōkō to shijō jittai: Gyakushintō maku, gengai roka maku, seimitsu roka maku, ion kōkan maku* (Development trend and market status of polymer porous membranes: Reverse osmosis, ultrafiltration, microfiltration, and ion exchange membranes). Okinawa: Sōgō Hōsō Publication.

Suzuki, A. (1972). Nihon ni okeru gyakushintō hō no kenkyū oyobi ōyo no genjō ni tsuite (On the present status of research and application of reverse osmosis method in Japan). *Bulletin of the Society of Sea Water Science Japan, 25*, 357–363.

Tokyo Kōgyō Shikenjo (The Government Chemical Industrial Research Institute, Tokyo). (1971). *Tokyo kōgyō shikenjo 70-nen shi (Tokyo National Industrial Research Institute: 70 Years of History)*. Tokyo.

Tokyo Kōgyō Shikenjo (The Government Chemical Industrial Research Institute, Tokyo). (1975). *Tokyo kōgyō shikenjo 75 shūnen kinenshi (The Government Chemical Industrial Research Institute, Tokyo 75th anniversary magazine)*. Tokyo.

Wada, Y. (2004). *Zōsui no Gijutsu (zōho-ban) (Technology of creating water (enlarged edition))*. Tokyo: Chijin Shokan.

Yarisawa, T. (1970). Kanagawa ken Chigasaki kaigan ni kensetsu sareta kaiusi tansuika rinkai kenkyū shisetsu ni tsuite (On the seawater desalination waterfront research facility constructed on the Chigasaki Coast of Kanagawa Prefecture). *Energy: Journal of the Japan Energy Association, 20*, 81–86.

Chapter 7
Toray: Development Aimed at Seawater Desalination

Abstract In this chapter, we describe the history of development and commercialization of RO membranes at Toray. Stimulated by DuPont's seminal achievement, Toray started its development activities for actualizing seawater desalination as a goal. We clarify that, in the middle of long-lasting development, the emergence of an ultra-pure water demand for semiconductor manufacturing provided economically viable markets and enabled Toray to continue its development. We also illustrate that a newly developed membrane, UTC-70, opened the way to the actualization of seawater desalination.

7.1 The Beginning of Development

7.1.1 Toray Versus DuPont

The first record indicating that Toray began exploratory research on RO membranes is dated May 23, 1968. In the previous year, DuPont succeeded in developing a polyamide hollow fiber RO membrane. This news had not only traveled around the United States but had also made its way into Toray.

For Toray, DuPont was both a role model and a target to aspire to. As far back as 1951, Toray integrated nylon technology from DuPont at a cost that exceeded its own capital at the time and succeeded in producing nylon domestically. It is no wonder that Toray, which had been constantly monitoring DuPont's technological development trends, promptly received news of the successful development of RO membranes.

Toray's first question was, "What exactly is Reverse Osmosis?" After receiving information from DuPont, Toray immediately began exploratory research on RO membranes. It officially began research on RO membranes in 1968. The development team explored every technical possibility. Membrane shapes were studied with both flat and hollow fiber types in mind. A variety of polymers were considered for the flat membrane type while nylon and acetate were used as materials for the hollow fiber type. As for the market, Toray envisioned desalination of seawater and brine as

M. Fujiwara and Y. Aoshima, *Mechanisms for Long-Term Innovation*,
Advances in Japanese Business and Economics 31,
https://doi.org/10.1007/978-981-19-4896-1_7

the main application while also considering solution concentration and wastewater treatment. At that time, Toray often used animal names for its development projects, and the RO membrane development project was called CAT.[1]

The development of RO membranes was carried out in three research organizations: the Industrial Materials Laboratory of the Fibers & Textiles Research Laboratories, the Engineering Research Laboratories, and the Applied Research Laboratory of the Central Research Laboratories.

In the Industrial Materials Laboratory, researcher Ryoichi Bairinji worked on the development of RO membranes using polyamide and cellulose acetate materials. Bairinji mainly developed the composition of the membrane to function as an RO membrane. The membranes developed by Bairinji were sent to the Engineering Research Laboratories, where they were developed into elements.

There were two teams involved in RO membranes in the Engineering Research Laboratories. One team, led by Naokatsu Kanamaru, was responsible for the development of machinery and equipment for membrane elementalization. The other team was responsible for the development of membrane production and evaluation technologies and the design of membrane modules. Tatsuo Kawabata, who would later play an important role in the development and commercialization of cellulose acetate RO membranes, joined this team after joining Toray in 1970 and conducting basic research on carbon fiber sintering for a year.

Meanwhile, in the Applied Research Laboratory of the Central Research Laboratories, researcher Kojuro Ikeda led the development of completely new membranes based on polyether- and polyamide-based materials, which began in 1970.

Thus, the development of RO membranes at Toray, triggered by DuPont's success, was carried out separately by three research groups belonging to different divisions. However, in 1971, as part of the company-wide integration of R&D functions, these research groups were consolidated into the Environmental Technology Research Laboratory of the Engineering Research Laboratories. Although consolidated, it was a small organization with only about 10 researchers.

7.1.2 Advanced Commercialization of Cellulose Acetate Membranes

Although Toray's developers considered polyamide synthetic membranes to be the most promising material, they decided to commercialize cellulose acetate (CA), a general-purpose material, first. This was because it would be easier to find a way to commercialize CA materials that had a proven track record in the United States.

For Toray, which had grown through synthetic fibers, business development with its own synthetic membrane would be more consistent and compatible with strategy.

---

[1] The project to develop carbon fiber, which was underway at the same time, was called CROW, after the color of the wet feathers of a crow.

However, to commercialize RO membranes, it needed to develop various complementary technologies including elements and modules. It was also necessary to establish technologies for day-to-day operation and maintenance. In order to expand the business, it is also important to build relationships with customers such as plant manufacturers and water supply companies. These technologies and information are not something that can be built or accumulated overnight.

Therefore, the developers decided to commercialize CA-based RO membranes, which are relatively easy to develop, and wait for the emergence of next-generation RO membranes made of completely new materials. They envisioned that they would establish these complementary technologies and customer relationships ahead of time, and then replace only the membrane material with a new synthetic material at a later date. Kawabata looks back on his role as a relay in the development of CA RO membranes as follows.

> With CA (cellulose acetate RO membranes), our goal was to enter the market as quickly as possible. There were things that can be learned and accumulated by commercializing membranes that had reached the level of being able to compete in the market. The idea was to replace the membrane with a different one (polyamide RO membranes) at the end of the game. The idea was to look at what was going on in the trenches and for everyone to be excited "about what amazing thing would be following down the pipeline next."
>
> What Toray is all about is "making things that no one in the world has ever made and making things that have the best performance worldwide." But when we first decided to try cellulose membranes, we found that they were just comparable to other products in the world. Toray has never done anything like this in its entire history. Normally, we are not allowed to do this. When I look back, I think, "How could they let us do this?" That was because they thought, "There's going to be a great product (polyamide RO membranes) coming next." ... There was potential to be able to create something that was unique in the world and of the highest standard. However, in terms of the total water business, we needed a lot of knowhow beyond just membranes, so (CA RO membranes) played the role of being the first to be implemented.[2]

At that time, Toray had a huge development team for carbon fiber development while the development team for RO membranes was very small in comparison. Despite this, one of the reasons for the approval of the relayed development, which was rare in Toray's history, was that the project leader, Shunji Kumazawa, was a strong proponent of the development of RO membranes. Masatoshi Ito, who joined Toray in 1948, and who had been the head of the R&D Laboratories since 1969 was also interested in the development of RO membranes while he supported the development of carbon fiber. Ito was later to serve as the company's president from 1981 to 1987.

---

[2] From the authors' interview with Tatsuo Kawabata at the Shiga office on April 25, 2014. The text in parentheses was added by the authors. Masaru Kurihara, who was working on the development of RO membranes based on polyamide materials, said the same thing about the cellulose-based materials development team: "If our boss sees the idea of developing cellulose as the only material, he would say, 'There is no way you can commercialize a material that is already being used in the world'. Thus, Kawabata and Bairinji (who were in charge of cellulose) were in a position to say, 'We have a fastball pitcher (Polyamide) here, so let us go ahead and do it." (The text in parentheses was added by the authors.)

Along with the membrane material, it was also important to determine the membrane shape. The developers mainly considered the hollow fiber shape and the flat membrane shape, and finally chose the flat membrane shape (spiral type in the element). The reasons for this choice are not completely clear, but the following reasons are generally considered to be the basis for that decision.

The first is differentiation from DuPont. As mentioned earlier, DuPont had developed a polyamide hollow fiber membrane. Since DuPont, which was a company to look up to at that time, used hollow fiber membranes, the chances of success in pursuing the same shape were thought to be low. Toray aimed to differentiate itself from DuPont by adopting the flat membrane shape. Second, it was said that hollow fiber membranes were easier to contaminate than flat ones. Third, hollow fiber engineers had been assigned to another business. Although Toray, as a fiber company, had excellent technology to make hollow fibers, its skilled hollow fiber related engineers were assigned to the development of membranes for artificial kidneys that was expected to be commercialized soon. Fourth, since Bairinji, who developed the membranes, belonged to the Industrial Materials Laboratory, he was able to select the shape of the membranes relatively freely without sticking to the spinning technology that had been cultivated in the textile business.[3] From these considerations, Toray's developers decided to proceed with the development of the flat membrane type.

7.1.3 The Search for New Materials: Launch of the PEC Group

While the development team for CA RO membranes was hurrying to commercialize the product, the development of membranes made of other new materials was also underway. Masaru Kurihara played a central role in this development. After joining the company in 1963, Kurihara spent five years at the Basic Research Laboratory and two years at the Central Research Laboratory before going to the United States to study at the University of Iowa in September 1970. Since he had consistently conducted research on heat-resistant polymers since joining the company, the initial purpose of his study abroad was not to develop RO membranes, but to research and develop heat-resistant polymers.

However, when Kurihara came to the United States, he wanted to do new research and asked his supervisor, John K. Stille, for a list of research themes. One of the topics was membranes that could be applied to seawater desalination and artificial kidneys. Intrigued, Kurihara decided to start research on membranes. Kurihara describes the situation at the time as follows.

> Foreign students (sent from Toray) at that time were highly enthusiastic, so they were eager to bring back work that was not being done Toray at all. I boldly asked him to list the research themes he was working on, and one of the fields was membranes. I said, "This sounds interesting," and the response was, membranes are great because they have multiple

[3] From the authors' interview with Masaru Kurihara on February 6, 2018 at Shiga Works.

> applications, like desalination and artificial kidneys. And polymers can be processed and then made into membranes. So, I said, "Please let me work on this theme."[4]

Kurihara's research theme was on piezodialysis membranes, which push salt out of water, unlike RO membranes, which push out water under pressure. Kurihara received research funds from the U.S. Department of the Interior's OSW and Toray to conduct this research.

At that time, researchers who received support from the OSW were required to report and disclose the results of their research. Kurihara sent these various reports on RO membranes to Bairinji and Ikeda, who were working on the development of RO membranes in Japan, and provided them with information. While studying abroad, Kurihara played an important role as a nexus of information to convey the status of RO membrane development in the United States to the development team in Japan.

After returning to Japan from the United States in September 1972, Kurihara returned to the Basic Research Laboratory from which he had originally been dispatched and was asked by the head of the laboratory whether he wanted to conduct research on artificial kidneys or RO membranes. Kurihara replied, "I want to develop RO membranes," and was transferred to the Environmental Technology Research Laboratory, where RO membrane development activities were concentrated. The development group to which Kurihara was assigned was later called the PEC Group.

At that time, the development of composite membranes was being actively promoted in the United States. In 1970, Cadotte developed NS-100 at North Star. And in 1972, when Kurihara returned to Japan and started membrane development, Riley obtained a patent for composite membranes at GA.[5] With the achievements of these individuals, the mainstream of development began to shift from asymmetric membranes to composite membranes. In line with this trend, Kurihara also proceeded with the development of composite membranes.

7.1.4 *Initial Development in Wastewater Treatment Applications*

The mission of the development team for the earlier CA system was to quickly acquire the complementary technologies and customer relationships needed for commercialization. In 1975, the development team, which had reached a certain level of success in developing membrane materials, was transferred from the Environmental Technology Research Laboratory to the Development Department with the aim of commercializing membrane elements.

The following year, in September 1976, Toray externally announced its intention to commercialize RO membranes. The following month, Toray participated in a

---

[4] From the authors' interview with Masaru Kurihara on August 30, 2010 at Shiga Works.

[5] US Patent 3,648,845 (filed 1969, obtained 1972). The material used at this time was a CA.

demonstration test for seawater desalination led by the WRPC established by the Ministry of International Trade and Industry. The 4-inch CA membrane module SC-5000 was used in the test. Later, the SC-5200 RO membrane module, which was expanded to 8 inches, was introduced.

However, it was believed within the company that it would be difficult to develop desalination applications for CA-based materials. Kawabata commented as follows.

> The ace (polyamide RO membrane) is for desalination. Those that are not ace (CA RO membranes) are used for slightly dirty or recycled water. You could desalinate seawater with CA, but the pressure resistance was a problem. (So,) you can use it for slightly salty brine, or for reclaimed water after secondary or tertiary treatment of waste water that has undergone primary treatment. And then, if you use ordinary water, you get very clean water, which is ultrapure water. Although it is possible to achieve a desalination rate of 99 (%) and some decimal values with cellulose membranes, little water comes out. Even when pressure is applied, the amount of water processed per membrane area is extremely small... Furthermore, when pressure is applied, the membrane is damaged, and the permeation volume decreases.[6]

CA membranes could not overcome the trade-off between desalination rate and water permeability performance. And in terms of cost, they could not compete with the evaporation method that was the mainstream method for seawater desalination applications at that time.

The first application of Toray's RO membrane business was for wastewater treatment. At the end of 1976, when Toray announced its commercialization, it succeeded in receiving an order for CA RO membrane elements from IBM's Yasu Plant.[7] The plant's wastewater purification capacity was 300 m^3/day. The membrane elements ordered at this time were manufactured at the Shiga Plant and delivered in 1977.

7.2 The Development of New Membranes and Struggles

7.2.1 Development of PEC-1000

While keeping an eye on the CA group that had achieved commercialization, Kurihara and his PEC Group were striving to develop membranes based on new materials. They used polyether-based materials to develop composite membranes by coating monomers on the surface of the support layer and polymerizing them to form a dense layer. The first cross-linked polyether composite membrane developed in this way was PEC-1000.

---

[6] From the authors' interview with Tatsuo Kawabata, from the same source referenced in footnote 2.

[7] At the time, Toray Engineering was in charge of the Toray Group's environmental engineering business. The company was looking for a market for the wastewater treatment technology using RO membranes developed by Toray itself, and this order was won after stiff competition. It should be noted that after these initial efforts, Toray Engineering made a full-scale entry into the RO membrane equipment business in October 1980.

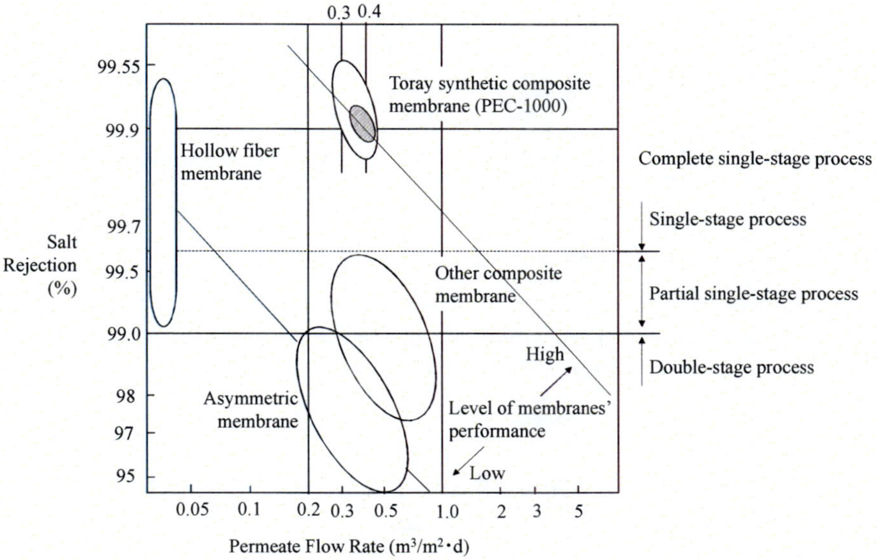

Fig. 7.1 Positioning of PEC-1000. *Source* Prepared by the author based on Kurihara (1983), p. 104

Figure 7.1 compares the performance of PEC-1000 with that of RO membranes made by other companies at the time. The horizontal axis shows water permeability and the vertical axis shows desalination rate. It can be seen that PEC-1000 succeeded in achieving a high level of both while other RO membranes were biased toward either the desalination rate or water permeability. 99.9% desalination performance, which PEC-1000 achieved, is actually comparable to that of today's RO membranes. In 1980, Kurihara and his colleagues in the PEC Group were transferred to the Development Department to promote the commercialization of PEC-1000.

In their 1980 paper on PEC-1000, Kurihara and his colleagues wrote, "Our next research target is to conduct long-term field tests of the 4-inch spiral PEC-1000 membrane element for desalination in a single-stage process that will provide high freshwater recovery under low pressure operation." In 1981, SP-120, an 8-inch membrane element using PEC-1000, was installed in an 800-meter^3-per-day desalination plant of the WRPC, and a series of demonstration tests were conducted.[8]

However, there were two problems with PEC-1000. The first was the weakness of chlorine resistance. This was an inevitable consequence of using polyether or polyamide materials. When RO membranes are used for seawater desalination, microorganisms thrive and clog the membranes. An inexpensive way to effectively prevent this is to use a chlorine disinfectant, but this treatment is not possible with membranes that have weak chlorine resistance.

[8] Kurihara et al. (1980) p. 20. The results of field tests on seawater desalination using PEC-1000 were reported in Kurihara et al. (1981).

The second, and more serious, problem was that PEC-1000 was extremely vulnerable to dissolved oxygen in seawater. This was an inherent and fatal problem of PEC-1000. Ultimately, Kurihara advised the company to stop selling PEC-1000. Kurihara looks back as follows.

> In the first stage, there was a membrane called PEC-1000 in the 1980s. This had good performance in elemental technology, but … it didn't go to Ehime (Plant). We were working on it in Shiga, and before it went to Ehime, I advised the head of research and development that it should be stopped. This kind of advice to stop was headed by researchers. I apologized (to my boss), knowing full well that I could have been ridiculed.[9]

While the market launch of PEC-1000 was abandoned, the CA RO membrane, which had been commercialized earlier, was blessed with a new market opportunity. This was the ultrapure water production application in the semiconductor industry.

7.2.2 The Rapid Expansion of Ultrapure Water Production Applications for Semiconductors

In the 1970s, when the semiconductor industry was rapidly developing, a higher level of purity was required for the water used to clean wafers. In the past, pure water was produced by the ion exchange method. But as miniaturization progressed, RO membranes with even finer separation performance became necessary.

Toray seized this new business opportunity with its CA RO membranes. Of course, the ion exchange method could not be replaced immediately. Semiconductor manufacturers were cautious about introducing the new technology, as they were afraid of stopping the production line due to failures.

In response to this, Toray approached the company with the idea of building a new pretreatment process and introducing RO membranes into it, rather than replacing the existing cleaning process. The idea was to build up a track record in the pre-treatment process while gaining customer understanding of the membrane's performance. This strategy proved to be successful, and the existing cleaning process gradually switched from the ion exchange method to RO.

The use of ultrapure water for the production of semiconductors proved to be a savior for Toray. Tadahiro Uemura, who joined Toray in 1974, and who was in charge of the development of polyamide RO membranes with Kurihara, recalls the significance of the demand for CA RO membranes in the semiconductor manufacturing process.

> I guess you could call it an amazing stroke of luck, this ultrapure water opportunity. When Japan's semiconductor business became number one in the world, one ton of ultrapure water was needed for each wafer, and they decided to produce ultrapure water even if it cost about 1,000 yen per ton.

---

[9] From the authors' interview with Masaru Kurihara, from the same source referenced above. Sales of PEC-1000 actually ended around 1993, after the development of UTC80 membrane described later.

> The price of RO membranes then must have been dozens of times higher than today's prices, but even after buying them, the semiconductor business was still viable. It was definitely an amazing stroke of luck. From the beginning, no one thought, "Let's start RO membranes because need for it is coming." It started to be used for ultrapure water, and although we could not sell a lot of products because they were used for producing ultrapure water, we were able to feed dozens of researchers (thanks to this demand). Then business opportunities came knocking, and the research and development just happened to line up with that.[10]

Toray's RO membrane business grew in tandem with the growth of the semiconductor industry. While not large compared to today's market, it was large enough to support the R&D activities of developers of new materials. 60–70% of RO membranes sold around the end of the 1970s were for industrial water such as water for the electronics industry and boilers.[11] The production of ultrapure water for semiconductors was included in this category, and it is clear that this was a major focus for Toray.[12] Toray was able to maintain a high share of this market, partly because it was ahead of its competitors.

While the business of CA-based RO membranes was expanding for industrial ultrapure water applications, commercialization of polyether- and polyamide-based materials still required more time. Kurihara faced the difficulty of implementing PEC-1000 and entrusted Uemura with the development of PEC-2000 as a next-generation membrane. PEC-2000 was a polyamide-based composite membrane produced by the polycondensation of polymer-based amine and acid chloride monomers.

7.2.3 *The Development and Abandonment of PEC-2000*

In 1982, many of the developers of PEC-2000 were transferred to the Development Department, where they moved toward mass production. With PEC-2000, materials were tailored for a variety of applications, and several variations, such as UTC-20 (an NF membrane), UTC-30 (an RO membrane for seawater desalination), and UTC-40 (an RO membrane for brine desalination), were commercialized.

Kurihara and Uemura were not transferred to the Development Department and became independent as the Functional Membrane Group with two or three other researchers. And in October 1982, they were transferred to the Functional Membrane Laboratory established in the Pioneering Research & Development Laboratories.[13]

---

[10] From the authors' interview with Tadahiro Uemura at Hitotsubashi University in Tokyo on July 15, 2010. The text in parentheses was added by the authors.

[11] Sōgō Hōsō Publication (1980) p. 106.

[12] As for other applications, 10% of the total was sterile water and purified water for hospitals and the confectionery industry, 5% each was for wastewater treatment and drinking water production, and 15% was for other applications.

[13] The Pioneering Research & Development Laboratories was the corporate research center and served as Toray's core.

From the Functional Membrane Laboratory, they supported the mass production of PEC-2000 and promoted the development of other new materials.

On April 2, 1984, with the mass production of PEC-2000 paving the way, Toray established the Membrane Business Division with RO as its main product. At that time, Toray announced a plan to increase the division's sales to 10 billion yen in three years.[14] This was an extremely ambitious plan since the division's annual sales in FY1983 were less than 2 billion yen.

The Ehime Plant was selected as the plant responsible for mass production. The manager of the Ehime Plant was Katsunosuke Maeda, who joined Toray in 1956, and who later became the president of Toray.[15] Maeda, who was impressed by the potential of the RO membrane business, strongly lobbied within the company for establishing the mass-production plant in Ehime instead of Shiga, and the RO business was transferred to the Ehime Plant in December 1984. In the following year, 1985, the Ehime Plant was made ready for mass production of PEC-2000, and the RO Production Department was established. Sales of the RO membrane business were 1.62 billion yen in FY1984, and further growth was needed to achieve the goal of 10 billion yen.

However, Kurihara again advocated the discontinuation of PEC-2000 because, as with PEC-1000, the chlorine resistance was weak. This time, however, a mass production plant had already been set up. It must have taken a great deal of determination for the developers to put a stop to sales at this stage. However, Kurihara was convinced that PEC-2000 would not be able to compete in the seawater desalination market because of its basic weaknesses based on its molecular structure. This was met with strong opposition from the development department and the plant, who asked why the company was giving up before it had even started. Kurihara described the situation at the time as follows.

> This was the part that was difficult. This (PEC-1000) was excusable, but (PEC-)2000 had already gone to Ehime and was running all over the place, and the development manager was in a situation where he could not back down. So when I said to him, "That is a material you should not use", the development manager and the director of the development laboratory quarreled with each other … Chemical and mechanical engineering experts have a tendency to think they can fix even low quality materials through how the materials are used. (So) their feeling was, "Why would you declare defeat when no one is even saying the product is weak yet?"[16]

The decision to abandon PEC-1000 and then PEC-2000 put a major shroud of doubt over the commercialization of RO membranes for seawater desalination. In the first place, the seawater desalination business is a risky business with large waves. Business inquiries come only every few years, and once an order is received, large-scale production activities are required. Preparing a large-scale mass-production

[14] Nikkei Sangyō Shimbun, July 29, 1983, p. 18; March 3, 1984, p. 9; April 12, 1984, p. 12. The Pharmaceutical Division and Ceramic Division were also established on April 2, 1984.

[15] Maeda served as president and CEO from 1987 to 1997, and then returned to the CEO position from 2002 to 2004.

[16] From the authors' interview with Masaru Kurihara, from the same source referenced above. The text in parentheses was added by the authors.

plant for a big wave that comes once every few years is not necessarily a rational choice for a division that needs to be profitable every year. The decision to abandon the project was made in the midst of such adverse conditions. Naturally, there was pressure from management to focus on the ultrapure water and brine applications that had been pioneered using CA membranes, rather than desalination applications using composite membranes. Uemura looks back on the situation as follows.

> At that time, desalination was not a thing that happened every year like it is now, but rather a situation in which a large facility with a capacity of tens of thousands of tons would be installed only once every few years. They said, "When you build a production facility for that purpose and there is nothing to do while there is no project, what are you supposed to do in the interim?" We originally developed PEC for desalination, and when we tried it, we only had one project every few years. (So I was told) "Desalination is not a good target. Instead, do something like ultrapure water or brine that can be sold continuously every year, even if it's just small items, and that has many customers and sells consistently."[17]

Although the development team did not abandon their primary goal of desalination, they decided to develop polyamide composite membranes for ultrapure water and brine as a practical decision. Specifically, they developed UTC-60 (NF membrane) by improving UTC-20, and commercialized it for use in brine and other applications.[18]

7.3 Breakthrough with the Development of UTC-70

7.3.1 The Search for New Materials

Although brine was targeted as a realistic application for the composite membrane for the time being, the development team did not give up on seawater desalination. To actualize desalination, the problem of chlorine resistance had to be solved at all costs.

The material for PEC-1000 was a monomer available mostly through commercial channels. Although some modifications were made to the chemical structure of PEC-2000, it was also made from existing monomers. Kurihara and his colleagues came to believe that it was necessary to synthesize material from a new monomer in order to fundamentally solve the chlorine resistance problem.

In 1984, Kurihara assigned the development theme of "the synthesis of an ideal membrane" to Yoshio Himeshima, a new researcher who had just been assigned to the Functional Membrane Laboratory. Himeshima, who had studied organic synthesis at university and graduate school, pushed forward with the search for a material that would exhibit high chlorine resistance while maintaining desalination performance.

---

[17] From the authors' interview with Tadahiro Uemura from the same source referenced above. The text in parentheses was added by the authors.

[18] UTC-20 is polymer-based, whereas the newly developed UTC-60 is monomer-based.

Based on past experience including that in other companies, the basic direction of development was clear for him. The idea was to form a polyamide layer by interfacial polycondensation of an aqueous amine solution and an organic solvent solution of acid chloride. The key point was to determine the specific molecular structure to be used. Himeshima mainly used as a benchmark the performance of the cross-linked polyamide membranes synthesized by interfacial polycondensation of a bifunctional aromatic amine and a trifunctional aromatic acid chloride, which was developed by Cadotte of FilmTec. The goal was to develop a membrane that surpassed this performance. Himeshima states the following.

> When you think of aromatic monomers, Cadotte was the obvious source to think of, so I devoured information about his patents and conference presentations. At that time, Cadotte was a giant. I wanted to catch up with him, and I wanted to make something good by noticing even one thing that they (he and other star scientists) hadn't already noticed.[19]

Through repeated trial and error, Himeshima narrowed down the materials while searching for the optimal parameters that would resolve the trade-off between performance. After synthesizing hundreds of monomers, he handed them over to an experienced researcher to form a membrane and have it evaluated. After about a year of repeating this process, they found triaminobenzene.

Following this discovery by Himeshima, the development team began investigating the use of triaminobenzene in RO membranes in May 1985. Triaminobenzene was not a new monomer, but it was not a commercially available material. Therefore, Himeshima decided to develop an original method to create triaminobenzene with high purity. In parallel with Himeshima's efforts, a group in the Functional Membrane Laboratory led by Uemura proceeded to identify the specific membrane structure. Finally, they developed UTC-70, a four-component cross-linked aromatic polyamide membrane.

When the pore diameter of the membrane is reduced to increase the desalination rate, the water permeability is usually reduced. However, UTC-70 was able to increase the amount of water produced while maintaining a dense mesh structure. Later, analysis revealed that when a composite membrane is formed by interfacial polycondensation, the surface of the membrane protrudes in the shape of numerous folds. These folds increase the surface area of the membrane, making it possible to achieve both a high desalination rate and a large amount of water production. In addition, the problem of chlorine resistance, which had been a concern, has been resolved to a certain extent to a level that can withstand commercialization.

7.3.2 Expanding into Seawater Desalination

In February 1987, the technology of UTC-70 was transferred to the RO Production Department. Toray commercialized it as SU-700 and delivered it to Toshiba's Oita

[19] From the authors' interview with Yoshio Himeshima at Toray's headquarters in Tokyo on August 19, 2010.

Plant in the same year. The plant was a 1 Mb DRAM factory that brought together the most advanced technologies of the time. UTC-70 achieved significant commercial success not only in the ultrapure water production application in semiconductor plants but also in the brine market.

In these applications, it is not necessary to increase the operating pressure as much as in the case of seawater desalination because the raw water is water with low salinity. Rather, the key is to achieve high desalination performance at the lowest possible pressure. If the system can be operated at a low pressure, electricity costs can be saved and desalination costs can be reduced. Therefore, the development team proceeded to reduce the pressure of the membrane by improving the membrane formation conditions, the reaction catalyst, and the control of the nanostructure. Table 7.1 shows the performance of Toray's RO membranes for brine applications, and it can be seen that the pressure reduction has progressed with each new membrane, from SU-720 (launched in 1987) to SUL-H20 (launched in 1999).

UTC-70 once again paved the way for Toray to pursue desalination applications. Knowing that the developers were very interested in desalination applications, the management of Toray took advantage of the success of UTC-70 development and encouraged its commercialization. The developers developed UTC-80 for seawater desalination, using triaminobenzene as in UTC-70, with an improved membrane to increase the desalination rate and an improved element structure. In 1991, UTC-80 was commercialized as SU-800 and delivered to Kita-Daito Island and Minami-Daito

Table 7.1 Development of RO membranes for brine desalination

		Low pressure		Ultralow pressure	Super-ultralow pressure
	Membrane element name	SU-720	SU-720L	SUL-G20	SUL-H20
	Launch year	1987	1988	1996	1999
Performance	Desalination rate (%)	99.4	99	99.4	99.4
	Water permeability (m^3/day)	26	22	26	26
Test conditions	Operating pressure (MPa)	1.5	1	0.75	0.5
	Temperature (°C)	25	25	25	25
	Feed concentration (mg/L)	1,500	1,500	1,500	1,500
	Brine flow rate (L/min.)	80	80	80	80

Source Prepared by the authors based on Uemura and Henmi (2008)

Island in Okinawa. It was also commercialized as SU-820 with an 8-inch element and introduced in 1996 when the desalination center in Chatan, Okinawa was partially completed (10,000 m^3/day). The performance of SU-820 was 6.5 MPa of operating pressure, a 99.75% desalination rate, and 16.0 m^3/day of desalinated water.[20]

The focus of Toray's RO membrane business shifted from CA membranes to polyamide membranes. However, Maeda, who had supported the RO membrane business as the general manager of the Ehime Plant and became president in 1987, was not satisfied with the situation. Business was still very small even though desalination applications were beginning to emerge. In 1994, Maeda gave strict orders to the developers to build a more efficient desalination system with the aim of commercializing it in earnest.

In response to his orders, the developers started developing a two-stage desalination system. In this system, the concentrated water discharged from the first stage of the desalination process is passed through another RO membrane to obtain fresh water twice and secure the amount of desalinated water. In the second stage, seawater with a higher concentration is passed through as raw water, so the operating pressure needs to be dramatically increased. Therefore, membranes that can withstand high pressure are required. In contrast to the development of membranes for brine desalination, which aimed to achieve low pressure, the development of membranes for seawater desalination here focused on improving pressure resistance.

SU-820BCM (desalination rate: 99.70%, desalination capacity: 16.0 m^3/day), which has a high pressure resistance of 9.0 MPa, was developed to overcome this problem. A highly efficient two-stage seawater desalination system was constructed using this membrane.[21] In this system called "Brine Conversion Seawater Reverse Osmosis Desalination System (BCS system)", the concentrated water discharged in the first stage is separated again in the second stage and 20% of freshwater is newly recovered, making it possible to obtain 60% of freshwater when combined with the 40% recovery rate of the first stage. It was calculated that the desalination cost would be reduced by about 20%.

The basic invention of this system was accomplished in 1994 when Maeda issued the order. In 1997, a pilot plant was successfully made operational in Ehime. And in 1999, the system was adopted as a commercial system in Maspalomas, Spain (3 plants, 4,500m^3/day each), Curaçao, the Caribbean (11,400m^3/day), and Trinidad and Tobago (136,000m^3/day; started operation in 2002).

---

[20] Kurihara et al. (2001).

[21] The fact that Katsunosuke Maeda's name appears on the patent for this system (Patent Publication 8-108,048; US Pat. 6,187,200) suggests that Maeda was strongly involved in the RO membrane business.

7.3.3 *Tackling New Technological Challenges*

As the practical application of seawater desalination began in earnest in the latter half of the 1990s, challenges also became apparent.

The first issue was to establish a stable mass production system. Conventional production at the Ehime Plant relied on human labor. In particular, the process of winding up the membrane when producing elements was completely dependent on skilled workers, which was a bottleneck in achieving mass production. Therefore, the members of the Ehime Plant introduced an automatic winding machine and made the membrane winding process unmanned. This quickly boosted production capacity. As a result of these efforts to automate production, the company was able to gradually lower production costs, which was necessary for market expansion.

The second challenge was to improve the membrane performance. There were two major technical issues to be overcome. The first is the removal of boron from seawater. In the 1990s, the World Health Organization (WHO) was concerned about the adverse effects of boron on the human body, set a provisional guideline of 0.3 mg/L for boron in drinking water in 1993, and later revised the value to 0.5 mg/L in 1998. In line with these guidelines, Toray developers had to develop membranes with excellent boron removal performance.[22]

The second technical challenge was to prevent fouling. Fouling is a general term for the contamination of membranes. In the process of obtaining fresh water by passing supplied water through the membrane, various substances adhere to the membrane. In particular, when the supply water is seawater, which contains a wide variety of substances, RO membranes become contaminated in various ways. This contamination deteriorates the performance of the membrane.

There are two ways to prevent fouling: one is to make the membrane itself difficult to foul, and the other is to make it easy to sterilize or clean the fouled membrane. For polyamide materials, it is most important to make the membrane resistant to fouling. Among the fouling problems, biofouling, in which microorganisms in the sea thrive and clog the membrane, is often a serious problem. The use of chlorine disinfectants is effective against biofouling, but polyamide materials have low chlorine resistance and cannot be injected with chlorine disinfectants. Therefore, it is critically important to develop a membrane that is as stain-resistant as possible. An RO membrane with low fouling was announced in 2002 as TML20-365, an RO membrane for sewage.[23]

[22] The WHO significantly relaxed the guideline value for boron to 2.4 mg/L in 2011.

[23] Inoue et al. (2002).

7.4 Business Expansion

7.4.1 Prioritized Business Development

In 2002, Sadayuki Sakakibara, who joined the company in 1967, took over the presidency from Maeda and announced "Project New Toray21" (Project NT21) as a management reform to transform the company into a "New Toray for the 21st Century". Three areas were positioned as strategic growth areas: Information Technology, Life Science, and Environment, Safety, and Amenity. The water treatment business including RO membranes was identified as an important business in the environment, safety, and amenities, and efforts to expand the business were now underway.

In order to expand its RO membrane business, Toray sought to go beyond components such as membranes and elements and enter into the downstream water treatment plant market. In 2002, Toray acquired a 20% stake in Suido Kiko, a well-established company in the public water treatment market, and further strengthened its business in the downstream. In 2004, the company announced Project NT-II, a follow-up to Project NT21, and reaffirmed the importance of the water treatment business. In the same year, Toray increased its stake in Suido Kiko and made it a subsidiary.

Even so, however, the company was not yet fully equipped to promote growth in overseas markets, the main market for seawater desalination. In addition, the introduction of high-performance energy recovery equipment in seawater desalination plants diminished the significance of using a two-stage system, which had been Toray's forte. It became possible to efficiently desalinate water in a single-stage system by reusing the recovered energy and repeating the desalination cycle.

The person who broke out of this situation and pushed forward with the global expansion of the business was Akihiro Nikkaku, who joined the company in 1973, and who became a director in 2002 and a managing director in 2004. In 2005, he served concurrently as the Head of Water Treatment Headquarters, and focused on cultivating and expanding a global-scale sales organization.[24]

In the U.S., Toray hired Randy Truby (see Chap. 5), who had long experience in the RO membrane industry, and used his contacts to form a global sales team of about 20 people. They established Toray Asia Pte. Ltd. (TAS) in Singapore to handle the Asia/Oceania region, Toray Bluestar Membrane Co. (TBMC) in China, and Toray Membrane USA (TMUS) in the U.S., and optimized regional sales activities.[25] Of these organizations, Truby was responsible for leading TMUS.

---

[24] As of 2016, Nikkaku is the president and representative director of Toray Industries, Inc.

[25] Tsumuji (2013) pp. 80–81 (http://president.jp/articles/-/10670?page=3, Accessed November 28, 2015) has a similar statement.

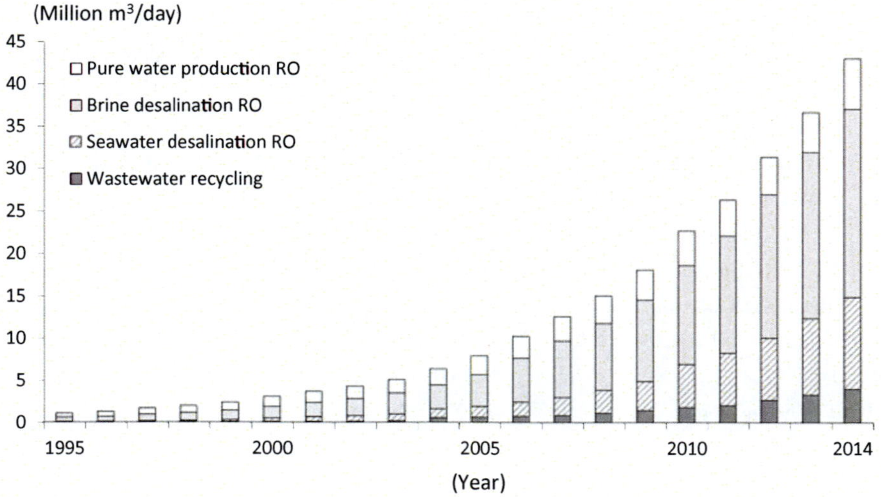

Fig. 7.2 Cumulative volume of water production. *Source* Partly estimated by the authors based on data provided by Toray Industries, Inc.

The water treatment business was also positioned as a priority business in the company-wide plan "Innovation Toray 2010" (IT-2010), which started in 2006 as a successor to Project NT-II. Toray expanded its global sales structure, increasing the number of its main sales bases from five to six, sales offices from five to 13, and sales agents from 11 to 13 between 2006 and 2008.[26]

These expansion measures have helped Toray to increase its presence in the global market. Figure 7.2 shows the cumulative volume of water produced by water treatment plants using Toray's membranes since 1995. The graph shows a particularly large increase since the mid-2000s. A breakdown by application shows that in the early stages, the weight of pure water production and desalination of brine was high, and since then, desalination of brine has grown significantly. The seawater desalination application appears in the graph from 2000 and has been expanding since around 2005.

In the 2000s, plants became larger in scale. Table 7.2 shows the top 12 large-scale plants using Toray's RO membrane. Except for Sulaibiya (Kuwait) and Changi (Singapore), which are wastewater treatment plants, all of these plants are desalination plants. The fact that all of these plants have been in operation since the 2000s shows that the scale of plants has increased during this period.

After assuming the post of president in 2010, Nikkaku invested 7.5 billion yen to build an integrated production plant for RO membranes in Beijing and quickly increased its production capacity by 1.5 times.[27] In addition to increasing the size

---

[26] Toray Industries, Inc. (2008) (http://www.toray.co.jp/ir/pdf/lib/lib_a270.pdf, Accessed November 28, 2015). Major sales offices refer to the sales offices directly responsible for each region.

[27] Nihon Keizai Shimbun (July 31, 2010, p. 12).

Table 7.2 Major plants using Toray's RO membranes

	Plant (Country)	Plant scale (m^3/day)	Beginning of Operation (Year)
1	Magtaa (Algeria)	500,000	2014
2	Sulabiya (Kuwait)	320,000	2005
3	Tual II (Singapore)	318,500	2013
4	Changi (Singapore)	228,000	2009
5	Al Dur (Bahrain)	218,000	2011
6	Hamma (Algeria)	200,000	2008
7	Shuaibah (Saudi Arabia)	150,000	2009
8	Fujairah exp 1 (UAE)	137,500	2014
9	Point Lisas (Trinidad and Tobago)	136,000	2002
10	Tuas (Singapore)	136,000	2005
11	Fujairah 2 (UAE)	136,000	2010
12	Shuwaikh (Kuwait)	136,000	2010

Source Data provided by Toray
Note As of the end of July 2015

of the plant, Nikkaku's consistent measures to expand its water treatment business have stimulated its rapid growth in the second half of the 2000s.

7.4.2 Monetization in the Face of Competition

Sales of RO membranes for seawater desalination have been growing steadily, supported by increasing demand in many parts of the world where droughts are advancing. However, while demand has grown, competition among companies has also intensified. The construction of seawater desalination plants is often a large-scale project, resulting in fierce competition among engineering companies for orders. RO membrane companies also tend to compete on price at the order stage, because if they can deliver the first element, they can expect replacement demand thereafter.

However, even if a product line is adopted in the first order, there is no guarantee that the line will dominate replacement demand later on. RO membrane elements for seawater desalination are virtually standardized to an 8-inch diameter and 40-inch length. There is no significant difference in performance between membrane manufacturers. Therefore, water treatment companies can choose a new membrane manufacturer at the time of replacement. In fact, the desalination center in Okinawa designated a membrane manufacturer for replacement in 1999 and 2001, but switched to a bidding system in 2002.[28] Membrane manufacturers are exposed to competition in replacement demand as well.

[28] Kuniyoshi (2007).

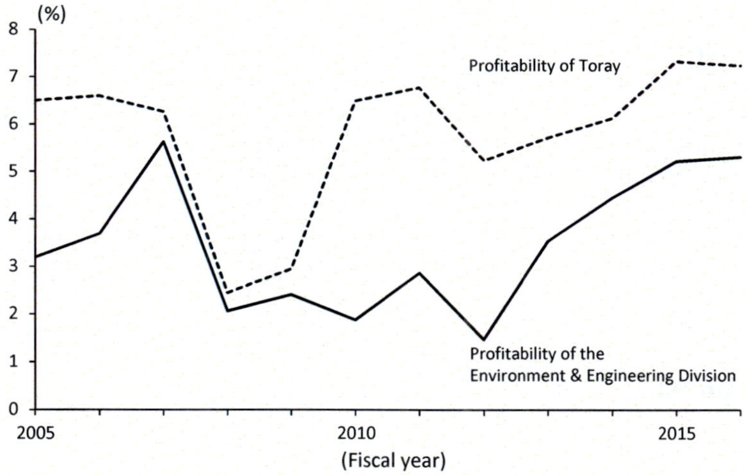

Fig. 7.3 Toray's profitability for the overall company and the Environment & Engineering Division. *Source* Toray "Securities Report" for each year

Of course, water treatment companies do not want to unnecessarily change membrane manufacturers and cause performance problems. Therefore, they tend to continue using the same manufacturer's membranes as long as there are no particular problems with operating conditions. However, the mere fact that the products are compatible with each other increases the bargaining power of the water treatment companies, which leads to downward pressure on prices.

The standardization of elements and the homogenization of membrane performance have made it difficult for the RO membrane business to achieve high profit margins. Figure 7.3 and Table 7.3 show the performance of the Environmental & Engineering Division, which includes the RO membrane business, over the last decade or so. Although the data cannot be taken at face value because it includes business other than RO membranes, it is clear that the division's profit margin has been lower than Toray's overall profit margin.

In recent years, Toray has set a major policy to become the world's number one company in the water treatment membrane business, including RO membranes, UF membranes, and Membrane Bioreactors (MBR). In line with this policy, it acquired Woongjin Chemical in 2013, which is strong in the field of RO membranes for home water purifiers, and placed it under its umbrella to expand its home water purifier business for emerging countries in Asia.[29] In addition, as already mentioned, the company is expanding into downstream areas, including plant engineering, through Toray Engineering and Suido Kiko.

[29] Nihon Keizai Shimbun (September 27, 2013) reported that Woongjin Chemical's sales for the fiscal year ending in December 2012 were 80 billion yen, and that the purchase price by Toray was about 40 billion yen.

In the "AP-G 2016" medium-term management project launched in 2014, the water treatment membrane business is positioned as an important component of the company-wide effort to expand its Green Innovation Business.[30] The plan is to strengthen the link between the membrane business and the equipment business and expand the business as a whole plant system. Standardized membrane elements tend to be subject to fierce price competition, and the scale of sales is small compared to plants. Therefore, there is a certain rationale for entering the downstream sector. However, there is also the concern that entering the downstream business means competition with other engineering companies that are Toray's customers.

7.5 Summary

From the time Toray officially began development in 1968 to the present, many of the companies that led the development of RO membranes have either withdrawn from the market or have been acquired by other companies. In the face of fierce competition, Toray has continued to develop and commercialize its own products and has established a solid position in the market.

The road to the present state of Toray's RO membrane business has not been smooth sailing. It was a journey of twists and turns. Initially, Toray's developers envisioned RO membranes being used mainly to desalinate seawater and brine. However, the technology was still far below the level required to achieve the performance and cost required for seawater desalination. Toray's developers, while focusing on composite membranes, decided to commercialize CA-based membranes, choosing to accumulate the complementary technologies and customer relationships necessary for commercialization.

In the 1980s, the rapid miniaturization of semiconductors created an opportunity for RO membranes to contribute to the purity of cleaning water and they seized this opportunity with their CA RO membranes. The RO membrane business achieved a certain level of business results, giving the company time to develop new composite membranes. This led to the successful development of UTC-70 that paved the way for seawater desalination applications. The emergence of semiconductor applications was, in a sense, an unexpected stroke of good fortune. However, the company's ability to seize this opportunity quickly is what supported the continuation of its business.

[30] http://www.toray.co.jp/ir/pdf/lib/lib_a385.pdf (Accessed December 1, 2015) Water treatment here includes RO membranes, MF membranes, UF membranes, MBRs, and household water purifiers. Based on these data, the authors estimate that the water treatment business generated sales of around 46 billion yen in FY2012.

In addition, in explaining the process by which the RO membrane business grew to its current scale despite unstable demand for a long time, it is important to mention the role of top management with strategic intentions. In the early stages, Kumazawa actively led development, and Ito, the head of the development laboratory who later became president, valued him. Despite the discontinuation of PEC-1000 and PEC-2000, the management never stopped their development. Kurihara recalls, "I never heard any harsh opinions from top management that forced us to dissolve the development of RO membranes".[31] The decision-making and support of top management in the commercialization phase seems to have played a major role in the development of Toray's RO membrane business, which has been established over the long period of nearly 50 years. The organizational transitions are shown in Fig. 7.4.

Table 7.3 Performance of the Environment & Engineering Division and its relationship with water treatment business

FY	Sales (mil yen)	Operating income (mil yen)	Operating profit ratio (%)	Report on the water treatment business (abstracted from financial results briefing materials)
2005	154.14	4.92	3.19	Revenue increased partly due to the acquisition of Suido Kiko as a subsidiary. The engineering subsidiary expanded sales and strengthened its structure, resulting in an increase in profit
2006	161.31	5.95	3.69	The water treatment business increased sales and improved profits due to strong exports of RO membranes
2007	173.21	9.75	5.63	Sales of RO membranes and household water purifiers were strong. With the establishment of a U.S. water treatment subsidiary, sales remained flat and profits improved due to an increase in sales through the subsidiary

(continued)

[31] From the author's interview with Masaru Kurihara on February 6, 2018, in Shiga.

Table 7.3 (continued)

FY	Sales (mil yen)	Operating income (mil yen)	Operating profit ratio (%)	Report on the water treatment business (abstracted from financial results briefing materials)
2008	160.21	3.30	2.06	In the water treatment membrane business, orders for RO membranes and MBRs increased mainly in Europe, the U.S., and the Middle East. Domestic sales of water purifiers for home use also remained strong. On the other hand, the strong yen reduced proceeds from export business, and expenses increased due to business expansion
2009	159.79	3.85	2.41	Shipment volume, especially of RO membranes, remained steady, but was affected by exchange rate fluctuations
2010	178.18	3.35	1.88	Sales of RO membranes for large-scale overseas projects expanded
2011	170.25	4.88	2.87	In the water treatment membrane business, sales declined due to no shipments of large-scale projects and the strong yen, but the company promoted activities to win orders for various water treatment membranes, including RO membranes, around the world
2012	178.36	2.63	1.47	Despite sluggish demand in the major markets of Europe, the U.S., the Middle East, and China, the company continues its efforts to expand global sales of various water treatment membranes, including RO membranes, and to reduce costs

(continued)

Table 7.3 (continued)

FY	Sales (mil yen)	Operating income (mil yen)	Operating profit ratio (%)	Report on the water treatment business (abstracted from financial results briefing materials)
2013	180.2	6.40	3.55	Although the market has yet to achieve a full-fledged recovery amid lingering uncertainty about the future of the global economy, the Company saw steady shipments of RO membranes and other products for the Middle East, as well as cost reduction efforts that have been underway since the strong yen began to take effect
2014	179.99	8.02	4.46	Despite weak global demand, shipments of RO membranes for seawater desalination and other products increased, and Toray Chemical Korea Inc., which became a consolidated subsidiary at the end of the previous fiscal year, contributed to earnings
2015	183.3	9.60	5.23	Against a backdrop of cost-cutting progress and yen depreciation, profit was improved for exporting RO membranes from Japan, etc. Additionally, in foreign countries, business performance in the U.S., China, and South Korea all transitioned to firm standing
2016	186.1	9.9	5.32	Sales of RO membranes and other products expanded, but exports from Japan were affected by the ongoing appreciation of the yen
2017	238.3	13.3	5.58	Demand for RO membranes and other products was generally firm in Japan and overseas

Source Toray financial results briefing materials (for each year)

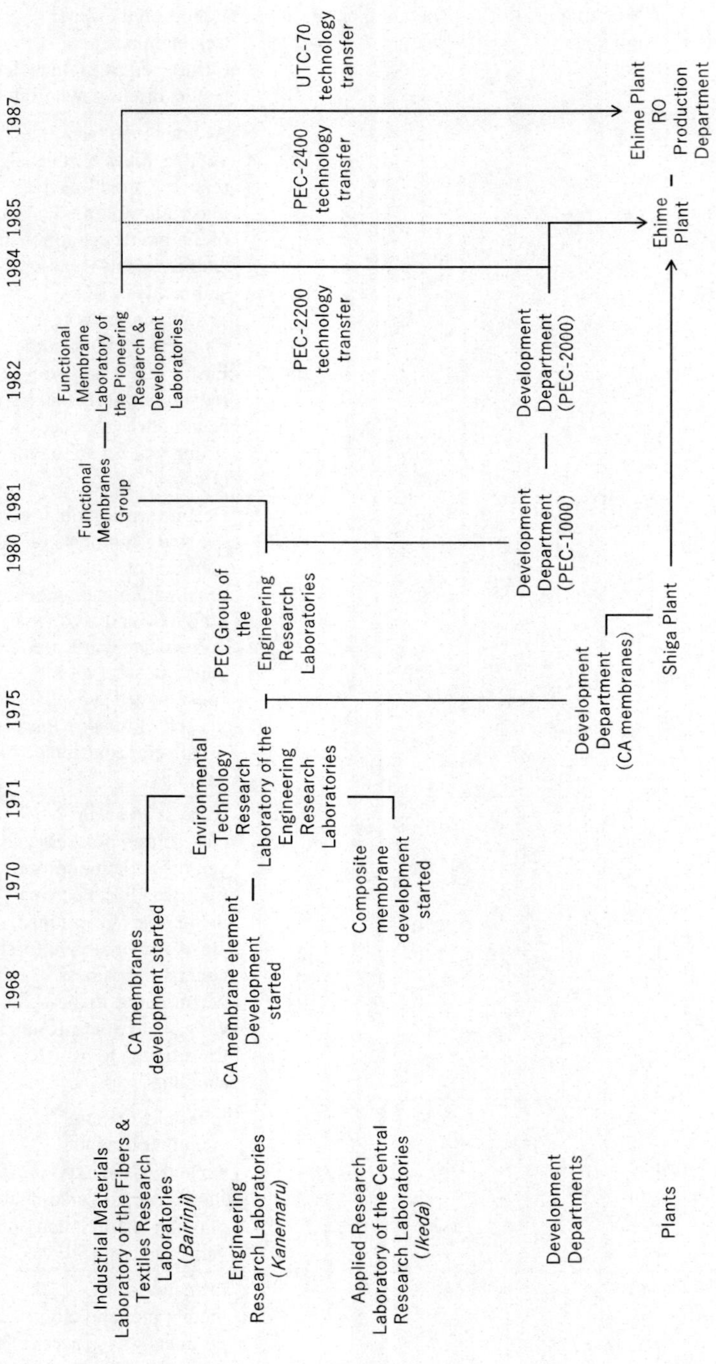

Fig. 7.4 The organizational transitions. *Source* Prepared by the authors based on interviews

References

Inoue, T., Sugita, K., Isaka, H., & Fusaoka, Y. (2002). Low fouling RO membranes. *Membrane, 27*, 209–212.

Kuniyoshi, N. (2007). Operation status of seawater desalination facility of Okinawa prefectural enterprise bureau. *Journal of Water Re-Use Technology, 31*, 35–39.

Kurihara, M. (1983). Proceedings: Concentration and recovery of dilute aqueous solutions containing valuable materials by synthetic composite reverse osmosis membrane. *Membrane, 8*, 97–112.

Kurihara, M., Kanamaru, N., Harumiya, N., Yoshimura, K., & Hagiwara S. (1980). Spiral-wound new thin film composite membrane for a single-stage seawater desalination by reverse osmosis. *Desalination, 32*, 13–23.

Kurihara, M., Harumiya, N., Kanamaru, N., Tonomura, T., & Nakasatomi, M. (1981). Development of the PEC-1000 composite membrane for single-stage seawater desalination and the concentration of dilute aqueous solutions containing valuable materials. *Desalination, 38*, 449–460.

Kurihara, M., Yamamura, H., Nakanishi, T., & Jinno, S. (2001). Operation and reliability of very high-recovery seawater desalination technologies by brine conversion two-stage RO desalination system. *Desalination, 138*, 191–199.

Nihon Keizai Shimbun. September 27, 2013; February 19, 2014; July 31, 2010; January 21, 2015; March 14, 2017.

Sōgō Hōsō Publication. (1980). Kōbunshi takōshitsu maku no kaihatsu dōkō to shijō jittai: Gyakushintō maku, gengai roka maku, seimitsu roka maku, ion kōkan maku (Development trend and market status of polymer porous membranes: Reverse osmosis, ultrafiltration, microfiltration, and ion exchange membranes). Sōgō Hōsō Publication.

Toray Industries, Inc. (2008). *Mizu shori jigyo no kakudai senryaku (Strategy for Expansion of Water Treatment Business)*. Accessed November 28, 2015, from http://www.toray.co.jp/ir/pdf/lib/lib_a270.pdf.

Tsumuji, T. (2013). Keieisha tachi no 40 dai" (Managers in their 40s): Nikkaku Akihiro. *President*, 80–81. Accessed November 28, 2015, from http://president.jp/articles/-/10670?page=3.

Uemura, T., & Henmi, M. (2008). Thin-film composite membranes for reverse osmosis. In N. Li, A. Fane, W. Ho, & T. Matsuura (Eds.), *Advanced Membrane Technology Applications*. John Wiley & Sons, Inc.

Chapter 8
Toyobo: Focused Development in Cellulose Acetate Hollow Fiber Membranes

Abstract In this chapter, we describe the history of development and commercialization of the RO membrane at Toyobo. Toyobo started RO membrane development with the aim to create new business opportunities utilizing its fiber technology amid the stagnation of its mainstay fiber business. Unlike many other companies that developed polyamide-based flat-type membranes, Toyobo adopted cellulose acetate-based hollow-fiber membranes and established a unique position in the Middle Eastern market. We also clarify how its RO membrane business and development were sustained under the extremely difficult business state of the company.

8.1 The Beginning of Development

8.1.1 Starting Out with Hollow Fiber Membranes

In November 1971, Toyobo began a full-scale investigation of various functional membranes, including RO membranes. At that time, when the textile industry was showing signs of decline, Toyobo set up a research and planning office at the Toyobo Research Institute (TRI) in Katada, Shiga Prefecture, and was searching for new business ideas.[1] The investigation of functional membranes was one of the many initiatives toward diversification. Among them, Tetsuo Ukai, who joined the company in 1962, and four other researchers at the TRI were assigned to investigate RO membranes. The following year, in April 1972, four of these five researchers moved on from their research to the search for specific technologies.

Initially, Ukai and his colleagues considered the application of RO membranes to recover wool grease from the wool washing process and sell it for use in cosmetics. However, as they proceeded with this project, they naturally began to consider the

[1] The name of the institute in Katada was changed to the Toyobo Research Institute in September 1976. However, from the viewpoint of readability, the name is referred to herein as "the Toyobo Research Institute" consistently, regardless of the timeline. Additionally, the year in which the Research Planning Office was established in the institute was 1971 (Toyobo, 1986, p. 369).

M. Fujiwara and Y. Aoshima, *Mechanisms for Long-Term Innovation*,
Advances in Japanese Business and Economics 31,
https://doi.org/10.1007/978-981-19-4896-1_8

possibility of applying the technology to seawater desalination. Ukai looks back on the project as follows.

> Around 1971-72, the focus was the recovery of valuable materials . . . We were searching for a way to concentrate and recover high-value materials. However, when we were tinkering with various membranes, we concluded that desalination was the way to go as far as membranes were concerned. We decided to stick to that, so we started gathering developers from around 1973 to 1974.[2]

By 1973, the number of developers of RO membranes in the TRI had increased to eight, and by the mid-1970s to 18. In addition to the staff at the TRI in Katada, skilled workers who already had direct experience in the manufacturing of hollow fibers at the Iwakuni Plant were involved.

At that time, DuPont and Dow had already succeeded in developing RO membranes, and introduced their products to the market. DuPont was developing its business with polyamide-based hollow-fiber membranes and Dow with CA-based hollow-fiber membranes. In contrast, Toyobo initially selected polyamide as their membrane material. However, when it was discovered that the material might infringe on DuPont's patent, the company decided in July 1975 to switch to CA. The use of CA, which was readily available in the market and had already been successfully commercialized by Dow, was not desirable from the standpoint of differentiation from competitors. However, it was an unavoidable choice due to concerns about patent problems.[3]

While there were twists and turns in the selection of materials, the development of the membrane shape was narrowed down to a hollow fiber type from an early stage. This was because the development team thought they could make use of their own fiber technology. Toyobo possessed wet spinning technology from its experience in manufacturing fibers such as rayon, acrylic, chinon, and spandex. Through repeated production of hollow futon cotton fiber and hollow rayon fiber, Toyobo had accumulated spinning technology for hollow fiber, which comprises much know-how. In particular, since wet spinning technology for hollow fiber is not something that can be easily imitated by other companies, it is desirable to utilize this technology

[2] From the authors' interview with Tetsuo Ukai at the head office of Toyobo on December 13, 2013.

[3] Nobuya Fujiwara, general manager of the Aqua Membrane Division (as of 2016), explains as follows how the development of RO membranes at Toyobo was narrowed down to the hollow fiber type at an early stage. "For the shape, I think Toyobo decided on the hollow fiber type from the beginning. For materials, there were other candidates, though. In addition to CA, there was another material called polyamide... But even for that, what we were working on was a polyamide hollow fiber. So, since the keyword was 'hollow fiber' from the beginning, we didn't think about making spirals like other companies did in the early stage, and I think Toyobo decided to use fibers, and hollow fibers at that time." From the authors' interview with Nobuya Fujiwara at Toyobo's headquarters on October 11, 2013.

when aiming for differentiation.[4] This choice was also in line with the company-wide policy of promoting diversification by utilizing the technologies they already possessed.

8.1.2 The Establishment of Production Technology

To create a hollow fiber membrane with both high desalination and water permeability performance, it was necessary to create a homogeneous ultra-fine filament structure with a hollow fiber cross-sectional area of 0.2 mm in diameter and a hole inner diameter of 0.1 mm. To establish such advanced production technology, abundant contributions were made by skilled engineers in the field who inherited the production knowhow, including tacit knowledge, cultivated in the textile business. For example, as early as the end of 1976, one of the skilled engineers, Kazuto Hamada, achieved desalination performance that satisfied the quality requirements of pure water,[5] which surprised Ukai and his colleagues. The importance of the production technology inherited from the textile business is described by Ukai as follows.

> (Toyobo) started working with acrylic after the end of World War II. After that, they created a new fiber called Chinon based on acrylic. Mr. Hamada and Mr. Takada (who later became involved in the spinning of hollow fiber membranes) were involved in this. At the same time, there was a group that was working on a fiber called spandex that was made by chemically removing solvents. This and the wet technology (of Chinon) merged to form the technology needed to make hollow fibers later. For example, spinning hollow fibers is a very amazing technology. It is pulled continuously, but it is pulled while being invested with performance. Especially in the case of membranes, if they are pulled too much, they will not have desalination performance.[6]

The development team of Toyobo devised an original method not only for making hollow fibers but also for weaving the hollow fibers into elements by bundling them. In contrast to the method used by Dow and DuPont, in which hollow fibers are arranged in parallel and bundled to form elements, Toyobo adopted a method called "cross winding" in which hollow fibers are braided so that they cross each other, as shown in Fig. 8.1. This method had the advantage of making the membrane element less dirty and easier to clean because it changed the way the supplied water flowed.

---

[4] As described in Chap. 3, there was a problem that unfamiliar materials and processes had to be developed when the spiral type was adopted. For example, in the spiral type, a member called a spacer is inserted between two flat membranes. The spacer is an important component that affects the way the feed water flows and the volumetric efficiency of the RO membrane. The development and manufacturing of spacers requires technology not found in the textile industry.

[5] From the authors' interview with Tetsuo Ukai, at the Research Institute for Advanced Studies on January 24, 2014.

[6] From the authors' interview with Tetsuo Ukai, from the same source referenced in footnote 2. The text in parentheses was added by the authors.

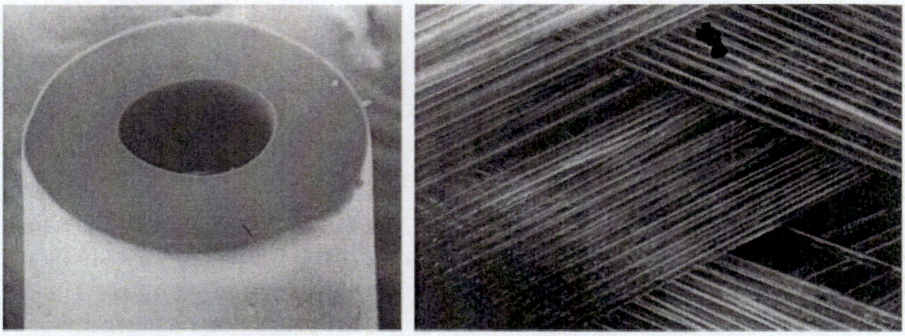

Cross-sectional view of hollow fiber membrane

Cross winding

Fig. 8.1 Hollow fiber membranes of Toyobo (as of 1980). *Source* Ukai et al. (1980), p. 4. *Note* The outer diameter and inner diameter of the membrane in the cross-sectional photograph are 225 μm and 110 μm, respectively

Cross winding also has the derivative effect of removing dirt from the hollow fiber membranes as they rub against each other.[7]

Cutting and sealing were also issues that arose in manufacturing the elements. First, it was necessary to cut the hollow fiber membranes without collapsing them to secure the necessary openings. Since the optimum cutting method varies slightly depending on the working environment, it took time to identify the optimum conditions. Next, the process of sealing both ends of the cut membrane element was also difficult. Since the performance of RO membranes degrades significantly when they dry out, the elements must be immersed in a liquid to keep them wet. However, it was not easy to make the sealant penetrate well into only ends of the element, not the entire element immersed in the liquid. Moreover, a high level of sealing that would not allow the supply water to leak out or deform the opening surface was required even when high pressure was applied.

The developers established the manufacturing process by solving these problems one by one through repeated trial and error with skilled workers at the plant. At the same time, they decided to externally procure the pressure vessel that is used to store the elements and make the modules. The hollow fiber RO membrane module developed in this way was named HOLLOSEP.

8.1.3 Feedback in the Demonstration Test

HOLLOSEP was first used in the desalination demonstration tests conducted by the WRPC established by the MITI. The first phase of the demonstration tests, which

[7] DuPont's hollow fiber membranes were arranged in parallel inside elements. In the parallel arrangement, the yarn had to be fixed by inserting a nonwoven fabric between each layer in order to arrange the spacing of hollow fiber membranes was constant (Matsui, 1980; Sekino, 1996).

lasted from 1974 to 1977, used foreign manufactured RO membranes from sources such as UOP, DuPont, and Dow, followed by Toyobo's HR5350, HR5350S, Toray's SC-5000A, SC-5000B, and Nitto's NRO-A, NRO-B. In the second phase of the demonstration tests, which started in 1978, Toyobo's HR8350/HR8650 and Toray's SC-5200 were used.[8]

Having sensed a positive response from the first phase of demonstration tests, Toyobo used cellulose triacetate (CTA), which was still in the basic research stage within the company at the time, as the membrane material for HR8350 and HR8650. The use of CTA had the advantage of increasing the pressure resistance of RO membranes and enabling them to withstand the high-pressure operation of seawater desalination. The developers at Toyobo clearly had seawater desalination in mind as an application of RO membranes. In addition, since CTA is less soluble in solvents than cellulose diacetate, unique knowhow was required to control the spinning process. This meant that if Toyobo could possess such knowhow, they would be able to open the way to seawater desalination and also aim at differentiation from other companies. Although both cellulose diacetate and CTA were easy to procure in the market, CTA was more effective at differentiating Toyobo because it was more difficult to handle.

In November 1978, six months after the start of the demonstration with CTA, Toyobo established the RO Business Development Department and stepped up its efforts to commercialize RO membranes. The development team also made efforts to develop a higher performance RO membrane. The result was HR8650 RO membrane module (desalination rate: 98.7%, flux: 20 m^3/day).

With the development of HR8650, Toyobo succeeded in realizing a single-stage seawater desalination process that had previously required a two-stage process. This attracted a great deal of attention from the industry. In addition, HR8650 was the first hollow fiber RO membrane module to incorporate two membrane elements, thereby increasing the amount of water desalinated and saving energy.

The following year, in July 1979, Toyobo installed a 10 m^3/day single-stage seawater desalination facility at the Iwakuni Plant. And in August 1979, the company delivered 52 RO membrane modules to a seawater desalination demonstration plant completed by the WRPC in Chigasaki with a desalination capacity of 800 m^3/per day.[9] Toyobo was responsible for the demonstration operation at this plant for the next 10 years, feeling a great response to the practical application of the technology.

In October, Toyobo announced that it would invest about 300 million yen to build a new RO membrane plant and commercialize it.[10] The former rayon building at the Iwakuni Plant was chosen as the production base because the research Institute in Katada responsible for starting up production had a limited production capacity. The

[8] The four digits on the Toyobo modules refer to the diameter, length (in feet), and operating pressure. This means that 8350 is a RO membrane module with a diameter of 8 inches, a length of 3 feet, and an operating pressure of 50 atm.

[9] Toray's CA spiral RO membrane (SC-5200) was also supplied for this plant and operated alternately in one-month intervals (Kunisada, 1981).

[10] Nikkei Sangyō Shimbun (October 9, 1979, p. 9). Based on the authors' interview with Ryuzo Sakamoto, June 16, 2016, Osaka.

Table 8.1 Early products

	Product name	Desalination rate (%)	Water permeation rate (m^3/day)	Module price	Application
Low pressure	HR3110	90	0.3	N.a	Water purification (Sterile water and ultrapure water)
	HR5110	90	1	N.a	
Medium pressure	HR5230	90	9	N.a	Water purification Brine water desalination Water recycling
	HR5330	90	15	Approx. 500 thousand yen	
	HR8330	90	40	Approx. 1 mil. Yen	
	HR8630	90	80	N.a	
High pressure	HR8350	98.7	10	N.a	Seawater desalination
	HR8650	98.7	20	N.a	

Source Prepared by authors based on Sōgō Hōsō Publication (1980)

Tufts Cell Division, which was familiar with cellulose, took over the job. Through technical training, production knowhow was transferred from the TRI to the plant, and production began in May 1980.

8.2 Commercialization

8.2.1 Initial Deployment

At that time, Toyobo's RO membrane modules were divided into three types according to their operating pressure: the low pressure type, medium pressure type, and high pressure type. The products corresponding to each type are shown in Table 8.1. The price was around 500,000 yen for the medium pressure type HR5330 and around 1,000,000 yen for HR8330.[11]

The HR series was often used in seawater desalination equipment for bonito fishing boats. Fishermen stored fish by diluting seawater with freshwater produced by the equipment to create water with an osmotic pressure equivalent to that of the water environment of the fish. It is desirable that the desalination equipment installed on fishing boats be as small as possible. In this respect, the hollow fiber type was advantageous because it could pack a large number of membranes into a module and had excellent volumetric efficiency. Hirohito Matsui, who was in charge of the

[11] Based on estimates by Sōgō Hōsō Publication (1980).

development of RO membranes at the TRI, stated the following in a lecture he gave in 1979.

> Although the amount of water permeated per unit area is one-third that of the spiral type, the membrane area per unit volume of the hollow fiber type is about 10 times that of the spiral type. So ultimately, the hollow fiber type has the largest amount of water treated per volume, about three times that of the spiral type and about 30 times that of the tubular type. Therefore, the hollow fiber type is very good for use in a limited area such as a ship.[12]

Small seawater desalination equipment for fishing boats was dominated by Toyobo. The introduction of desalination equipment to fishing boats began in 1978. By 1981, more than 600 fishing boats in Japan had been equipped with the equipment. Because of their important role in maintaining the freshness of fish, seawater desalination equipment sold well despite its high price of 2 to 3 million yen per set. Later, Toyobo expanded its RO membrane business to include the production of pure water for use in hospitals, such as hand-washing water for surgery.

However, whether it was for fishing boats or hospitals, the modules needed were small, and the scale of the business was not large enough. In terms of the cumulative sales volume until around 1980, domestic sales accounted for 80% and overseas sales only 20%.[13] The project needed to expand the scale of its business to demonstrate its presence both inside and outside of the company. And as a means of achieving this, the development of overseas markets became an important consideration. In particular, the company aimed to win orders for large-scale seawater desalination plants overseas. Tatsuo Nakahara, who served as general manager of the Technology Export Department, actively advocated overseas expansion and encouraged developers.[14]

8.2.2 *Expansion into Saudi Arabia*

Toyobo focused its attention on Saudi Arabia as a destination for overseas expansion. Saudi Arabia was suffering from a serious drought. To take advantage of its abundant crude oil, it had been obtaining fresh water by using the evaporation method that involves distilling seawater. DuPont was the first company to introduce the RO method there. Toyobo saw this move and followed suit.

However, recognition of the RO method and confidence in the technology were still low in Saudi Arabia, and the market was far from accepting. Therefore, in January 1981, Toyobo formed a caravan team to demonstrate seawater desalination in various cities to raise awareness of the technology. The first caravan conducted demonstrations in Jeddah, which overlooks the Red Sea on the western side of the country, for a half-year from March to August 1981. After gaining a certain level of recognition in Jeddah, the caravan traveled about 1,000 km east across the Arabian Peninsula to

[12] Matsui (1980) p. 39. Unit area refers to the membrane area and unit volume refers to the module volume.

[13] Sōgō Hōsō Publication (1980) p. 110.

[14] From the authors' interview with Tetsuo Ukai, from the same source referenced in footnote 2.

Al Khobar, which overlooks the Arabian Gulf, and conducted demonstrations from October 1981 to March of the following year.

The caravan conducted demonstrations on both the east and west coasts because of the different nature of seawater in the Red Sea and the Arabian Gulf. Normal seawater has a salinity of 3.5% and a maximum seawater temperature of about 30 degrees Celsius, whereas in the Red Sea, the salinity is 4.3% and the seawater temperature rises to about 35 degrees Celsius. In the Arabian Gulf, the salinity is even higher at 4.6% and the temperature rises to about 40 °C. Since differences in salinity and seawater temperature have a significant impact on the basic performance and durability of RO membranes, it was important to demonstrate them on both shores. Both demonstrations produced good results and succeeded in increasing awareness of the RO method in Saudi Arabia.

In 1982, Toyobo announced a three-year plan aimed at expanding the RO membrane business. This plan was ambitious in that it aimed to increase sales from approximately 500 million yen in FY1981 to 2 billion yen in FY1984.[15] In response to this announcement, the production line at the Iwakuni Plant was expanded to two lines, and its production capacity was increased to 70,000 tons per year in terms of water production. In this way, Toyobo embarked on the full-scale development of large-scale seawater desalination facilities in the Middle East market.

It was in 1983, when these series of expansion plans were underway, that Nobuya Fujiwara (general manager of Aqua Membrane Division as of 2016) joined the company and was assigned to the TRI where he and about five colleagues were in charge of RO membranes. By this time, the basic development of CA RO membranes had been completed, and the company was steadily setting up organizations for commercialization. In May of the same year, the Functional Membrane Division was newly established, and the RO membrane business became an independent business entity as an organization. In November, a new functional membrane plant was opened in Iwakuni, and production activities began in earnest.

It was not only RO membranes that were produced at the Iwakuni Functional Membrane Plant. The Artificial Kidney Hollow Fiber (AKH), which was developed in March 1978 and commercialized in 1981, has also been produced at the same plant. AKH products were developed using CTA, following the lead of RO membranes. In 1985, the company set a goal of producing 500,000 units per AKH month and focused on exporting it overseas.[16]

[15] Nikkei Sangyō Shimbun (June 10, 1982, p. 7). Nitto Denko's sales of membranes at that time were said to be about 600 million yen (Nikkei Sangyō Shimbun, October 1, 1982, p. 15).

[16] Nikkei Sangyō Shimbun (June 6, 1983, p. 11). This was a product jointly developed with Nissho, a medical equipment manufacturer.

8.2.3 *Orders for Large Projects and Stagnation*

In 1984, Toyobo was awarded a contract to supply RO membrane modules for seawater desalination plants in Haql and Duba, Saudi Arabia, partly due to successful demonstrations. Toyobo delivered the RO membrane modules through the intermediary of the ITOCHU Corporation to Envirogenics Systems, a U.S. plant engineering company, which was awarded a contract by the SWCC. Of the 27-billion-yen Envirogenics Systems bid, Toyobo received an order for about 1.5 billion yen.[17] The total desalination capacity of the two facilities reached 11,000 m^3/day (6,600 m^3/day for Haql and 4,400 m^3/day for Duba). This was the first large-scale project for Toyobo. Seizing this opportunity, Fujiwara was transferred from the TRI to the Functional Membrane Division where he was responsible for improving the structure of the membrane module and promoting its practical application.

Orders for large-scale projects continued to be received in 1986. Mitsubishi Heavy Industries (MHI) won the bid for Jeddah 1 Phase 1, the world's largest seawater desalination plant at the time with a capacity of 56,800 m^3/day. MHI adopted RO membrane modules from Toyobo. It was reported that MHI's bid was worth about 6 billion yen, of which Toyobo's order was worth about 1.5 billion yen.[18] In response to this order, Toyobo developed HM10255 RO membrane module, which has an increased water permeability by enlarging its diameter to 10 inches, and delivered 1,480 units (2,960 elements).

Although the company succeeded in winning a large-scale project in Saudi Arabia, this did not mean that the business was secure. In the case of large-scale seawater desalination projects, once the supply of large numbers of modules to coincide with the start-up of a facility is completed, demand ceases until the time when the elements are replaced, and the business enters a period of stagnation.

Figure 8.2, which shows the cumulative desalination capacity of RO membrane modules for seawater desalination, reveals this situation. After receiving a large order, maintaining the factory operating rate became a serious problem.

In addition to the problems associated with production fluctuations, Toyobo was also hit by problems that occurred at "Jeddah 1 Phase 1" in the 1990s. When commercial operations began in May 1989, it was thought that "Jeddah 1 Phase 1" would continue to operate smoothly.[19] However, after about a year and a half, the desalination performance began to deteriorate rapidly. And in 1991, the condition became so serious that it could not be overlooked.

---

[17] Nihon Keizai Shimbun (March 2, 1984, p. 10).

[18] Nikkei Sangyō Shimbun (January 20, 1987, p. 17). An evaporation method plant had been constructed at this site in 1970. In order to update and expand the desalination capacity of the plant, the construction of a seawater desalination facility using the RO method was planned to replace the evaporation method.

[19] Hideo Iwahashi and Masahiko Nagai of MHI investigated the operation of the plant for eight months from April through November 1989, and reported that the plant was operating smoothly (Iwahashi and Nagai, 1990).

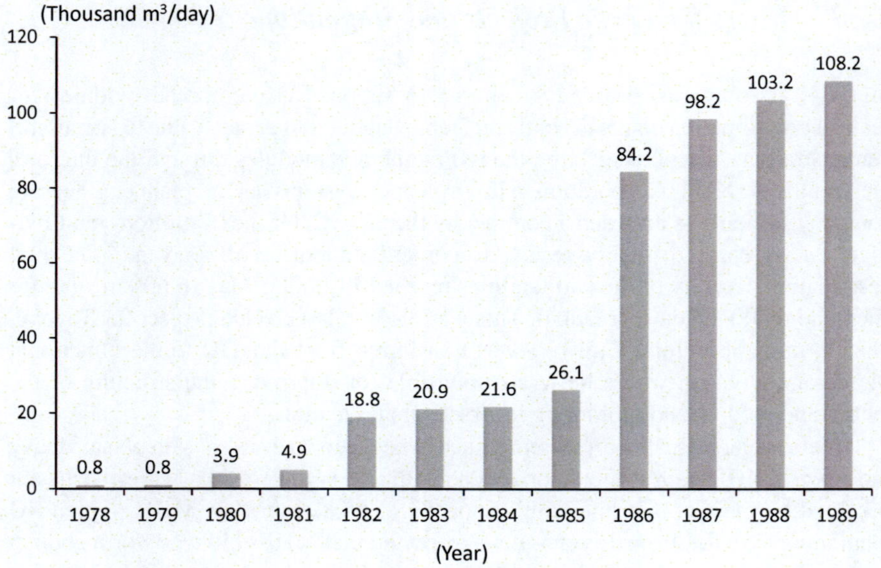

Fig. 8.2 Cumulative capacity of the RO membrane system for seawater desalination at Toyobo. *Source* Prepared by the authors based on Kumano (1992). *Note* Contract-based values

8.3 Repeated Struggles

8.3.1 Troubleshooting in Jeddah

The trouble with Jeddah 1 Phase 1 caused a great deal of concern for the SWCC. This was because Jeddah 1 was set to be followed by construction for Phase 2 (56,800 m^3/day). The contract for Phase 2, signed in 1991, called for plant engineering by MHI and the delivery of 1,480 HM10255 membranes by Toyobo, as in Phase 1. There was concern that the problems encountered in Phase 1 would have an impact on Phase 2.

In November 1991, Toyobo and MHI formed an emergency response team with about 20 members to investigate the cause and hasten to solve the problem. As a result, they discovered that there was a problem with chlorine sterilization. At Jeddah 1, the continuous chlorine disinfection method had been adopted. The frequency of chlorine disinfection was increased to keep up with the contamination of the Red Sea. However, many heavy metals such as copper (Cu) and cobalt (Co) were dissolved in the Red Sea near Jeddah, and they caused the catalytic effect of oxidizing and cleaving the main chain of CTA, which significantly deteriorated the desalination performance.

After identifying the cause of the problem, Toyobo devised a method called the intermittent chlorine injection (ICI) method and solved the problem by moderately reducing the frequency of chlorine injection into the membrane module.[20] This method of operation enabled the membranes to be sufficiently sterilized while reducing the burden of chlorine on the membranes, and thus high desalination performance could be maintained for a long time.[21] Toyobo solved the problem in May 1992, only six months after the emergency response team was organized. After careful field tests, the ICI method was officially adopted in February 1994.

Based on this achievement, the ICI method was introduced into Phase 2 from the start of commissioning in March 1994. Toyobo's prompt response to the trouble brought relief to the SWCC and gained their trust. Fujiwara commented as follows.

> There is no doubt that the resolution of this trouble between November 1991 and May 1992 was an important event. The owner of this plant is the Saudi government. We gained the trust of the Saudi government. We gained their trust because we solved the trouble in a short period of time. This is another reason why we became strong in Saudi Arabia in addition to purely technical reasons.

In 1993, MHI won the bid to construct a seawater desalination plant (planned desalination capacity: 128,000 m^3/day) in Yanbu, a city facing the Red Sea, to pump water to the industrial city of Medina. Toyobo delivered 3,210 HM10255 membranes. With the addition of the Yanbu plant, the total volume of water produced by the SWCC using RO was 405,500 m^3/day. According to the calculation, about 60% of this would be produced using Toyobo membranes.[22] In response to this series of orders, Toyobo invested 500 million yen to increase the production capacity of its Iwakuni Plant by 50%, which can produce 300 membranes per month.[23]

Although Toyobo earned the trust of the SWCC by quickly resolving problems, the SWCC continued to be cautious. For example, unlike in Phase 1, it demanded that MHI and Toyobo guarantee the membrane replacement rate in the Phase 2 and Yanbu projects. Specifically, in Phase 2, MHI and Toyobo were required to guarantee a membrane replacement rate of 0% for the first year and 10% or less per year for the remaining four years of operation.[24] For the Yanbu plant, the cumulative membrane replacement rate was required to be kept below 60% ($15\% \times 4 = 60\%$) while meeting the WHO water quality standards for the first five years of operation (Iwahashi, 1996). Despite these high assurance demands, the ICI method was effective and Toyobo was not troubled by the requirements.

However, extending the life of the membrane modules was not necessarily desirable in terms of business development, as it led to a decrease in replacement demand.

---

[20] It might be possible to sterilize membranes continuously by controlling the chlorine concentration below 0.1 mg/L. However, in reality, it is difficult to keep controlling it, so Toyobo devised a method to sterilize membranes moderately by controlling the time when chlorine passes through the membrane module. This method is also thought to have the advantage of preventing microorganisms from becoming resistant.

[21] Nada et al. (1994); Kumano and Tanaka (2013).

[22] Nikkei Sangyō Shimbun (June 7, 1993, p. 11). Percentages are based on desalination capacity.

[23] Nikkei Sangyō Shimbun (June 21, 1993, p. 21).

[24] Iwahashi (1996).

In addition, the Saudi Arabian government was experiencing financial difficulties at the time, and the operation of the Yanbu plant was frozen until 1998. The RO membrane business had been unable to seize the opportunity for growth.

8.3.2 The Search for New Materials

While promoting the business development of CA-based RO membranes mainly for seawater desalination, Toyobo also developed membranes using new materials in parallel from 1981. In light of developments in other companies' polyamide-based membranes that were suffering from low chlorine resistance, Ukai designated the developers to "conduct research on 'Innovative RO' that is resistant to chlorine."[25]

One of the developers who received this designation was Shigetsugu Konagaya, who joined the company in 1974 and a member of the TRI. Konagaya formed a synthesis group with Osamu Watanabe, Haruhiko Narusawa, and Masaya Tokai. Then, they started a literature survey on the molecular structure of membranes with high basic performance and excellent chlorine resistance as well as the evaluation of chlorine resistance.

Based on the literature survey, they attempted to synthesize a wide range of polymers and first identified about 900 types of new materials as candidates. From there, they divided the search among the materials and narrowed their list down to about 600 types. Konagaya was in charge of aromatic polyethers and epoxy resin crosslinked materials containing aromatic rings, Watanabe and Tokai were in charge of aromatic polyamides, and Narusawa was in charge of aromatic polyamide-imides and aromatic polyimides. After that, they evaluated about 200 types for their chlorine resistance, about 150 types for their membrane formability, and about 100 types for their desalination rate and water permeability.

In this process, they thought that the low chlorine resistance of the aromatic polyamide film was caused by the formation of N-chloroamide, and they proceeded to develop a method to prevent the formation reaction.[26] They narrowed down their search to about 100 candidates and finally developed an aromatic copolyamide using diaminodiphenyl sulfone, which has an electron-withdrawing group.

The new asymmetric membrane made of aromatic copolyamide was planned to be used for seawater desalination, but unfortunately, its pressure resistance was insufficient. A pressure of about 60 atm is said to be necessary for desalination. When this level was applied, the inner diameter of the hollow fiber deformed, and sufficient performance could not be achieved. Therefore, the developers had to change their target to brine desalination that works at low pressure.

In 1985, only Konagaya was transferred from the Synthesis Group to the Film Formation Group. In the Group, he conducted experiments with Yoichi Matsui,

---

[25] From the authors' interview with Shigetsugu Konagaya at Nagoya University on January 15, 2015.

[26] Nita (1993) p. 372.

Kazuhide Nita, Eiji Kuzumoto, Morio Miyagi, and Atsuo Kumano. After conducting a series of experiments on a beaker scale, Konagaya visited the Takefu Plant of Toyo Kasei Kogyo every month to improve his synthesis skills to the mass production level. It was not until 1989 that mass production level film manufacturing knowhow was established.

Although the performance of this membrane was good, the manufacturing cost was very high. In addition, a large amount of money was required to treat the waste liquid generated in the manufacturing process.

8.3.3 Struggles in Ultrapure Water Production for Semiconductors

Around the time that Konagaya and his colleagues changed the application of the new membrane to brine desalination, the use of ultrapure water production for semiconductors was rapidly expanding.

From the late 1970s to the 1980s, as the miniaturization of semiconductors continued at a rapid pace, there was a demand for ultrapure water with organic matter and fine particles removed as much as possible from the cleaning water used in manufacturing. The RO method, which was different from the conventional ion exchange method, attracted attention. Toray and Nitto Denko were among the first to respond to this demand. While the seawater desalination market failed to expand, the emergence of the market for semiconductor manufacturing was a precious gift.

However, this market was not one in which Toyobo's technological strengths could be utilized. The market for ultrapure water for semiconductors requires a high water permeability and desalination rate under low pressure operation. Unfortunately, CTA hollow fiber membranes have a fundamental problem in that their desalination rate drops significantly when the operating pressure is lowered. Under low pressure operation, it could not compete with polyamide composite membranes in terms of water permeability. Therefore, in the 1980s, Konagaya was hardly even considering applications for semiconductors.

It was not until the 1990s that Toyobo's developers began to clearly target the market for semiconductor manufacturing. According to Satoshi Tanaka, who joined Toyobo in 1990 and was involved in the development of polyamide RO membranes at the TRI, new membranes not suitable for seawater desalination due to pressure resistance problems, were launched for semiconductor manufacturing application.[27]

In June 1991, Toyobo announced that it would introduce polyamide hollow fiber membranes to the ultrapure water market for semiconductor manufacturing in the fall of the same year.[28] The chlorine resistance of this new membrane was reported to be six times higher than that of polyamide RO membranes from other companies. Even after 8,000 h of operation, the new membrane could maintain the same

[27] From the authors' interview with Satoshi Tanaka, February 1, 2016, in Jeddah.

[28] Nikkei Sangyō Shimbun (June 5, 1991, p. 22).

desalination rate as Toyobo's CTA RO membrane HA series. However, the market launch did not go smoothly. In the fall of 1991 when the product was planned to be launched, the original announcement was revised to say that the product would be introduced in 1993, another two years later.[29] The products were actually introduced in 1993. Toyobo sold two types of HS series, low-pressure HS-3110 and HS-5110, and medium-pressure HS-5230 and HS-5330.

The exploration of domestic industrial applications mainly for the production of ultrapure water for semiconductors was especially important for Toyobo's RO membrane business. Orders from seawater desalination plants overseas were extremely volatile. Therefore, to ensure the stability of the business, it was important to incorporate industrial applications, which generate demand in a short cycle, into the business portfolio. Fujiwara recalls his approach to business development at the time as follows.

> For the overseas market, we focused almost exclusively on seawater desalination. But in Japan, the market for pure water production was still large, so we were working on both. Overseas, desalination is a project, so it's not a business that earns money every day. In addition, that alone was not enough to make it a viable business, so we also had to work on domestic businesses that were close to home. Desalination was our main focus, but we also wanted to develop pure water as a business that could earn money every day.[30]

However, the polyamide hollow fiber membranes developed by Matsui, Nita, Tanaka, and their colleagues were asymmetric membranes and could not outperform the composite membranes of other companies in water permeability. In addition, the problem was that the desalination rate did not increase under low-pressure operation. Toyobo was unable to penetrate the ultrapure water market dominated by Toray and Nitto Denko.

To solve the technical problem of asymmetric membranes, Atsuo Kumano (joined Toyobo in 1983) and Akihiro Ariji (joined Toyobo in 1994) took up the challenge of developing polyamide composite membranes at the TRI. In order to manufacture a composite structure with high water permeability efficiently and at low cost, it was necessary to establish an interfacial polymerization method. However, it was extremely difficult to form hollow fiber membranes by interfacial polymerization. For example, if yarn touched the equipment even slightly during the spinning process, the dense separation layer would be damaged. In response to this problem, they developed a new technology for making hollow fiber composite membranes, but were unable to increase productivity as much as they wanted. In the end, Toyobo was unable to produce a polyamide hollow fiber membrane with a composite structure.

Struggles in the semiconductor manufacturing market led Toyobo to re-focus on its traditional CTA hollow fiber membranes. Toyobo's developers concluded that they should concentrate on markets where the technical strengths of CTA hollow fiber membranes could be exploited. They targeted the seawater desalination market, especially in areas where microorganisms tend to thrive, as the market where chlorine resistance that is the decisive advantage of CTA hollow fiber membranes would be

[29] Nikkei Sangyō Shimbun (October 9, 1991, p. 1).

[30] From the authors' interview with Nobuya Fujiwara at Toyobo's head office on October 11, 2013.

most effective. Saudi Arabia was the ideal market, as it faces such an area and has a large demand for water.

8.4 Business Expansion

8.4.1 The Impact of Company-Wide Structural Reform

The 1990s was a period when Toyobo fell into a very difficult financial situation not only in the RO membrane business but also company-wide. As shown in Fig. 8.3, the return on sales continued to decline from FY1990, falling to 1.3% in FY1995. The need for reform to improve profitability was strongly recognized across the company.

The pressure on the RO membrane business naturally intensified, and it was no longer permissible for the business to continue to operate at a loss. Therefore, the RO membrane business reduced the number of employees in the division to about three in order to reduce personnel costs, and in FY1998 the division somehow managed to turn a profit for the year. At that time, Fujiwara, who was actually leading the RO membrane business, tried to keep the business alive by continuing to report profits.

Although the RO membrane business managed to return to profitability, it did not have a significant impact on the recovery of the company's performance. In June 1999, Junji Tsumura, who had been newly appointed president, publicly stated both

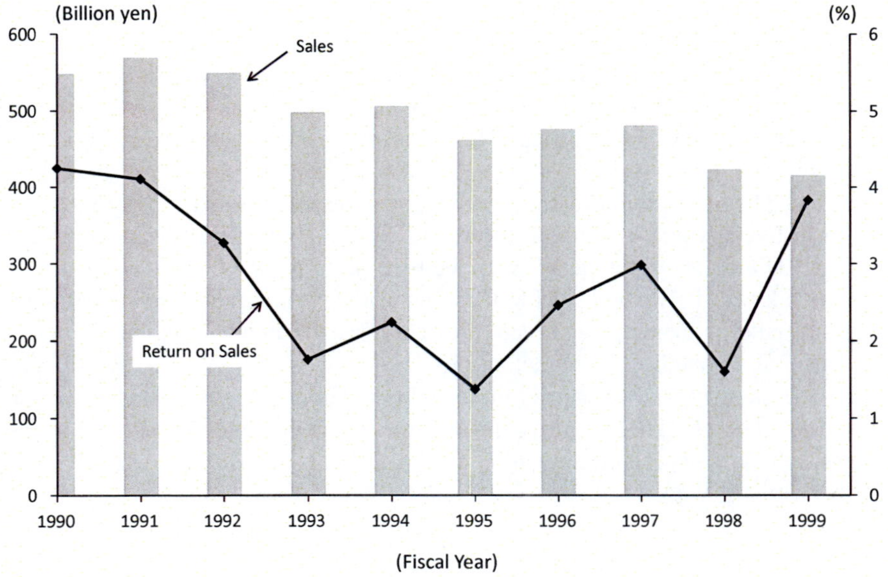

Fig. 8.3 Sales and return on sales of Toyobo. *Source* Toyobo "Securities Report" for each year. *Note* The left axis shows the bar graph and the right axis shows the line graph

inside and outside the company his philosophy that "Losses are bad and profits are good" and proceeded with thorough restructuring.[31] Naturally, the RO membrane business also came under further pressure to improve profitability.

Fortunately, since the AKH business was doing well, the performance of the functional membrane business as a whole was good. Although AKH and RO membranes were viewed as independent entities, and their earnings were managed separately, RO and AKH shared the same technological base of hollow fiber membranes, which meant that they were able to exchange personnel not only in R&D but also at their plants. By sharing fixed costs with well-performing AKH business, the development and commercialization of RO membranes continued, albeit at a slower pace. In December 1999, the Iwakuni Plant began a project to develop a higher performance RO membrane that had been underway since 1996.

While pressing hard to improve the profitability of each business, Toyobo's management was strongly promoting the development of new businesses to shift from the traditional textile business. The company promoted a shift to what it called "specialty businesses", which were businesses that could meet market needs and establish a long-term competitive advantage through proprietary technologies. The RO membrane business was expected to be one of the businesses that could play a leading role in the future, as it would meet the social need for water shortages by utilizing the company's hollow fiber technology.

8.4.2 Expansion in the 2000s

In May 2000, Toyobo officially signed a contract to supply RO membrane modules to a seawater desalination plant (desalination capacity: 50,000 m^3/day) to be constructed in Fukuoka, Japan. The adoption of the modules for Japan's largest seawater desalination plant attracted a great deal of attention and expectation within the company.

In December of the same year, the development members from the TRI in Katada moved to Iwakuni and established the FROP Group, a development project dedicated to RO membranes. Together with the high-performance RO project that was already underway in Iwakuni, the FROP Group worked to make the membranes more porous, smaller in diameter, and more pressure resistant. In addition, they changed the conventional method of extracting produced water from only one side of the membrane element, and actualized the opening of the membrane element at both ends. The result was HB10255FI, a 10-inch membrane module that was delivered in June 2005 to a seawater desalination plant in Fukuoka, Japan, where 1,000 units (2,000 elements) were installed.

Business opportunities also arose overseas. When DuPont announced its withdrawal from the RO membrane market in 1999, the possibility emerged of capturing

---

[31] For more information on the company-wide structural reform process, see Fujiwara and Aoshima (2016).

replacement demand for the RO membrane elements they had delivered. In particular, since Toyobo was the only company that sold the same hollow fiber membrane modules as DuPont, it had the potential to monopolize replacement demand after the withdrawal. Table 8.2 shows the list of contractors and adopted membrane manufacturers for large desalination plants around 1999, showing that DuPont membranes were used in six out of ten plants. This indicates that there emerged a good business opportunity for Toyobo to enter markets from which DuPont had retreated.

The fact that some customers were dissatisfied with DuPont's membranes was also a tailwind for Toyobo. Since DuPont's polyamide-based membranes were not chlorine-resistant and could not be directly sterilized by chlorine, they frequently clogged due to the growth of microorganisms. In addition, some DuPont membrane elements were not sealed properly, causing water leakage in some cases. The dissatisfaction of these customers was also conveyed to Toyobo. As Toyobo became more deeply involved in the Saudi Arabian market, it began to receive a variety of information from the SWCC and other seawater desalination plant operators.

Having grasped the dissatisfaction of its customers, Toyobo aggressively pursued sales activities to replace DuPont. After the withdrawal of DuPont, the company launched HB9155 in 2002 and HJ9155 in 2005, both of which were compatible with DuPont's membrane elements. This made it possible to replace only the inner

Table 8.2 Top 10 desalination plants around 1999

Plant name	Country	Capacity (m^3/day)	Prime contractor	Beginning of operation	Membrane manufacturer
Medina-Yanbu	Saudi Arabia	128,000	Mitsubishi Heavy Industries	1998	Toyobo
Al Jubail	Saudi Arabia	90,900	Preussag	1999	DuPont
Jeddah 1 Phase 1	Saudi Arabia	56,800	Mitsubishi Heavy Industries	1989	Toyobo
Jeddah 1 Phase 2	Saudi Arabia	56,800	Mitsubishi Heavy Industries	1994	Toyobo
Marbella	Spain	56,400	DECOSOL	1998	DuPont
Pembroke	Malta	54,000	Volumetrics	1993/1994	DuPont
Al Dur	Bahrain	45,000	Weirwest Gathge	1989	DuPont
Mallorca	Spain	42,000	De Gramont	1997	DuPont
Okinawa	Japan	40,000	Kurita Water Industries and others	1996/1997	Toray Nitto Denko
Dhekelia II	Cyprus	40,000	Cadagua	1998	DuPont

Source Sekino and Fujiwara (1999) p. 440

membrane elements with Toyobo's products without changing the pressure vessel already delivered by DuPont.

Toyobo continued to explain that the "high chlorine resistance" of CA materials lowers the cost of desalination. If the chlorine resistance is high, inexpensive and general-purpose chlorine disinfectants can be used, and the number of cleaning cycles can be reduced. Fewer cleanings not only mean lower costs for the chemicals used, but also less downtime to shut down equipment for cleaning. For example, the desalination plant in Jeddah only needs to be cleaned twice a year. On the other hand, when polyamide RO membranes are used, it is said that they need to be cleaned once every two to three months.[32] Therefore, although Toyobo's membrane modules are more expensive than those of other companies, the desalination cost per unit is not necessarily higher. Fujiwara points out the following in this regard.

> In a desalination plant, the cost of producing water is the key to success. For example, in the case of other companies, it takes a lot of time to clean the membranes, or they have to clean them many times. If you compare them with actual plants, you can see that although our membranes are expensive, the cost of producing water is low. This is because membranes account for only 7% or so of the construction cost of a plant. Even if the cost of membranes is high, it doesn't have much impact on the plant construction cost itself. When you consider the cost of cleaning the membrane, downtime, and the life of the membrane, it is cheaper to use our membranes. It's not a comparison of individual membranes, but rather the cost of producing water, which is of course thoroughly calculated and reviewed by the end-user.[33]

In the 2000s, long-awaited demand for membrane replacement began to emerge in Saudi Arabia. And in FY2004, Toyobo's RO membrane business finally eliminated its accumulated deficits.

8.4.3 Business Performance

As the price of crude oil soared in the 2000s, Saudi Arabia's finances were boosted, and the government promoted the construction of new and larger seawater desalination plants. A large-scale plant with a desalination capacity of 218,000 m^3/day was planned for Rabigh on the Red Sea coast, for which Toyobo received an order for membrane modules in 2005. In response to the increased order volume, Toyobo expanded the production capacity at the Iwakuni Plant by 1.5 times from 2005 to 2006.[34] Large orders continued to come in after Rabigh, with Shuqaiq-II (240,000 m^3/day) in 2008, Jeddah RO 3 (260,000 m^3/day) in 2010, and Ras Al-Khair (345,000 m^3/day) in 2012.

Behind these series of orders was the growth strategy of Ryuzo Sakamoto who took over the baton from Tsumura in 2005 and became president. In 2008, Sakamoto

[32] From the authors' interview with Yaser Zaki Al-Jehani at the SWCC on February 3, 2016.

[33] From the authors' interview with Nobuya Fujiwara at Toyobo's headquarters on November 15, 2013.

[34] Nihon Keizai Shimbun (November 29, 2005, p. 13). It reports that the monthly production would be expanded from 20 to 30 tons.

Table 8.3 Major large seawater desalination plants that use Toyobo's RO membrane modules

Plant (Country)	Capacity (m^3/day)	Beginning of operation (Year)
Ras Al-Khair (Saudi Arabia)	345,000	2014
Jeddah RO3 (Saudi Arabia)	260,000	2013
Shuqaiq II (Saudi Arabia)	240,000	2010
Rabigh (Saudi Arabia)	218,000	2008
Yanbu (Saudi Arabia)	128,000	1998
Rabigh 2 (Saudi Arabia)	109,000	2015
Jubail (Saudi Arabia)	85,000	2007
Jeddah RO1 (Saudi Arabia)	56,800	1989
Jeddah RO2 (Saudi Arabia)	56,800	1994
Marafiq Yanbu (Saudi Arabia)	50,400	2005
Fukuoka (Japan)	50,000	2005
Al Dur (Bahrain)	45,500	2005
Manifa (Saudi Arabia)	27,000	2012
Florida (U.S.A.)	11,400	2005
Tanjung Jati B (Indonesia)	10,800	2005

Source Prepared by the authors based on http://www.toyobo.co.jp/news/2014/release_4747.html (Accessed January 17, 2019)

completed the structural reforms that he had inherited from Tsumura, and pushed ahead with investment activities aimed at new growth. In 2010, the company, together with ACWA of Saudi Arabia and the ITOCHU Corporation, established the Arabian Japanese Membrane Company, LLC (AJMC) in Rabigh and invested 700 million yen to build a plant there. This plant, which started operation in 2012, is gradually transferring manufacturing technology from Japan to the local area, and the scope of local production is gradually expanding.[35]

Partly as a result of these investment activities, Toyobo succeeded in deeply penetrating the Saudi Arabian market. Table 8.3 shows 15 major desalination plants that use Toyobo's membrane modules. The top 10 plants are located in Saudi Arabia, indicating the concentration of the Saudi Arabian market.

As of 2016, Toyobo's RO membrane business is estimated to have generated sales of around four to five billion yen.[36] Although the sales itself is not large, a profit margin is thought to be high. As shown in Fig. 8.4, while the profit margin of overall Toyobo has been around 5%, the profit margin of the business segment including RO membranes has been around 15%. It is thought that the company has secured high profit margins by focusing on specific areas where its technological superiority can be utilized and by occupying the replacement demand for those areas.

[35] Based on the authors' interview at AJMC on February 2, 2016.

[36] The authors' estimation from various reports and public information.

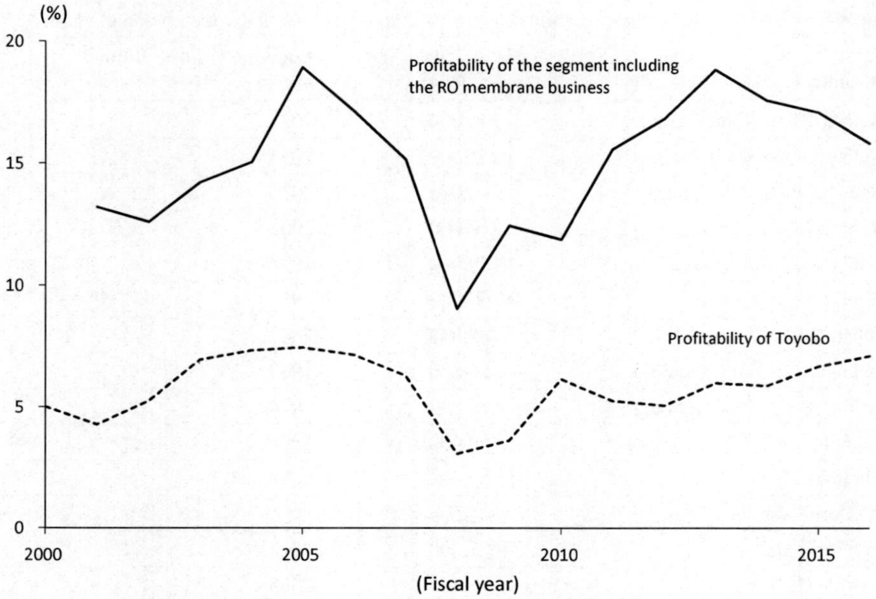

Fig. 8.4 Toyobo overall and segment profit margin. *Source* Prepared by the authors based on Toyobo's "Securities Report" for each year. *Note* Segment profit ratio values are values for the following divisions.
FY2001 – FY2004: Bio/Medical/Functional Products
FY2005 – FY2014: Life Science, FY2015 onward: Healthcare

8.5 Summary

Toyobo's motivation for launching RO membranes was to establish a new business. The developers focused on RO membranes in their search for a business opportunity to utilize the technology and knowhow they had accumulated in the textile business. To convert their own technology, they selected a hollow fiber membrane that would allow them to utilize their spinning technology and adopted CA materials that they were familiar with. The company's goal was to desalinate seawater, but it also developed small-scale applications such as fishing boats and hospitals for which it could see demand.

In the 1980s, the company turned its attention to Saudi Arabia with the aim of entering the market for seawater desalination. After gaining the trust of the SWCC through its prompt response to the problems that occurred at Jeddah Phase 1 in the early 1990s, Toyobo was able to advance its business development in Saudi Arabia.

In the 1980s, the ultrapure water market for semiconductor manufacturing expanded rapidly in Japan. However, Toyobo's CA-based membranes were unable to compete with competitors in the market for ultrapure water production, where performance under low-pressure operation is important. Therefore, developers at Toyobo developed a hollow fiber RO membrane using polyamide materials. However, this could still not surpass the polyamide composite membrane in terms of both water permeability and desalination performance. Finally, they were forced to withdraw from this market.

With the withdrawal from the ultrapure water market and the lack of growth in demand for seawater desalination, Toyobo's RO membrane business fell into a predicament that could be described as a crisis of survival starting in the mid-1990s. However, despite this, the business was able to survive. One of the reasons for this was a thorough downsizing of the organization to reduce losses. Fujiwara expected that demand for replacement membranes would surely increase in the 2000s, and he chose to endure the situation for the time being. The fact that the RO membrane business was highly socially responsible may have been another reason why some losses were tolerated. It was also positioned as one of the new specialty businesses in the company-wide structural reform process. In addition, strong sales of hollow-fiber membranes for artificial kidneys, for which the company shares a technological base, provided a tailwind.

After overcoming the crisis of the late 1990s, the business entered a period of expansion in the 2000s. Toyobo focused on the large-scale seawater desalination market in Saudi Arabia, where the high chlorine resistance of CTA hollow fiber membranes could be utilized. There, the company highlighted to customers and plants that the unit price of desalinated water could be lowered by saving cleaning costs and shortening downtime. Saudi Arabia has a large number of desalination plants based on the evaporation method, which consumes a large amount of crude oil. In order to conserve oil as a strategic export commodity, it is expected that Saudi Arabia will continue to replace its evaporative desalination plants with RO. This replacement demand alone is still considered to have sufficient growth potential. The list of Toyobo's RO membrane modules for seawater desalination is shown in Table 8.4.

If we were to describe the character of Toyobo's RO membrane business in one phrase, it would be "narrowly focused". The company has thoroughly narrowed down its technologies and markets, and competes only in areas where it can make the most of its strengths. As a result, although the scale of sales is small, the company has secured high profitability. However, this strategy of narrowing down was not necessarily intended from the outset. It was formed in the process of learning from mistakes, such as the defeat in the ultrapure water market for semiconductor manufacturing and the troubles at Jeddah. It could be said that the source of the company's competitive advantage lies in this series of emergent learning capabilities.

Table 8.4 List of Toyobo's RO membrane modules for seawater desalination

Series name	Product name	Launch (Year)	Membrane module external dimensions (Outer diameter mm x length mm)	Water permeation rate (m^3/d)	Salt removal rate* (%)	Pressure measurement requirement (MPa)	Operation pressure range (MPa)	Opening format (One end or both ends)	Notes
HR series	HR3155	1976	104 × 400	0.4	99.6	5.39	Not over 5.9	One end	
	HR5155	1976	153 × 444	1.2	99.6	5.39	Not over 5.9	One end	
	HR5255	1977	153 × 825	3	99.6	5.39	Not over 5.9	One end	
	HR5355	1977	153 × 1230	5	99.6	5.39	Not over 5.9	One end	
	HR8355	1977	305 × 1330	12	99.6	5.39	Not over 5.9	One end	
HM series	HM8255	1978	298 × 2602	27.5	99.6	5.39	Not over 6.4	One end	
	HM9255	1984	344 × 2613	35	99.6	5.39	Not over 6.86	One end	
	HM10255	1987	390 × 2863	45	99.6	5.39	Not over 6.86	One end	
HB series	HB9155	2002	278 × 1498	15	99.6	5.39	Not over 8.23	One end	Exchangeable with DuPont modules
	HB10255	2003	400 × 3097	62	99.6	5.39	Not over 8.23	One end	
	HB10255FI	2004	400 × 2875	67	99.6	5.39	Not over 8.23	Both ends	
HJ series	HJ9155	2005	294 × 2051	34	99.6	5.39	Not over 8.23	Both ends	Exchangeable with DuPont modules
HL series	HL10255	2006	380 × 4433	95	99.6	5.39	Not over 6.86	Both ends	
HU series	HU10255	2008	396 × 4720	95	99.6	5.39	Not over 6.86	Both ends	

Source Prepared by the authors based on the Toyobo website (accessed on December 25, 2017) and the authors' interviews
*Salt removal rate = (1 – permeate salt concentration / feed salt concentration) × 100 (%)

References

Fujiwara, M., & Aoshima, Y. (2016). Toyobo: Promoting fundamental corporate reforms. *Hitotsubashi Business Review, 64*, 124–141.

Iwahashi, H. (1996). Aspects of the sea water desalination technology. *Bulletin of the Society of Sea Water Science Japan, 50*, 250–256.

Iwahashi, H., & Nagai, M. (1990). Operation of the largest seawater reverse osmosis desalination plant in the world. *Bulletin of the Society of Sea Water Science Japan, 44*, 146–151.

Kumano, A. (1992). Recent trends in water desalination technology by reverse osmosis. *Journal of the Society of Fiber Science and Technology Japan, 48*, 28–34.

Kumano, A., & Tanaka, T. (2013). Development of hollow fiber reverse osmosis membrane for seawater desalination and actual operating plants. *Bulletin of the Society of Sea Water Science Japan, 67*, 264–272.

Kunisada, Y. (1981). Technological development of reverse osmosis seawater desalination in Chigasaki. *Bulletin of the Society of Sea Water Science Japan, 35*, 82–92.

Matsui, H. (1980). Chūkū sen-igata gyakushintō module ni tsuite (On hollow fiber type reverse osmosis modules). *Membrane, 5*, 33–49.

Nada, N., Iwahashi, H., & Umemori, F. (1994). Test result of the intermittent chlorine injection method in Jeddah 1 plant. *Desalination, 96*, 283–290.

Nihon Keizai Shimbun. March 2, 1984; November 29, 2005.

Nikkei Sangyō Shimbun (Nikkei Business Daily). October 9, 1979; June 10, 1982; October 1, 1982; June 6, 1983; January 20, 1987; June 5, 1991; October 9, 1991; June 7, 1993; June 21, 1993

Nita, K. (1993). New chlorine resistant copolyamide RO membrane with hollow fiber configuration. *Membrane, 18*, 371–374.

Sekino, M., & Fujiwara, N. (1999). Newly developed economical sea water desalination system by hollow fiber reverse osmosis module. *Bulletin of the Society of Sea Water Science Japan, 53*, 439–444.

Sekino, M. (1996). Aspects of the sea water desalination technology: Hollow fiber reverse osmosis module. *Bulletin of the Society of Sea Water Science, Japan, 50*, 231–239.

Sōgō Hōsō Publication. (1980). *Kōbunshi takōshitsu maku no kaihatsu dōkō to shijō jittai: Gyakushintō maku, gengai roka maku, seimitsu roka maku, ion kōkan maku* (Development trend and market status of polymer porous membranes: Reverse osmosis, ultrafiltration, microfiltration, and ion exchange membranes). Okinawa: Sōgō Hōsō Publication.

Toyobo Corporation. (1986). Toyobo 100-nen shi: jou/ge (100-Year History of Toyobo: Volume 1 and 2). Osaka.

Ukai, T., Nimura, Y., & Matsui, H. (1980). Gyakushintō hō no saikin no dōkō (Recent trends in the reverse osmosis method). *Nenryō oyobi nenshō (Fuel or Burn), 47*, 1–16.

Chapter 9
Nitto Denko: Development Under Earnings Pressure

Abstract In this chapter, we describe the history of development and commercialization of RO membranes at Nitto Denko. The president of the company in the 1970s envisioned a corporate growth strategy that defines the membrane business as one of its main pillars. The company thus developed not only RO membranes, but also various other membranes for a wide variety of applications. We illustrate that, in the 1980s, the company expanded its business to overseas markets by acquiring Hydranautics. We also depict that Nitto Denko began full-scale business development in the seawater desalination market after overcoming a patent dispute with Dow Chemical.

9.1 The Beginning of Development

9.1.1 Launch of Membrane Business

Nitto Denko's first contact with the world of RO membranes was in the 1960s. It occurred when Kobe Steel, which was commercializing tubular RO membrane modules, offered the company to supply parts for pressure vessels. Since Kobe Steel did not treat water treatment membranes as its core business, it soon began to ask Nitto Denko's Kameyama Works to produce RO membranes itself. In this way, Nitto Denko moved from the periphery of the RO membrane business to the core. Research on RO membranes did not officially begin within Nitto Denko until 1973.[1]

The following year, in 1974, Saburo Hijikata took over as president and promoted a company-wide policy of focusing on the three new business areas of electronics, medical care, and membranes as the pillars of future growth. However, RO membranes alone would not be enough to make the membrane business a pillar of growth. Therefore, the company decided to develop a wide variety of separation membranes under the concept of a "department store of membranes". The development of RO membranes was one of these endeavors.

[1] Suzuki (2004), p. 14.

M. Fujiwara and Y. Aoshima, *Mechanisms for Long-Term Innovation*,
Advances in Japanese Business and Economics 31,
https://doi.org/10.1007/978-981-19-4896-1_9

In exploring the possibility of commercialization, Nitto Denko decided to explore a variety of applications for RO membranes. Although seawater desalination was in their sights, it was not a clear goal. Nitto Denko was determined to establish its membrane business, and it was not necessary to adhere to RO membranes alone. RO membranes were just one of many separation membranes, and desalination was just one of many possible applications.

Nitto Denko paid a license fee to Kobe Steel and began production of CA tubular RO membrane modules. It delivered 900 modules for the plant treating starch waste liquids in Hokkaido. The order was placed with Nitto Denko, a member of the Hitachi Group at the time, since Hitachi, Ltd. received an order for the design and construction of this facility. However, this was the only order received by Nitto Denko for the treatment of starch waste liquids. These applications, such as liquid waste treatment and the recovery of valuable materials, did not expand as much as expected.

Nitto Denko also slightly explored desalination as one potential application. In 1976, Nitto Denko participated in a seawater desalination demonstration test that the WRPC had begun in Chigasaki. However, rather than actively participating with the aim of desalinating seawater, the company participated half-heartedly in response to an offer from the WRPC. It was clear that it would be difficult to achieve good results with tubular RO membrane modules that have inferior volumetric efficiency. In fact, the company withdrew from the experiment after about a year. The company then set up a small-scale seawater desalination plant at a fishing port to test the technology, but for the time being there was no hope of developing the business.

The 50 Nitto Denko developers were divided into groups to develop not only RO membranes but also UF, NF, MF, and other liquid treatment membranes, as well as gas separation membranes and gas–liquid separation membranes. For RO membranes, all kinds of materials, structures, and shapes were tested, including polyamide composite membranes as well as tubular CA membranes. Sensing the limitations of the tubular type that had poor volumetric efficiency, Nitto Denko began to develop a new kind of membrane module.

In order to develop a variety of membrane businesses, the company actively introduced technologies from overseas. One such technology was polyamide hollow fiber UF membranes introduced from Berghof in Germany. These UF membranes were delivered to a Japanese semiconductor manufacturer in 1978 for the production of ultrapure water. Although this membrane had high water permeability, it had the drawback of low chlorine resistance. Nitto Denko experienced difficulties in receiving complaints from customers due to unexpected clogging. Later, a polysulfone UF membrane was developed as an alternative.

Despite these efforts, the membrane business did not produce the results that were expected. In order for membranes to become one of the pillars of new business as envisioned by Hijikata, they needed a major boost.

9.1.2 Top-Led Development

In April 1978, Hijikata appointed Hideki Yamamoto, then deputy general manager of the company's planning division, as the promoter of the membrane business and designated him to develop businesses with monthly sales of about 2 billion yen from the combination of medical and membrane business.[2] Yamamoto first interviewed each of the membrane developers to find out what goals and future visions they had in mind. As a result, he found that no one was developing membranes with a clear plan or hope for the future of the membrane business. Yamamoto was keenly aware that while the name "membrane" sounded good, the direction and goals of the business were not unified within the organization. Yamamoto described the situation at the time as follows.

> No one was studying it with any solid certainty about what direction they should go . . . No one knew how big the market would become. No one knew what applications would grow in the future. It turned out to be a project that simply had the word "membrane" in the name with almost no future plans or dreams in sight . . . And yes, membrane is an attractive name, isn't it? It could be used for anything. I thought it was a technology with a versatile character that could be used for anything, such as desalination, wastewater treatment, or enrichment processes, but no one knew what field it would be based on, and no one was trying to find out.[3]

At that time, the development of membranes was carried out at the Technical Research Laboratory in Ibaraki city in Osaka, but in two separate buildings across the street from each other. In one building, the development of tubular RO membrane modules and UF membrane modules, which had been introduced for the treatment of starch liquid waste, was being carried out. In the other building, there were several teams in charge of the development of hydrophilic polyolefin UF and MF membranes, polysulfone UF membranes, and polyimide RO membranes. Hijikata expected Yamamoto to bring together the membrane technologies scattered throughout the company and develop them into a single area of business, and instructed him to integrate the membrane development teams.

In April 1979, the membrane developers, who had been working in two separate buildings, were integrated into a single location. In the same year, a company-wide membrane project was launched. The membrane business was positioned as a strategic area for the future based on the strong will of top management.

9.1.3 Development of the Spiral Module

Keisuke Nakagome was responsible for the development of the flat RO membranes and spiral modules that would later support Nitto Denko's RO membrane business.

[2] From the authors' interview with Hideki Yamamoto at Nitto Denko's head office on August 29, 2014.

[3] From the authors' interview with Hideki Yamamoto referenced above.

Nakagome was an engineer who joined Kurita Water Industries (Kurita) in 1959 and then moved to Nitto Denko in 1969. After joining Nitto Denko, Nakagome was involved in the development of insulating varnishes and heat-resistant films, and the development and mass production of Flexible Printed Circuits (FPCs) in Kameyama. Then, in August 1978, he became involved with water treatment again when he was transferred to the Technology Research Laboratory where he was instructed to develop flat RO membranes with a composite structure and spiral modules. This was the year before the two separate development teams were integrated. Nakagome's development efforts began on a small scale in the building where the development teams for tubular RO membranes and UF membrane modules were located.

The impetus for the development of the flat membrane was a suggestion from Baker of Bend Research, with whom Nitto Denko had been working for some time. In the U.S., North Star had succeeded in developing a flat composite membrane called NS-100 in 1970, and Baker suggested that they follow this example. Another reason for choosing the flat membrane type was that the technical directors believed that the sheet technology that Nitto Denko had accumulated up to that time could be utilized in the manufacturing process.

The development group for the tubular type was far larger than the group of the flat composite membrane conducted by Nakagome alone. 1-m-wide membrane manufacturing equipment based on Baker's sketches was already available in the company, but there was no budget to make a prototype using it. And even if that was tried, the equipment was not capable of producing RO membranes. Nakagome read the literature published by the OSW and concentrated on membrane production at the laboratory level first. Based on his own experience in coating technology, he developed a method to form sheet-type composite membranes with high precision even on a small scale, and achieved a desalination rate of 99% for a 75-mm-diameter membrane. One more person was added to the staff on this occasion.

Furthermore, in April 1979, membrane development was integrated into one place, and Yoshiyasu Kamiyama, who joined the company in 1978, and Noriaki Yoshioka, who joined the company in 1977, participated in Nakagome's membrane development group. In addition, three developers joined the group for the development of flat membrane modules. This was still a small development team with a total of seven people. NS-100, their performance target, was said to have the level of performance applicable to brine and seawater desalination. For brine, high permeability was required, and for desalination, a salt rejection rate of 99% or higher at a salt concentration of 3.5% was required. They concentrated on synthesizing a composite membrane that could achieve both water permeability and desalination rate at a high level, which are generally in a trade-off relationship.

At this stage, Nakagome and his team did not have a specific application such as brine or seawater desalination in mind. Nakagome and Kamiyama describe the difficulty of foreseeing a clear application as follows.

> It was mentioned in the literature, but I didn't know what the specific use was . . . There was no market to speak of, and there was no one who could definitively say how things should be. It was seed-oriented, and that's the way it is in the introduction period. Maybe I was

> able to continue because I was just stupid. When the results come out, everyone always says something that sounds better.[4]

> I didn't have a clear goal . . . NS-100 was one performance goal. But there was no one who said what would happen (in terms of the application) after it was developed. Almost no one was talking about the desalination (of brine or seawater). If you read the reports, they say that there is, "a potential market for this kind of thing," but that never inspired certainty beyond the level of, "Well, okay, I guess so." They didn't tell us exactly what to do or by when we should do it.[5]

Even at the laboratory level, it was not easy to form a uniform and defect-free membrane. In those days, developers had to make even the equipment for measuring desalination performance by themselves, not to mention the equipment for making flat membranes. Even when they thought they had made membranes under the exact same conditions, the performance varied greatly. Under the slogan, "Don't be fooled by average values," they focused on the highest value obtained in the experiment and tried to optimize it.

After confirming the level of feasibility through trial and error, they moved on to practical development using a 1-m-wide membrane forming system, but they faced many problems. For example, even a 500-m roll of taffeta or nonwoven fabric, which is used as the base material for the membrane, had several seams that were caught and torn apart during the manufacturing process. This problem was gradually solved with the cooperation of paper manufacturers. In the winding process of the elements, if a large number of sheets were not shifted evenly, wrinkles would form and damage the membrane surface, making it impossible to ensure the required membrane performance. This problem was solved by newly developing their own winding machine.

The two basic patents held by UOP were potential obstacles to the adoption of the flat membrane type. Toray, which also adopted the flat membrane type, had devised and commercialized its original element structure. However, Nakagome and his team confirmed that the patent held by UOP was about to expire in one year and decided to adopt the UOP method, which has a simple structure, judging that a complicated structure would cause many problems in terms of water flow and dynamics.[6]

In 1980, the Membrane Module Development Department was established in the Technical Research Laboratory. In the same year, Nitto Denko succeeded in developing the composite membranes NTR-7197 and NTR-7199 using polyurea-based materials. The NT in NTR stands for Nitto, the R for Reverse Osmosis, and the last two digits refer to the desalination rate. In other words, the desalination rates of these membranes were 97% and 99%, respectively. Regarding the water permeability, both membranes achieved a water production rate of 1 m^3/day per 1 m^2 of membrane area. The developers set up a small plant at the company's waterfront test site on

---

[4] From the authors' interview with Keisuke Nakagome on April 11, 2014 at Hitotsubashi University (interview) and September 23, 2014 (e-mail).

[5] From the authors' interview with Yoshiyasu Kamiyama at Okayama University on August 25, 2014. The text in parentheses was added by the authors.

[6] The patent was invalidated by objection from Kurita.

the Atsumi Peninsula and conducted demonstration tests of seawater desalination for about a year. In this experiment, a desalination rate of 99% was achieved despite insufficient pretreatment.

However, these membranes were not chlorine-resistant and could not be used for desalination that requires frequent sterilization and cleaning of the membranes. Therefore, the developers aimed to develop the technology in the food industry. For example, when it was used to concentrate grape juice in the winemaking process, it tasted almost like noble rot wine and was well received. Here, Nitto Denko's characteristic of developing various types of separation membranes for various markets can also be seen. Nakagome describes this approach as follows.

> Since we didn't know what kind of applications the membranes would be used for, we decided to add RO membranes to the product number list if they had any notable characteristics, such as high permeate velocity (water permeability), chlorine resistance, heat resistance, and fouling resistance even if the salt rejection rate was low. If they could find applications for it later, they could use it. Of course, we wanted to have a high rejection rate and a high permeate flux. But, we also had to make it easy to use with chlorine resistance, usable pH range, heat resistance, degradation resistance, and fouling resistance.[7]

9.2 Exploration of Applications and Development into the Ultrapure Water Market

9.2.1 The Search for Applications

In 1981, following the successful development of NTR-7197 and NTR-7199, the Membrane Module Development Department became independent as the Membrane Module Business Promotion Department. This meant that a great deal of commercialization pressure would be placed on the developers. The department consisted of about 60 people of which about 40 were engineers. Yamamoto was the first head of the Department. As the pressure to commercialize increased, Yamamoto always took two things into consideration.

One was the technological capability to create new RO membranes. At that time, Nitto Denko was selling only tubular RO membrane modules and UF membrane modules. However, Yamamoto got a feel of being able to introduce a new RO membrane module in the near future as he proceeded with development under the integrated development organization.

The other was the future potential and ripple effect of the application market. Yamamoto and his colleagues continued to examine the future potential and ripple effect of membranes while searching for various applications. To develop potential food applications, they participated in the "Membrane Research Association in Food Industry" (MRAFI), which was established in 1982 with the support of the Ministry of

---

[7] From the authors' interview and e-mail with Keisuke Nakagome. The text in parentheses was added by the authors.

Agriculture, Forestry and Fisheries. In this initiative, food companies and membrane manufacturers paired up to tackle a variety of application themes. Nitto Denko teamed up with Ajinomoto.

As a result of joint development, Nitto Denko succeeded in replacing the ion exchange resin used by Ajinomoto to decolorize its flavor solutions with its NF membrane NTR-7450. NTR-7450 was more resistant to chlorine than NTR-7197 and NTR-7199. This was the only project that was put to practical use in the Association. Kenichi Ikeda, who joined the company in 1977 and developed NTR-7450 at the Technology Research Laboratory, describes the development as follows.

> My boss told me, "Make a strong membrane." And that was the sole theme. I started by trying to find the answer to the question, "What is a strong membrane?" The result was the 7450, which is now called nanofiltration, but at the time we called it loose RO. That membrane was used in food products to filter soy sauce. . .
>
> In the food industry, there are many substances that contaminate membranes. If you can't wash the membrane with chlorine when it gets dirty, it's not good. You can't get rid of the dirt. In the case of 7450, however, you can wash it with chlorine when it becomes dirty. Then, the system can be operated very stably . . .
>
> But at that time (when we were developing it), we didn't know what we were going to use it for at all. It just so happened that Ajinomoto approached us at that time and said, "We'd like to do something like this." We were working on a project for the Ministry of Agriculture, Forestry and Fisheries, and we gave it a try and found it surprisingly interesting.[8]

Fields that contribute to the natural environment were also considered. For example, as a measure to prevent pollution in the Seto Inland Sea, Nitto Denko's tubular UF membrane modules were used to treat pulp waste liquid at a major paper manufacturer in Shikoku. Another example is the use of Nitto Denko's tubular UF membrane modules in the field of water reuse (grey water) for large buildings and factories. This application has the potential to spill over into the fields of water reuse at water purification plants and brine and seawater desalination.

However, they were still small and not a big business. Even if Nitto Denko developed RO membranes in collaboration with customers in the food industry, the membranes would only be custom-made products for that particular customer. Even if they were able to expand their business to other companies, this would be limited to one or two companies at most, and Nitto Denko would not be able to recover their investment in development. It was able to develop applications such as wine grape concentration and fruit juice flavor concentration, but not on a scale that would support the business.

Although tubular membrane modules were used in the fields of electrodeposition coating, oil–water separation of wastewater, and manure treatment, this market was also limited. Nitto Denko once made a brine desalination system using 20 to 30 NTR-7197 4-inch elements and exported it to Indonesia. However, since Nitto Denko competed with its customers, water treatment equipment manufacturers, with this system, its deployment was limited to special cases. It also participated in the

[8] From the authors' interview with Kenichi Ikeda at Research Institute of Innovative Technology for the Earth (RITE) on December 27, 2013. The text in parentheses was added by the authors.

"Research and Development of Comprehensive Water Reclamation and Utilization System" (Aqua-Renaissance'90 Project) launched in 1985 at the initiative of the MITI, but did not achieve any significant results.

9.2.2 *"Sanshin Katsudo" (Three New Activities)*

The development and commercialization of RO membranes at Nitto Denko was characterized by the fact that developers worked on anything related to separation membranes and cultivated a variety of markets without limiting targets of technology and market. Nakagome likened separation membranes to a "sieve" and hoped to find some applications by having a large number of membranes with various features.[9] Seawater desalination was just one of the many potential applications. Since the performance required for seawater desalination was far beyond the level of technology that Nitto Denko had achieved at the time, it was not expected that they would be able to develop this application immediately.

In addition to concentrating on the development activities themselves, the developers were also required to visit customers frequently. This was not only to explore a variety of applications, but also to gain a deep understanding of the customer's use environment for membranes and modules, including compatibility with equipment, daily operations, cleaning, and maintenance. A series of exploratory activities also led to the supply of pressure vessels for storing membrane modules to Toyobo that was developing RO membranes at the same time.

These efforts to explore a variety of applications reflected the company's new business development philosophy of "Sanshin Katsudo" (Three New Activities). "Sanshin" means three new things: new product development, new application development, and new demand creation. At Nitto Denko, developers are required to pursue these three directions whenever they create new businesses. Nakagome describes "Sanshin Katsudo" in membrane development as follows.

> Our corporate culture was based on the Sanshin Katsudo. Thus, if there is a possibility that the membrane can be used, we pick up whatever we can without regard to the business potential, because each separation membrane market is small. If we refused to do something that the sales staff wanted to do, they would put strong pressure on us. I don't remember that we were told to stop. That is why it is necessary to keep a stock of various membranes, and it is extremely rare to throw them away.[10]

The developers' attitudes toward Sanshin Katsudo were evident in the development of NTR-7250. Although NTR-7250 achieved its targets in terms of chlorine

---

9 From the authors' interview and e-mail with Keisuke Nakagome, as referenced above.

10 From the authors' interview with Keisuke Nakagome at Hitotsubashi University on July 23, 2014. The quality and flow rate of the permeate water depends on the quality of the feed water, module configuration, and operating conditions. Based on his experience during his water treatment days, Nakagome promoted the development of software, which is important for designing RO membrane systems, and was one of the first to provide it to water treatment equipment manufacturers.

resistance and water permeability, the desalination rate was remarkably low at around 60%, making the membrane completely useless for seawater and brine desalination. However, the company had previously introduced an RO membrane with a low desalination rate to the market under the name "loose RO membrane". Based on this experience, the developers thought that even if the desalination rate was low, some feature might open up an application somewhere. NTR-7250 had such a feature.

There are various types of salt, including monovalent salts such as sodium chloride and divalent salts such as magnesium sulfate. Although NTR-7250 had only about a 60% desalting rate for sodium chloride, which is a common salt, it achieved a 99% rate for magnesium sulfate that is a divalent salt. Noticing this feature, the developers introduced the new membrane to various customers for the purpose of softening hard water.

At that time, Yamamoto had a strong relationship with the top management of Kurita that was engaged in the production of ultrapure water for semiconductors. The two companies regularly held technical meetings to promote mutual exchange between their developers. In 1981, when Yamamoto introduced NTR-7250 at the meetings, Kurita showed strong interest in the product. NTR-7250 was superior in its ability to remove certain total organic carbons (TOC). Yoshioka recalls the development of NTR-7250 as follows.

> Anyway, Kamiyama and I knew that we had to have a high interception rate (for salt). We said, "It has to be at least 90 (%). "Toray's membranes were already in the market, so we said, "It's no good unless we get 95 (%)." We selected materials that were chlorine resistant, that could withstand chlorine. While we were working on this, we suddenly came up with (NTR-7250 membrane). It produced a lot of water. But we said, "The desalination rate is low," and "There's nothing we can do about that." At that time, Kamiyama often visited Kurita, and when he introduced NTR-7250, they discovered its value.[11]

9.2.3 Capturing the Demand for Ultrapure Water

At that time, Toray dominated the RO membrane market for the production of ultrapure water for semiconductors in Japan. Toray's CA RO membrane was introduced in the process before ion exchange resin. In contrast, NTR-7250 was newly introduced into the process after the ion exchange resin. Kurita took advantage of the features of NTR-7250 and devised a new multi-step separation system.

In 1983, an order for NTR-7250 was delivered for the first time. Japanese semiconductor manufacturers, who had abundant financial resources, actively integrated Kurita's equipment. Kurita sought an exclusive supply contract with Nitto Denko for NTR-7250, which Nitto Denko agreed to for three years. As a result, Kurita, which had been in fierce competition with Organo in the ultrapure water production equipment market, rapidly increased its market share. Nitto Denko continued to maintain

[11] From the authors' interview with Noriaki Yoshioka at Nitto Denko's head office on August 29, 2014. The text in parentheses was added by the authors.

a good relationship with Kurita, and was able to obtain information on membrane applications and customer needs.

The success of NTR-7250 sparked momentum in Nitto Denko's membrane business. With the expansion of the semiconductor industry, sales in the membrane business, which had been 600 million yen in FY1982, reached 1 billion yen in FY1983.[12] In 1984, the company announced a grand plan to increase sales in the membrane business to 6 billion yen by FY1987.[13] In line with this, it launched a project to build the Shiga Plant, developing all the manufacturing equipment for the plant in-house.

In order to adapt the system to semiconductor applications, Nitto Denko introduced a semiconductor plant-level ultrapure water system (100 m^3/h) to evaluate the membrane module. Being able to evaluate the system at the mass production level in advance allowed the customer to start producing ultrapure water without having to spend time making adjustments through on-site trial runs. This reduced the waste of water and was well received by the customer.

The development team for RO membranes had consisted mostly of engineers in polymer chemistry. However, as the use of RO membranes expanded, it was expanded to include engineers to strengthen its engineering capabilities. In 1986, the Membrane Sector was established and the Shiga Plant began operations. Yamamoto, who had been promoted to Director while serving as general manager of the Business Promotion Department, became the first general manager of the Membrane Sector and also served as the manager of the Shiga Plant. Nakagome served as deputy plant manager. In the same year, Nitto Denko introduced NTR-729, the successor to NTR-7250, to expand sales in the market for ultrapure water production for semiconductors.[14] Figure 9.1 shows the machinery used in the manufacture of flat RO membranes at the Shiga Plant. By this time, Nitto Denko's market share was in close competition with that of Toray, and the scale of its business was around 100 million yen per month, although it did not reach the planned level.

Unfortunately, the semiconductor recession hit in 1986 when the Shiga Plant started operations. Nitto Denko's RO membrane business was also affected by the recession, and there were even months when sales were only about 20 million yen. Since there was nothing to produce, all they did for the first six months was trial production. The rest of the time, they were painting and weeding. During that time, people could learn the operation of the new equipment, but the business was still in a difficult situation. Nevertheless, the top management continued to support the membrane business. Goro Kamai, who had succeeded Hijikata as president in 1985, followed the same line.

Although he had the support of top management, Yamamoto had a strong sense of crisis that the business would never grow if it remained in Japan. In order to grow, he

---

[12] Nikkei Sangyō Shimbun (October 1, 1982, p. 15; August 7, 1984, p. 7).

[13] Nikkei Sangyō Shimbun (August 7, 1984, p. 7).

[14] This was because when NTR-7250 was produced at the Shiga Plant, the desalination rate increased to 90%, and the name was changed to NTR-729. This episode illustrates the difficulty of manufacturing RO membranes and the importance of establishing membrane production technology. The importance of this can be inferred from the fact that Hydranautics was not able to make products with its high-performance composite membrane, as described below.

Fig. 9.1 Production equipment for flat RO membranes (Shiga Plant). *Source* Nitto Denko (1986)

believed that overseas expansion was necessary. In 1985, he established Nitto Denko Technical Corporation (NDT)[15] in San Francisco, California, and set up a membrane division.[16] This was done with the aim of exporting electroplating tubular membrane modules to automobile manufacturers in the U.S. market.

Furthermore, Yamamoto was also looking to acquire overseas companies. Nitto Denko, which had been performing well thanks to the tailwind of the domestic economy, was looking for new investment targets, but had not yet decided on any. Furthermore, the appreciation of the yen following the Plaza Accord meant that foreign companies were rapidly becoming undervalued.

[15] At that time, Yoshioka was assigned as a researcher to Membrane Technology and Research (MTR), which had been established by Baker, and later moved to NDT.

[16] The president at the time of establishment was Katsuhiko Murata.

9.3 Progress in Commercialization

9.3.1 Acquisition of Hydranautics

At that time, the New York branch of Sanwa Bank (at the time) informed Nitto Denko that Hydranautics was for sale. Hydranautics was a subsidiary of Rohm and Haas, the second largest chemical company in the United States after Dow Chemical. Yamamoto was very interested in this situation. However, since it was Nitto Denko's first M&A, he did not jump at it immediately. Yamamoto sent Nakagome and others to investigate the company and requested an evaluation by Stanford Research Institute (SRI), an independent research organization.

The SRI's evaluation result was "neither good nor bad". In terms of technology, the company had the ability to develop and manufacture CA spiral RO membrane elements, while composite membranes were still in the research stage. Although the company's reputation was not particularly high, to expand the membrane business overseas, it was necessary to have a company that possessed comprehensive development, manufacturing, and sales functions. Yamamoto describes how he came to think about overseas expansion as follows.

> There was a limit to turn membranes into a business as long as we stay in Japan. I thought that it would never take off in the way that people expected. So, I pondered the question of what we should do to actualize membrane business expectations. The acquisition of Hydranautics in 1987 was the answer. I thought, "We don't have to stick to Japan. Let's seek markets in the world, and we let's instead see how well we can compete amidst the technology of the world." In order to do so, we moved on to the idea of partnering with or acquiring overseas companies.[17]

In November 1987, Nitto Denko acquired Hydranautics. Of course, not all of this acquisition was worthwhile. For example, Hydranautics, a vertically integrated business, was involved in the engineering of desalination facilities in addition to its membrane business. Hydranautics was so good at this that it was even in the business of installing and operating its own desalination plants at nuclear power plants on the West Coast and selling the desalinated water. However, the profitability of this plant business was low, and there was a risk that competing plant manufacturers would shy away from Nitto Denko and Hydranautics' membranes. Hydranautics thus withdrew from the desalination plant engineering business after the acquisition.

The CA RO membranes owned by Hydranautics also did not produce the desired effect. They were imported to Japan with the expectation that they would be used in the production of ultrapure water for semiconductors in Japan, but it was difficult to establish a competitive advantage for the pre-process applications that Toray had already dominated.

However, the acquisition had significant benefits. Hydranautics had sales channels and contacts all over the world and had gained a certain level of recognition from its

[17] From the authors' interview with Hideki Yamamoto referenced above. The text in parentheses was added by the authors.

customers based on its delivery record to date. As a result, various information from users in various countries began to flow into Nitto Denko. The door to the global market was opened wide.

9.3.2 NTR-759

Another achievement was the acquisition of polyamide composite membrane technology. This was a major benefit that Nitto Denko had not anticipated at the time of the acquisition. Nitto Denko knew that Hydranautics was researching composite membranes, but it was not aware of the details of the technology and did not have high expectations. However, the composite membrane turned out to be an extremely high-performance membrane with both a high desalination rate and water permeability. At the time, FT-30 developed by Cadotte at FilmTec was the talk of the market, and the membrane developed by John E. Tomaschke in the Hydranautics development department was comparable in performance to FT-30.[18] However, Hydranautics did not have the production technology to commercialize this membrane.

Realizing the potential of this new membrane, the development team at Nitto Denko made every effort to commercialize it by utilizing all the manufacturing technologies they had developed. As a result, in January 1988, only three months after the acquisition of Hydranautics, Nitto Denko was able to create a mass-produced product that demonstrated an acceptable level of performance. This new composite membrane was named NTR-759.

In May 1988, just as the development and commercialization of NTR-759 were underway, a major problem occurred on the Japanese side of the venture. The problem was that the permeability of NTR-739, which had been introduced for the ultrapure water production in Japan, suddenly dropped. Although the cause of the problem was hastily investigated, there was not enough time to clarify all the causes and take appropriate countermeasures. As a result of this problem, the gap in market share between Nitto Denko and Toray opened up to roughly 3:7.

A considerable amount of NTR-739 was already on the market. It was impossible to recover customer trust unless those membranes were collected and replaced with new membranes. It was NTR-759 that saved the company from this crisis. The commercialization of NTR-759 was promoted by all the engineers, and it was a success. Nitto Denko was able to deliver NTR-759 to Kurita in August, three months after the problem was discovered. Because of this prompt response, Kurita, which would have experienced immense inconvenience, wound up instead being grateful and praising Nitto Denko.

[18] U.S. Patent 4,872,984; 4,948,507.

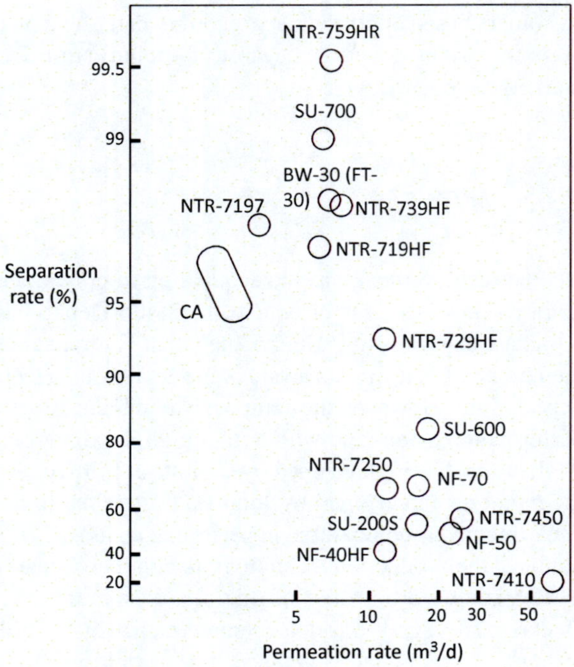

Fig. 9.2 Performance comparison between NTR-759HR and other RO membranes *Source* Prepared by the authors based on Kamiyama (1990), p. 526. *Note* φ4 × 40 (L) inch element, 0.15% NaCl, 1.5 MPa (15 kgf/cm^2), 25 °C

Figure 9.2 shows the performance of NTR-759 and other RO membranes.[19] What is noteworthy in this figure is that NTR-759 has the same water permeability as NTR-739, but achieves an overwhelmingly high desalination rate. Nitto Denko was able to replace NTR-739 with a higher-performance membrane. Since ultrapure water production equipment companies such as Kurita and Organo purchase from multiple membrane manufacturers to avoid risk and reduce costs, it is not usual to see a large difference in market share between membrane manufacturers. However, with the introduction of NTR-759, Nitto Denko succeeded in regaining the market share it had lost due to problems.

NTR-759 was also highly popular in overseas markets. NTR-759 penetrated the U. S. market where the only available membrane module was that of FilmTec despite their increasing popularity for the desalination of brine for drinking water and agricultural water. In 1989, this momentum led to the relocation of company properties from their previous dispersion of headquarters, plants, and laboratories between Santa Barbara and San Diego to consolidation in San Diego alone, along with the relocation of the headquarters there. At the same time, state-of-the-art equipment was introduced. NTR-759 was a hit both in Japan and overseas, and sales for Nitto Denko's membrane business expanded rapidly.

[19] Each performance is based on a product catalog, and the operating conditions differ slightly. Therefore, intercompany comparisons should be interpreted as a rough positional relationship.

9.3.3 Patent Disputes

In this way, NTR-759 was an opportunity for Nitto Denko to make a great leap forward in the RO membrane market, but at the same time, it brought the company the severe challenge of patent litigation.

In May 1990, a letter arrived from Dow to Hydranautics. The letter stated that NTR-759 infringed its patent (commonly known as the '344 patent) on FT-30 developed by FilmTec, a subsidiary of Dow. Of course, Nitto Denko must have fully considered the patent issues before the acquisition. It had obtained assurances from Rohm & Haas that Hydranautics technology did not violate any patents, including the '344 patent, and from its lawyers and researchers that the Hydranautics composite membranes were novel and did not infringe the '344 patent.

The '344 patent was the key patent that underpinned the revolutionary FT-30 composite membrane developed by Cadotte. It was because of this that Dow, which had acquired FilmTec at a high price, took a hardline stance to protect its rights. Nitto Denko's RO membrane business, which had begun to revive with NTR-759, suddenly found itself in a crisis of survival. Yamamoto, who had been appointed a Managing Director in 1989, took the lead in this crisis and stood as a witness in court.

In May 1991, in anticipation of losing the lawsuit, Yamamoto launched an emergency project in Japan. At the start of the project, Yamamoto designated Kamiyama to "produce a membrane to replace NTR-759 in three months." The challenge was to develop a membrane with the same performance as NTR-759 by August when the district court would render its decision. To solve this difficult problem, a total of 40 to 50 developers from all over the company gathered, with Kamiyama serving as the leader. Yoshioka was also instrumental in assisting in this urgent project. The developers were divided into a number of groups and proceeded at a rapid pace to develop a completely new composite membrane that did not use trimesic acid chloride, which had been a pending issue.

During this time, on August 31, Nitto Denko lost its first trial in the case. The court ruled that all membranes using trimesic acid chloride as a skin layer material infringed the '344 patent. As a result, Hydranautics was forced to stop manufacturing composite membranes in the United States. The urgent project in Japan had not been able to produce a new membrane to replace NTR-759. Therefore, Nitto Denko decided to switch to produce CA membranes in the U.S., and to assemble the composite membranes manufactured in Japan into modules in Mexico for sale in markets other than the United States. Responding to such actions, Hydranautics implemented layoffs at its plant in San Diego.

Fortunately, in March 1993, Nitto Denko's claim was accepted in the second trial for the case, which ruled that the '344 patent had been developed by Cadotte with research funds obtained from the U.S. government and that the U.S. government, not FilmTec, was the rightful owner of the patent. As a result, the urgent project in Japan was canceled.

With the case won, Nitto Denko immediately got the license of the '344 patent and resumed production at Hydranautics. Nitto Denko had continued to install state-of-the-art equipment even while production was halted due to the loss of the first trial and that contributed greatly to the rapid recovery in business performance. The employees who had been laid off were brought back, production began at the new facilities, and sales began to grow again.

It was in 1994 that Nitto Denko's membrane business became profitable for the first time on a single fiscal year basis. More than 20 years had passed since the start of development.

This patent litigation consumed a great deal of energy and time for Nitto Denko. At the same time, however, the developers were able to gain a great deal of knowledge through this process. In a patent lawsuit, it is necessary to conduct detailed analysis, scrutiny, and evaluation beyond the usual levels to prepare objective evidence with technical support. This led to the accumulation of new knowledge.

For example, when the developers observed the surface of NTR-759 with an electron microscope, they found that NTR-759 had a folded structure. The folds increased the surface area of the membrane and enhanced its water permeability.[20] The subsidiary, Nitto Analytical Techno-Center (NTC), also played a major role in the analysis and evaluation of the membrane. In the process, NTC's analysis and evaluation technology advanced, leading to the establishment of molecular structure simulation and evaluation technology, which were utilized in the subsequent development of RO membranes.

9.4 The Road to Seawater Desalination

9.4.1 Harnessing Low Pressure Performance

NTR-759, which achieved a dramatic improvement in performance, contributed greatly to the development of Nitto Denko's membrane business. However, they could not afford to become complacent. The development of new membranes had come to a halt after winning the patent lawsuit. Against this backdrop, Kamiyama, a Development Director in the Membrane Sector, felt the need to set a higher goal.[21]

Kamiyama gave an order to the developers to "do three times better than the 759." Specifically, he instructed them to develop a new membrane that would achieve the same water permeability and desalination rate as NTR-759 at one-third the pressure. In contrast to North Star's NS-100, which was Kamiyama's target when he first started developing composite membranes, the later introduced FilmTec's FT-30 had reduced its operating pressure to 1/3. Therefore, Kamiyama used the expressions "3

[20] Ikeda and Tomaschke (1994).

[21] Kamiyama was appointed the head of the Department in April 1992.

times" and "1/3" to mean that he was aiming for the next groundbreaking innovation after FT-30.

The developers were divided into three teams. However, no road map for a threefold improvement in performance was presented. Although this was an extremely difficult task that was unlikely to be achieved, Kamiyama never said it was okay to stop. In contrast, he kept telling his customer, Kurita, every year that a low-pressure membrane would soon be released. Kamiyama describes the situation at the time as follows.

> I instructed them to do ("3 times"), but they could hardly realize it. Because I didn't tell the developers to stop, they couldn't stop . . . because they are very serious and don't disobey their superiors. So, I think they struggled a lot. I visited Kurita and others and tell them every year that low-pressure membranes would be released. So, I think it was hard for the developers.[22]

About two years later, one of the teams reported data that looked like an outlier. The data showed dramatically higher water permeability and desalination rates than those of conventional membranes. A closer examination revealed that the membrane surface folds were deeper and the surface area was larger than that of NTR-759. However, if the folds are too deep, the concentrated high-salinity water does not diffuse on the membrane surface and the osmotic pressure increases, resulting in a decrease in permeability. In addition, there is a risk that fouling substances may stay on the membrane surface. Therefore, optimization of the fold structure was an important issue in the development of the membrane. After a trial-and-error process, the developers succeeded in establishing the technology to control the fold structure appropriately.

A new membrane developed in 1996 was named ES10, where ES stands for Energy Saving. The reason for attaching a different name from the existing products was to emphasize the low-pressure characteristics of the membrane. Although Nitto Denko had already been emphasizing the low-pressure characteristics of its membranes, with the introduction of ES10, the company became even more conscious of this as a clear sales point. As shown in Fig. 9.3, the fold structure deepens from NTR-7199 to NTR-759 and ES10. According to Hirose and Ito (1996), the skin layer of NTR-7199 has flat surface and no height whereas that of NTR-759HR has approximately 0.2 μm and ES10 has about 0.4 μm.

If the amount of permeable water could be maintained unchanged with 1/3 of the operating pressure, a significant savings in energy consumption could be achieved. To separate substances with RO membranes, a high-pressure pump must be used to apply a pressure greater than the osmotic pressure. The electricity cost to drive this high-pressure pump accounts for a large portion of the desalination cost that had been a source of concern for customers. Furthermore, if the pump could be driven at a low pressure, the piping could be made thinner. All of these factors lead to lower desalination costs, which is an effective selling point for equipment manufacturers to their customers. For example, if a customer wants to save on electricity,

[22] From the authors' interview with Yoshiyasu Kamiyama at Okayama University on August 25, 2014.

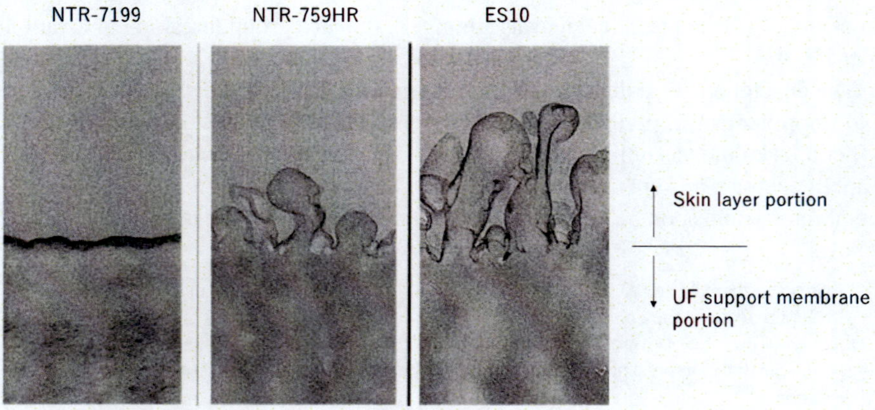

Fig. 9.3 Change in the structure of folds. *Source* Hirose and Ito (1996), p. 40

the company can offer "savings by lowering the pressure." If a customer wants to increase the permeability, the company can offer "three times the permeability at the same pressure."

ES10, called an ultra-low-pressure RO, was first introduced for the desalination of brine. At the time, the developers envisioned the use of this technology exclusively for the desalination of brine for the production of ultrapure water for semiconductors. In the 1990s, the company's annual report mentioned its activities of ultrapure water production for the semiconductor industry. The semiconductor industry accounted for a large proportion of their RO membrane business.

9.4.2 Results in Okinawa and Fukuoka

Prior to ES10, Hydranautics had delivered RO membrane modules for desalination applications through its own sales channels, but only on a small scale.[23]

At the time when Nitto Denko began developing ES10, the construction of seawater desalination facilities was underway in Okinawa, Japan. Okinawa's topography makes water intake from rivers unstable, and securing a stable supply of water was a longstanding issue. In 1977, when the number of days of water supply restriction reached 169 days in Okinawa, the "Okinawa Island Seawater Desalination Project Study (Phase 1)" was started. Construction of a seawater desalination facility began at the Chatan Purification Plant in 1993 and was completed in 1997.

[23] Nitto Denko's first RO membrane module based on cross-linked fully aromatic polyamide was launched in 1991. In 1992, Nitto Denko began operation in the Spanish Canary Islands at a capacity of 6,000m^3/day.

The desalination capacity of this facility was 40,000m^3/day, the largest in Japan at the time and the fifth largest in the world.[24] Nitto Denko received an order for 10,000 m^3/day of this capacity and delivered NTR-70SW, which was a modified version of NTR-759 for seawater desalination. Nitto Denko subsequently expanded the order to 15,000 m^3/day. As a result of this large order, the desalination application became the focus of attention within the company.

In the 2000s, the seawater desalination market began to expand in earnest. Following Okinawa, another desalination plant (50,000 m^3/day) was completed in Fukuoka in 2005. Like Okinawa, Fukuoka had also suffered from water shortages for many years, and after a major drought in 1994, the introduction of desalination facilities began to be considered. In July 1999, a call for proposals for plant and water intake facilities was issued.

As a result of this public solicitation, a plant proposal using Toyobo's hollow fiber membrane modules was selected, and Toyobo was to deliver 2,000 10-inch hollow fiber membrane elements. However, because there was still concern about the removal of impurities such as boron using the single-stage desalination method of Toyobo, 1,200 units of Nitto Denko's ES20-D8 RO membrane elements were installed for post-treatment. In addition, 3,060 Nitto Denko's PVDF-based UF membrane elements were also installed for pre-treatment.

One of Nitto Denko's differentiating factors in seawater desalination applications was its ability to remove boron at low pressure.[25] Since the boron removal process is carried out after seawater is desalinated, the salt concentration is low and high operating pressure is not required. Therefore, it is desirable to remove boron at low pressures with as little energy consumption as possible. The ES20B-D8 RO membrane element installed in Fukuoka used ES20, which was developed in response to the flow of ES10 ultra-low-pressure RO, and had the significant feature of low pressure in addition to boron removal capacity.

Another differentiating factor that the company liked to highlight was its long-term stability and longevity. Although, for a composite membrane, initial basic performance in terms of desalination and water permeability is of course important, so is the length of time that the performance values can be maintained as the membrane becomes contaminated during desalination operations. In particular, for membrane manufacturers that use polyamide composite membranes, which are difficult to disinfect with chlorine, the key is to create membranes that are resistant to the growth of microorganisms. This is the development of so-called low fouling. Nitto Denko

---

[24] Based on Ryūkyū Shimpō (April 10, 1997). The total project cost is about 34.6 billion yen, of which 85% (about 29.4 billion yen) is subsidized by the government (Okinawa-ken Kankyō Seikatsubu Seikatsu Eiseika (Okinawa Prefecture Department of Environment and Living Standards, Living Sanitation Division), 2012).

[25] It should be noted here that low pressure is not necessarily a strong differentiating factor in the seawater desalination process itself. This is because a high osmotic pressure of about 2.5 MPa is applied between seawater and freshwater, and it is necessary to apply a higher pressure to cause RO phenomenon. In general, the operating pressure in desalination must be much higher than 5.5 MPa. Therefore, since the operating pressure is high, even if there is a slight difference between companies, it is unlikely to be a strong differentiating factor.

developed and released LF-10 in 1997, and has been developing the series to this day. LF is the abbreviation for Low Fouling.

9.4.3 *The Full-Scale Development of Seawater Desalination*

In the 2000s, the global seawater desalination market began to expand in earnest and Nitto Denko also began to grow by catching the wave. The most important support was the global sales channels of Hydranautics. Kamiyama explains the importance of the Hydranautics sales channels as follows.

> After acquiring Hydranautics, I was really impressed that there are so many applications for RO membrane spirals, and (in desalination applications) the number of membrane modules per case is very large. In Japan, even if you build an ultrapure water facility, there are only a few hundred modules at most. It's not even a thousand. On the other hand, Hydranautics' facilities (at that time) already had a thousand units. It's a whole new level. I thought, "This is a completely different world." I mean, if you asked me whether we in Japan could go to such places to sell our products, the answer would be, we could not. We can't compete with plant manufacturers in the United States or overseas. It's completely different.[26]

One of the strengths of Hydranautics, which has customers all over the world, was that it could easily conduct field tests for seawater desalination. Until then, even if a company was able to develop a new composite membrane, it would have faced difficulties in conducting its own field tests. Hydranautics, however, made it easy to send test results. In this way, the development team of Nitto Denko was able to proceed with development more effectively than before by quickly obtaining useful feedback information.

Supported by Hydranautics' sales channels and field tests, the RO membrane business expanded overseas. For example, in the first half of the 2000s, Hydranautics succeeded in winning orders for projects in Carboneras, Spain (120,000 m^3/day; started operation in 2001), Fujairah, UAE (170,000 m^3/day; started operation in 2003), and Tampa, U.S.A. (100,000 m^3/day; order value approximately 1 billion yen).[27] As a result, sales in the membrane business steadily increased from 7.9 billion yen in FY2002 to over 13 billion yen in FY2004, 14 billion yen in FY2005, and 17.7 billion yen in FY2006.[28]

In anticipation of growth in the seawater desalination market, the company also worked to expand its production capacity. In 2006, the company announced a plan to increase sales in the membrane business to 28 billion yen by FY2010, and invested 6 billion yen in the first phase of expansion work at its Shiga Plant. The company had made a number of improvements to its membranes and introduced the SWC5

---

[26] From the authors' interview with Yoshiyasu Kamiyama referenced above. The text in parentheses was added by the authors.

[27] Nihon Keizai Shimbun (February 15, 2003, p. 13).

[28] Nikkei Sangyō Shimbun (February 24, 2003, p. 6; October 19, 2005, p. 3; September 14, 2006, p. 12) and Nihon Keizai Shimbun (May 18, 2007, p. 11). Note that 7.9 billion yen in FY2002 is equivalent to 28% of the global RO membrane market of 28.2 billion yen.

and SWC5 MAX to the market in 2007 and 2009, respectively.[29] The first phase of the expansion of the Shiga Plant was completed in 2009.

Overseas orders continued from the late 2000s onward. In Australia, in particular, the company won orders for the Gold Coast (130,000 m^3/day; started operations in 2009), Adelaide (280,000 m^3/day; started operations in 2013), and Melbourne (440,000 m^3/day; started operations in 2013). In 2010, the company received an order for a project in Israel with a water production capacity of 410,000 m^3/day, one of the largest in the world. In this project, the company received the world's first order for a large 16-inch diameter seawater desalination membrane module.

In China, the company began assembling RO membrane modules in Shanghai in 2002 to meet the rapidly growing demand for industrial water, and also developed a new RO membrane for the Chinese market. In 2006, for example, the company launched the PROC10 RO membrane element for pure water production, followed by the improved PROC20 in 2009.[30]

9.5 Performance of the Membrane Business

Over a long history of more than 40 years, Nitto Denko developed RO membranes into a business of its own. In the process of overcoming many technical hurdles, exploring various application markets and making bold investments, the business has grown significantly, despite the twists and turns.

The 1990s was a dizzying decade, with a patent dispute with Dow, the development of ES10 ultra-low-pressure membrane, and the development of desalination applications. Figure 9.4 shows the performance of the membrane business as a whole. Although this figure is not specific to RO membranes, the general trend can be traced.

According to the figure, although sales declined for a while from FY1991, they began to expand in FY1994. This expansion was supported by the boom in the semiconductor industry that accompanied the spread of personal computers. However, in the second half of the 1990s, the semiconductor industry fell into a serious recession as a reaction to the personal computer boom, and Nitto Denko's membrane business was also affected by this, resulting in a decline in sales. The reason the sales amount peaked despite semiconductor recession beginning around 1996 was that the Okinawa seawater desalination project compensated for the slump in demand for semiconductors. The commercial significance of the large-scale seawater desalination projects was so great that the term "seawater desalination" appeared for the first time in the company's annual securities report for the year ending in March 1997, and the company reported growth in this business.

From the latter half of the 2000s, the seawater desalination market came to support Nitto Denko's membrane business. In their annual reports from the fiscal year ending

[29] SWC5 MAX's sales target for 2009 was 2 billion yen.

[30] The sales targets for each element were 2 billion yen for PROC10 (FY2008) and 500 million yen for PROC20 (FY2009).

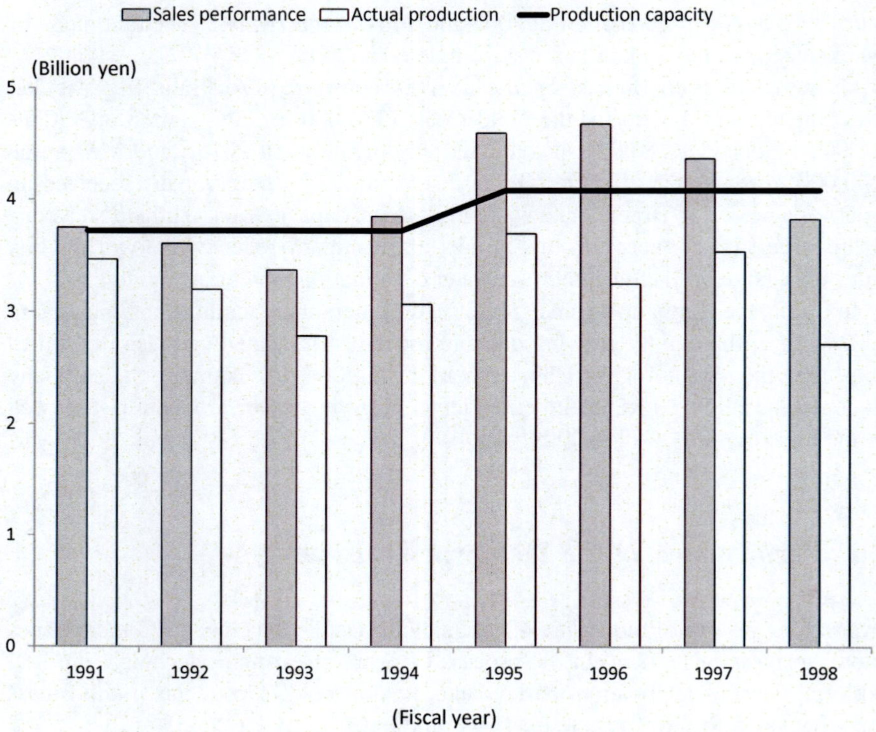

Fig. 9.4 Performance of the membrane business division in the 1990s. *Source* Nitto Denko "Securities Report" for each year

on March 31, 2008 onward, the term "ultrapure water production for semiconductors" almost disappeared and instead the results in the seawater desalination market were described. The performance of the membrane business in the 2000s is not clear as it was incorporated into the Functional Materials segment that has an order of magnitude larger sales. In the 2010s, the membrane business was again being reported. As shown in Fig. 9.5, the performance of the membrane business has been steadily expanding since FY2011 in line with the growth of the seawater desalination market.

On the other hand, there are some issues to be addressed if the membrane business is to become a pillar of earnings. There is little difference in the basic performance among competitors of polyamide spiral RO membranes. In addition, the shape of the module is virtually standardized, so even if a company wins the first order, there is no guarantee that it will be able to monopolize demand for replacement membranes. These factors have led to fierce price competition, which puts pressure on the profitability of each company. Figure 9.5 shows that the profitability of the membrane and medical divisions are significantly lower than that of the Nitto Denko overall. These data should be interpreted with caution, as they do not show the membrane business on a stand-alone basis, but they do suggest the difficulty of obtaining profits in the business.

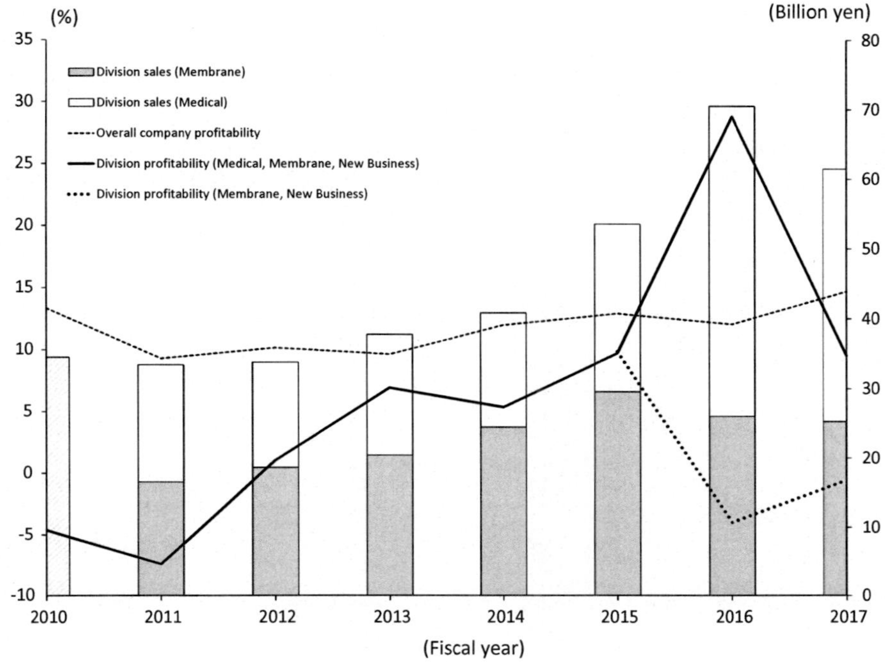

Fig. 9.5 Performance of the company overall and membrane-related divisions. *Source* Nitto Denko "Securities Report" for each year. *Note* Division sales in FY2010 are the sum of Medical and Membrane. The Medical business was separated in FY2016 by merging it with other related businesses, and became an independent segment as the Life Sciences business. To maintain continuity of the time series in this figure, even after the division was separated as the Life Sciences division, sales are also shown as the Medical division. The solid line shows the profitability for the Medical segment, Membranes, and New Businesses combined, whereas the profitability of the division excluding the Medical business is shown as a dashed line. The divisional profit margins for Membrane and New Business for FY2016 are calculated based on the Annual Securities Report of Nitto Denko. The left axis shows the line graphs and the right axis shows the bar graph

9.6 Summary

Nitto Denko did not develop RO membranes with the explicit goal of seawater desalination. RO membranes were only one of many separation membranes, and desalination was only one of many applications.

The history of Nitto Denko's RO membrane business is a history of the determined efforts of the top management to develop separation membranes as a pillar of the business, and the unrelenting efforts of the developers to commercialize them. While Hijikata determined to make the membrane business a new pillar of the company, Yamamoto pushed forward with the expansion of the business without the slightest thought of withdrawing. The engineers responded to his wishes by trying all kinds of technologies and searching for application markets. Each one of them was not

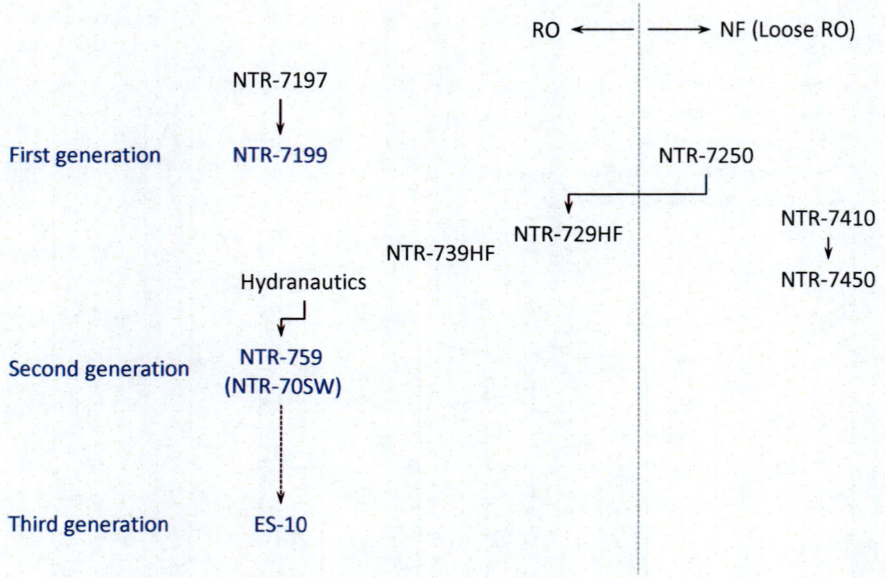

Fig. 9.6 The time flow of RO membranes at Nitto Denko. *Source* Prepared by the authors

necessarily a success as a business. However, they seemed to be useful in justifying the continuation of the business for about 20 years until the business got back on track in the early 1990s. In June 1996, when the business was back on track, Yamamoto was appointed president of Nitto Denko and led the company for just under five years until March 2001. Figure 9.6 summarizes the time flow of RO membranes introduced to the market by Nitto Denko.

Nitto Denko's researchers and engineers pursued a principled understanding of the technology and were also responsible for ensuring that the product functions properly at the customer site and creates economic value. Therefore, rather than blindly improving membrane performance, emphasis has been placed on designing products with a view to maximizing their value to the customer. This point is also related to the organizational structure of development at Nitto Denko. Looking back at the history of membrane development at Nitto Denko, research and development were integrated or mixed. In addition, engineering and sales ideas from the customer's perspective were also incorporated.

The engineers have learned a lot during the course of development from the customer's point of view. For example, from applications in the food market, they have gained knowledge about bacteria control, cleaning issues, and module design. From the semiconductor market, they learned about particulate evaluation and microbial problems. In particular, the establishment of analysis and evaluation technology for these problems had a major impact on the subsequent development of RO membranes. The analysis and evaluation technology established by Nitto Denko was so successful that their customers, water treatment equipment manufacturers, came to learn about it. The search for diverse application markets was not limited to learning about membranes themselves, but also included much about membrane module design and cleaning processes. This is why Nitto Denko's membrane engineers are also knowledgeable about peripheral technologies other than membranes. The Membrane Sector of Nitto Denko was proactive in applying for patents on the results of such technologies. It is said that the themes to be applied for were budgeted for each fiscal year and assigned to the engineers, and many applications were filed.

Various functions and performances are incorporated into membranes. It is the customer who extracts the value of these functions and performance. Nitto Denko appears to have learned a great deal from its customers by having its engineers come into direct contact with a wide variety of customers. It was Kurita, a customer, that showed Nitto Denko that the characteristics of NTR-7250 were suitable for the semiconductor manufacturing market. Looking back over the Nitto Denko history, we believe that by sharing the customer's point of view, Nitto Denko's engineers have discovered the value of their own technology, which has led to the growth of their business.

Appendix: Patent litigation concerning RO membranes (FilmTec vs. Hydranautics)

In May 1990, Nitto Denko was troubled by a patent infringement lawsuit in the U.S. District Court for the Southern District of California, in which FilmTec sued Hydranautics for patent infringement. Since FilmTec had been acquired by Dow in 1985 and Hydranautics had been acquired by Nitto Denko in 1987, the case was essentially Dow versus Nitto Denko.

FilmTec claimed that Hydranautics was infringing on patent 4,277,344 (filed Feb. 22, 1979, published July 7, 1981), owned by its co-founder Cadotte. FilmTec also sued another RO company, Allied Signal, for patent infringement. Cadotte's patent, commonly known as the '344 patent because of its patent number, was important to RO companies working with polyamide-based materials. And if it was found to have been infringed upon, this would have been a lawsuit gravely affecting not only Hydranautics and Nitto Denko, but other RO companies as well.

In August 1991, the U.S. District Court for the Southern District of California ruled in FilmTec's favor (Case No. 90–563 GT(M)). Hydranautics immediately appealed to the United States Court of Appeals for the Federal Circuit. In the appeal, Hydranautics challenged the origin of the invention reported in the '344 patent, which, although filed in 1979, was based on experiments conducted on February 23, 1978, which, in turn, were similar to experiments conducted on November 17, 1977.[31]

The date November 17, 1977, was significant because it was before Cadotte moved to FilmTec. The experiments conducted in November 1977 were conducted at the

[31] The only difference was that the temperature at which they were dried was lowered from 130 °C to 100 °C.

Midwest Research Institute (MRI), where Cadotte had previously worked. Cadotte left MRI on December 31, 1977.

In receiving government support for RO research, MRI had entered into a contract with the Office of Water Research and Technology in 1976. Under that agreement, inventions made by MRI between July 15, 1976 and January 15, 1978 belonged to the government. In other words, Hydranautics argued that the experiment from which the inventions of the '344 patent originated was performed on MRI on November 17, 1977, and that the '344 patent therefore belonged to the U.S. government, not FilmTec.

FilmTec's objection was to the difference in performance of the RO membranes obtained in the two experiments: the performance of the RO membranes obtained in the November 1977 experiment was 12.7 gfd and 9.5 gfd in water permeability and 95.0% and 88.5% in desalination. In contrast, the performance obtained in the February 1978 experiment was improved to 13.1 gfd and 11.0 gfd in permeability and 98% in desalination. FilmTec argued that this performance difference indicated a discontinuity between the two experiments. The arguments of the two sides were in direct conflict. The industry's attention was focused on the outcome of the decision.

On March 18, 1993, the U.S. The Court of Appeals for the Federal Circuit affirmed the continuity of the experiments and ruled in favor of Hydranautics. As a result, the '344 patent became the property of the U.S. government, which was a great relief not only to Hydranautics and Nitto Denko, but also to many other RO companies.

The series of lawsuits over the '344 patent were serious lawsuits that had the potential to wipe out many RO membrane companies. While the companies watched with bated breath, the Hydranautics/Nitto Denko filed an appeal, and the case finally ended with the reversal that the patent belonged to the U.S. government.

Patent infringement litigation ruling:

Case No. 92–1091 (982 F.2d 1546); Case No. 98–55274 (204 F.3d 880).

References

Denko, N. (1986). *Shiga kōjōannai pamphlet (Shiga plant information brochure).*

Hirose, M., & Ito, H. (1996). Ultra-low pressure type composite reverse osmosis membrane module 'ES series.' *Nitto Technical Report (nitto Gihō), 34*, 38–47.

Ikeda, K., & Tomaschke, J. (1994). Noble reverse osmosis composite membrane. *Desalination, 96*, 113–118.

Kamiyama, Y. (1990). Gyakushintō maku (Reverse Osmosis Membranes). *Sen-i Gakkai shi (Journal of the Textile Machinery Society of Japan), 43*, 523–532.

Nihon Keizai Shimbun. February 15, 2003; May 18, 2007.

Nikkei Sangyō Shimbun. (Nikkei Business Daily). October 1, 1982; August 7, 1984; February 24, 2003; October 19, 2005; September 14, 2006

Okinawa-ken Kankyō Seikatsubu Seikatsu Eiseika (Okinawa Prefecture Department of Environment and Living Standards, Living Sanitation Division). 2012. *Okinawa ken no suidō gaiyō: Heisei 23-nendoban (Outline of Okinawa prefecture's water supply: FY2011 edition).*

Suzuki, F. (2004). Sekai ni 'Anzen de anka na mizu' o teikyō suru membrane jigyō (Membrane business providing 'safe and affordable water' to the world). *Nitto Gihō (Nitto Technical Report), 42*, 85, 11–15.

Part IV
Analysis

Chapter 10
Policy Support and Spillover Effects: Initiation of Development in the United States and Japan

Abstract In this chapter, we analyze how development was launched in the United States and Japan by focusing on the role of "policy support". We argue that in the United States, a large amount of public support was generated by the confluence of different policy intentions, which supported early development. In Japan, development was triggered as a follow-up action for US predecessors such as DuPont and Dow. There, it becomes clear that US government policy indirectly influenced Japanese companies through the actions of US public institutions and private companies.

10.1 Two Questions

10.1.1 Why Did Development Start in Japan, a Country with Abundant Water Resources?

In this chapter, we summarize the role of the U.S. government in supporting the start-up of the development of RO membranes, and also clarify how the effects of public support in the U.S. indirectly spread to Japan.

In Chaps. 6–9, we clarified the process by which Japanese companies started to develop RO membranes and became highly competitive in the market. However, we must ask the following questions: why did the development of RO membranes start in Japan, where there were no serious water shortages, and how did these companies become globally competitive? By global standards, Japan is a country with abundant rainfall and water resources. Although some insisted that the amount of rainfall per capita was inferior to that of other countries (Honma, 1968) and that the flow rate of rivers fluctuated throughout the year, the 1960s was the decade in which the largest number of dams were constructed in Japan and a stable water supply infrastructure was developed.[1] While industrialization led to a threefold increase in demand for

[1] According to the National Institute for Land and Infrastructure Management (2006), the number of dams constructed in the 1960s was 356 (based on the year of completion).

M. Fujiwara and Y. Aoshima, *Mechanisms for Long-Term Innovation*, Advances in Japanese Business and Economics 31,
https://doi.org/10.1007/978-981-19-4896-1_10

industrial water between 1965 and 2000, the amount of water withdrawn from rivers and underground sources peaked in 1973 and has been declining ever since, thanks to progress in recycled water use.[2] Although weather conditions often cause droughts in some areas, it is unlikely that the development of RO membranes was driven as strongly by the need to address water shortages at the national level as in the United States.

Why, then, has the development of RO membranes become so active in Japan? This is the first question this chapter addresses. To answer this question, this chapter focuses on the spillover effect between the public and private sectors across national borders. As described in Chap. 4, the development of the industry was triggered by the U.S. government's support for the development of seawater desalination technology in response to concerns about serious water shortages. Resources supplied from a public perspective were often necessary for the development of a new technology that was not economically viable. With this support, many companies and research institutes in the United States began to develop RO membranes. This trend spread far across the ocean to Japan. Despite the fact that there was not enough demand for seawater desalination in Japan, the development of RO membranes by leading U.S. companies such as DuPont and Dow, which Japanese companies looked up to as role models, led Japanese chemical and textile companies to imitate their actions and begin their development.

It cannot be said that the reason why Japanese public institutions began to support the development of desalination technology was directly due to national measures against water shortages. It can be noted that it was desalination technology that attracted attention in the United States at just the right time when public research projects in Japan developing salt production technology were losing their way due to the emergence of alternative technologies. The project for developing salt production technology was redirected to desalination technology development and continued. This led to a joint public–private demonstration project at the WRPC in Chigasaki.

10.1.2 Why Was a Large Amount of Government Support Invested in the United States?

The development of the RO membrane industry driven by Japanese and U.S. companies can be traced back to the large-scale support provided by the U.S. government. It is precisely because of the massive budget invested at the national level that many U.S. companies started development, and the influence of this support spread across the ocean to Japan. Therefore, why did the U.S. government invest such a large amount of money in the first place? This is the second question addressed in this chapter.

[2] Ministry of Land, Infrastructure, Transport and Tourism (2004), "Current Status of Water Resources in the World and in Japan" (http://www.mlit.go.jp/common/001020285.pdf, accessed February 12, 2018).

At the time, there was certainly concern in the United States about future water shortages. In California, as a result of the urbanization of Los Angeles, water shortages had come to be recognized as a serious problem. While the impact of the situation in California was certainly significant, it could also be said that it was a regional problem. So, why was the national budget invested in this over a long period of time, and why was so much support given to companies, universities, and research institutions throughout the United States?

To answer this question, this chapter focuses on the point that the support for desalination research was justified by another policy intention of striving for the peaceful use of nuclear energy. Furthermore, we discuss the possibility that the budget for desalination research was increased by the large budget for space development promoted at the same time.

10.2 The Significance of Government Support

In general, it is not easy for private companies to start full-scale commercialization and development of budding new technologies. This is because there is a high degree of technological uncertainty and the prospect of practical application is often unclear. In the case of a new technology whose main purpose is to solve a social problem, it is even more difficult to start early because of the added uncertainty of the market. Since there is not sufficient capacity or incentive to pay where there is a social need, it is difficult to directly link the new technology to commercial results despite a strong need for awaiting its arrival. For example, the introduction of solar panels and storage batteries would certainly be useful in poor areas without electricity, but the people in those areas cannot bear the cost of installing them.

Even if a company can find a niche market with high solvency, its public mission makes it difficult for it to focus on specific customers and sell at high prices to secure profits. For example, if a pharmaceutical company were to supply cancer drugs only to the wealthy, it would not be able to avoid a certain amount of social criticism. That is why the national insurance system compensates for this. There would also be social resistance to the monopolization of such technologies by certain companies.

Thus, if pure profitability as a business is considered, it is not an easy task for private companies competing in the market to promote the development of technologies that solve social problems. As a result, there is often social underinvestment in this type of technological development (Jaffe et al., 2005). Public support from the government is needed to solve this problem.

The development of desalination technology corresponds to just such an example. Looking around the world, many regions are suffering from water shortages. Technology to produce drinking water from seawater and brine is a key technology in solving such water shortages. However, many of the regions suffering from water shortages are also poor. Even if the scale of potential users is large, they do not have sufficient capacity to pay the cost. In addition, it is socially unacceptable for a company that has established desalination technology to strategically raise the price of water in a drought area, focusing on profitability. Of course, if the main supplier of water is a municipality, the municipality will make efforts to supply water equally to people of all social strata at a reasonable price. However, it is difficult for society to accept a company that tries to gain monopoly profit by providing desalination services to the municipality at a high price. Under these circumstances, it is difficult to foresee significant commercial results beyond the establishment of a new desalination technology. This would reduce the motivation for companies to develop new technologies. That is why the U.S. government provided strong support for RO membranes in the early stages of their development.

Public support not only contributes to the process of knowledge creation, but also removes the initial uncertainty in the market and leads to commercialization, which is the only way to obtain benefits commensurate with investment.[3] In other words, whether or not an industry develops as a result of public support depends on whether it can act as a "catalyst" to induce investment by private firms, or whether it can trigger spillover effects. As will be discussed below, this spillover effect played a major role in the progress of RO membrane development.

The development of RO membranes in the 1970s and 1980s was actively pursued not only in the United States but also in Japan. Japanese companies have played a major role in the development of this industry. The development of RO membranes by Japanese companies was influenced by the increased development activities of private companies in the United States, which were triggered by public support in the United States. In short, public support in the U.S. indirectly spread across the ocean to Japan and supported the development of the industry, especially in the early stages.

Policies such as the U.S. government's support for the development of desalination technology are naturally intended for the development of domestic industries and companies since taxpayer money is invested in them. However, contrary to such intentions, the policies of the home country may not only contribute to domestic firms but also indirectly stimulate the behavior of firms in other countries and induce development activities. This may result in a decrease in the competitiveness of domestic firms, but may contribute to the development of the industry as a whole by inducing international competition and collaboration among many firms. The early RO membrane industry seems to have developed precisely through such a "policy spillover effect".

Figure 10.1 shows the overall picture of the discussion in this chapter. Section 10.3 analyzes the backgrounds for the development of desalination technology initiated by Japanese private companies and public institutions, and clarifies how the U.S. policy support spilled over to Japan. It is shown that the independent development activities of private companies and public organizations under the influence of the U.S. merged through the demonstration test at the WRPC in Chigasaki and contributed to

[3] Guan and Chen (2012) analyzed the efficiency of each national innovation system in 22 OECD countries by dividing it into the upstream knowledge creation stage and the downstream commercialization stage. They pointed out that there was no clear relationship between the efficiency of the knowledge creation stage and the commercialization stage, and that the overall innovation efficiency depended mainly on the commercialization process.

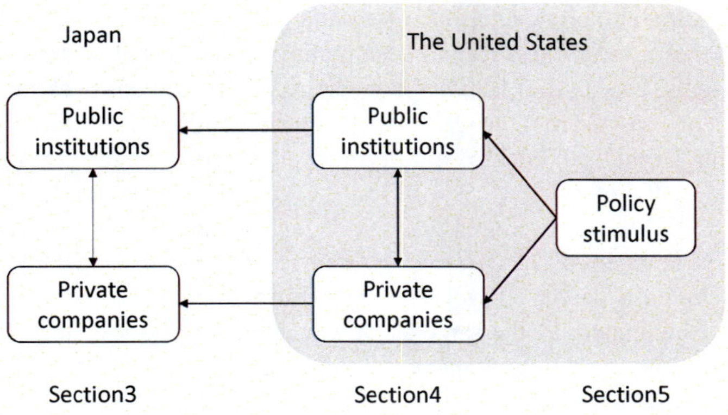

Fig. 10.1 The scope of analysis in this chapter. *Source* Prepared by the authors

the formation of the foundation for subsequent development activities by Japanese companies.

Section 10.4 clarifies that the support by the U.S. government bolstered the development of RO membranes by private companies and research institutes in the United States. We argue that many of the results of the development were widely disseminated as public goods, which led to the creation of many entrepreneurial businesses and the formation of industrial clusters, and also indirectly benefited Japanese companies. In Sect. 10.5, we explain the reasons for the continuation of the massive public support for such research. In addition to the social need to solve water shortages, we analyze the indirect influence of the peaceful use of nuclear energy during the Cold War and the space development projects led by NASA on the policy support for the development of RO membranes.

10.3 Why Did Development Start in Japan, a Country Rich in Water Resources?

10.3.1 Competitive Imitation by Japanese Firms

Why did the development of RO membranes begin in Japan, a country rich in water resources? To answer this question, we will take a look back at how the three major Japanese companies began developing RO membranes.

Toray, which had been seeking for new businesses, was influenced by the RO membrane Permasep presented by DuPont at the "International Conference on Water for Peace" in May 1967, and officially began the development of RO membranes in 1968. Thereafter, the central role in the development activities at Toray was played by Masaru Kurihara, who came to the United States for two years starting in 1970 to conduct research activities at the University of Iowa. Kurihara initially came to the U.S. to research and develop heat-resistant polymers, but he chose to develop RO membranes after witnessing the desalination research that was being actively conducted in the U.S. In the U.S., Kurihara provided his contacts back in Japan with information on reports from the OSW. After returning to Japan, Kurihara continued to inject stimulation from the U.S. into Japan by actively developing exchanges with Cadotte, Riley, and others.[4]

Like Toray, Toyobo also sought to diversify, and as part of this, began developing RO membranes in 1971. In doing so, it referred to DuPont and Dow. To avoid infringing DuPont's patents, Toyobo gave up using polyamide materials and chose cellulose acetate materials, the same as Dow's, and chose hollow fiber membranes to make use of the technology it had cultivated in the textile business.

At the time, DuPont and Dow were role models for Japanese textile companies. Japanese companies carefully collected information on the technology development trends and business strategies of U.S. companies, and were eager to incorporate these technologies from early on. For example, Toray signed a technical cooperation agreement with DuPont for nylon yarn in 1951. The development of RO membranes by Toray and Toyobo was another example of their efforts to follow the lead of U.S. companies. The fact that Teijin also began developing RO membranes early on also indicates that Japanese textile manufacturers were influenced by the trends of DuPont and Dow.

Although Nitto Denko was not a textile company, it was also strongly influenced by developments in the United States and began developing membranes. President Hijikata, who had visited the U.S., sensed the potential of the membrane business there and clearly positioned membranes as one of the pillars of the business in 1974, which became the basis for the company's membrane development. This policy of top management played an important role in the continuation of RO membrane development at Nitto Denko even though commercial success was difficult to achieve. It was at the suggestion of Baker, who was working for Bend Research in the U.S., that Nitto Denko shifted its resources to the development of flat composite membranes in the late 1970s when it integrated membrane development throughout the company.

As seen from these circumstances, the development of RO membranes by Japanese companies did not start in response to any social demand to solve domestic water shortages. Nor was it to put to practical use the results of basic research conducted by universities and public institutions. Rather, it was a search for future business opportunities for companies, in which one possibility was found amidst the actions of U.S. companies. The development of membranes by Japanese companies began as a competitive imitation of the U.S. companies that had been carefully benchmarked.

Thus, there is no evidence that the early development of RO membranes by Japanese companies was influenced by domestic public support or the R&D results of public institutions. However, this does not mean that no desalination research was conducted by Japanese public institutions, nor that they were slow to start development.

[4] Based on the authors' interview with Masaru Kurihara on February 6, 2018 in Shiga.

In fact, research on seawater desalination was already underway at the Tokyo Kōgyō Shikenjo (Government Chemical Industrial Research Institute, Tokyo (GCIRIT)). The institute began desalination research for freshwater production in **April 1961, seven years earlier than Toray's official start of development in 1968. The** Kaiyō Shigen Bukai Kaisui Riyō Shō Iinkai (Subcommittee on Seawater Utilization, Marine Resources Division) for Shigen Chōsa Kai (Natural Resources Research) of the Kagaku Gijutsu Chō (Science and Technology Agency) was established in 1963, and a report on the technological development of seawater desalination was compiled in 1967. Considering that Kennedy enacted Public Law 87-295 in September 1961 to support desalination research on a massive scale, research by Japanese public institutions was thought to be early.

Why did public institutions, especially the GCIRIT, start desalination research in the early 1960s when the country was not suffering from drought? To preemptively conclude, desalination research was emerging in the U.S. at a time when salt production research at the institute in Japan was in a deadlock, and desalination research was an area of research in which knowledge from salt production research could be put to good use.

10.3.2 Exploitation of Accumulated Research in Public Institutions and Imitation

Seiichi Ishizaka, who joined the GCIRIT in 1947, played a leading role in desalination research at the institute. Of course, there were many collaborators, but the promotion of desalination research projects at public institutions was largely due to the efforts of Ishizaka himself.

In 1947, the GCIRIT launched a research project to establish a method to efficiently extract dissolved resources such as salt from seawater to solve the extreme shortage of salt after the war.[5] Ishizaka, who had majored in electrochemistry at university and had conducted research on the production of magnesium chloride from magnesium hydroxide produced from seawater, was involved in this research together with Koichi Tahara who was a researcher at the same institute. They worked under the supervision of Kenji Tanaka, the director of the Fifth Department. The research project received strong support from Ryuichi Sano, chairperson of the Japan Soda Industry Association, and Harunari Inoue, director of the Sangyō Gijutsu Sōgō Kenkyujo (Agency of Industrial Science and Technology (AIST)). A large budget was invested in the project.

However, as a result of the availability of inexpensive salt on the market and the advancement of alternative technologies, their research was never commercialized in earnest. In an interview with the Chemical Society of Japan, Ishizaka recalled that the research did not produce the results they had initially envisioned despite the large amount of money invested.

[5] Tokyo Kōgyō Shikenjo (1960) pp. 75–86.

The GCIRIT invested a large amount of money in this research. But the results were not ready for commercialization. Salt would become cheaper, electricity would become more expensive, and the diaphragm method might be replaced by the ion exchange membrane method. Then the research wouldn't be economically viable, and there would be technical problems. It was eventually wondered why we had spent so much money on research for such a useless project, which put the director in a very difficult position. I don't think that was the cause of his death, but Kenji Tanaka died of asthma. It left a strong impression on me, and I made a firm decision at that time that we had to succeed in seawater utilization at all costs. Many years later, I came to the conclusion that we should desalinate seawater to produce water.

Anyway, I had to fight a battle of condolence for my late boss, so I couldn't get away from seawater. I was also experimenting with the freezing method as a way to make up for it.[6]

Research on seawater desalination for the purpose of extracting freshwater began in 1961. In the "Research Plan of the GCIRIT" published by the AIST at the time, it was stated that "Special research on seawater utilization was conducted from FY1952 to 1954 and 1956. Special research on the effective use of natural gas brine was conducted from FY1958 to 1960, and research on the refrigeration and concentration method was conducted from FY1962. In addition, research on solvent extraction was carried out in connection with special research on uranium enrichment. We are going to conduct research on seawater desalination (for the purpose of securing industrial water) by integrating these technologies."[7] The special research on seawater utilization referred to here is the research that Ishizaka and his colleagues were engaged in. Seawater desalination was a ray of sunshine that was discovered when research on extracting dissolved substances such as salt from seawater was half-abandoned.

Desalination was more than being able to utilize accumulated salt-producing technology. Desalination was an advanced research and development theme that the U.S. government had established the OSW to support on a national level. Ishizaka worked at the Japanese Embassy in the U.S. as a scientific attaché from 1958 to 1962, and had the opportunity to visit many research institutes established by the OSW. There, he witnessed the early days of desalination research supported by the Kennedy administration. Ishizaka returned to Japan with this experience and soon after decided to lead desalination research. Ishizaka reminisces about this process as follows.

Since I was stationed there for four years, I was blessed with opportunities to visit various places in the United States. In the field of seawater utilization, which was my specialty, the Department of the Interior established the OSW and built various research facilities throughout the United States, so I was able to visit many of them. In 1962, after four years in the U.S., I returned to the GCIRIT. My first job was as a section chief of the analysis section. After a while, I became a section chief of the inorganic manufacturing section that I had belonged to before, and began to focus on research regarding the desalination of salt water, which I had investigated in the United States. This research was an extension of my previous work on seawater utilization.[8]

[6] Ishizaka (2009) p. 25.

[7] Kōgyō Gijutsuin (1962), p. 37; The text in parentheses was added by the authors.

[8] Kitanihon Shimbun (2010) pp.233–234.

As already mentioned, the GCIRIT had been conducting research on seawater utilization for byproduct collection, not desalination. Therefore, when the U.S.-Japan Conference on Development and Utilization of Natural Resources (UJNR) was established in 1964 and desalination research became a theme, Japan contributed to the exchange of information in the field of byproduct utilization and focused on absorbing desalination technology from the United States. One such technology was the RO method that GCIRIT began to handle in earnest. Ishizaka looks back on this point as follows.

> In 1964, the U.S.-Japan Conference on Development and Utilization of Natural Resources, called UJNR, was established. The UJNR was an umbrella under which panels were formed. One of the panels was called "Desalting", which means desalination. We had only just begun to work on this. In the U.S., they were spending about 10 billion yen a year, depending on the exchange rate of the dollar. They did a lot of research and published reports every year.
>
> If Japan wanted to work together with them, it would not be possible to maintain a balance. So, Japan decided to work together with the byproducts. We formed a panel to discuss not only desalination but also byproducts, and every year we alternated between the two countries. This provided us with a lot of benefits for Japan's desalination technology. We received a lot of information from the other side. The RO project started as we were stimulated by what the other side said.[9]

For the researchers on salt production in Japanese public institutions, desalination research was a new path. Since desalination was a major research theme of the U.S. OSW, it was a good target not only for the survival of the research group but also for the possibility of new research development. They obtained reports published by the OSW, absorbed the latest information, and followed the desalination research that was being conducted in the United States.[10] Ishizaka himself reported on the trend of RO membrane technology in the U.S. in the academic journal "Polymer" in 1968 (Ishizaka, 1968).

The research group, led by Ishizaka as a leader and Tahara as a sub-leader, applied to the Ministry of International Trade and Industry (MITI) for a Large-Scale Project that had started around that time, citing the Ministry of Construction's concern about future water shortages.[11] The application was successfully approved, and a research project entitled "Seawater Desalination and Byproduct Utilization" was conducted from FY1969 to FY1977. Research on the RO method was also promoted in this project. Ishizaka had the following to say about the project.

> This is important to note that we were able to finance the beginning of the research on the RO method using the research fund for this project. It was just around the time the U.S. started working on the RO method and related reports came out, so we started working on it in Japan as well … in short, for membrane research, there were also in part achievements of the researchers who worked in this Large-Scale Project.[12]

---

[9] Ishizaka (2009) pp. 25–26.

[10] The Japanese academia was also linked to this trend. In 1947, when the AIST started a research project to establish an efficient salt manufacturing method, the Salt Technology Research Group was established. In 1950, it evolved from a research group into the Society of Salt Science, Japan. In 1965, the society was renamed the Society of Sea Water Science, Japan.

[11] Kensetsu Shō, Kasen Kyoku (Ministry of Construction, River Bureau) (1968).

[12] Ishizaka (2009) p. 20.

As desalination research progressed under the leadership of Ishizaka, the WRPC was established in 1973, and demonstration tests of the evaporation and RO methods were conducted in Chigasaki. The private companies that participated in these tests acquired information about operation of desalination through the operation of the demonstration plants and made progress in their efforts to establish desalination methods. Furthermore, during this period, the price of crude oil soared due to the effects of the oil shock, which further increased the focus on the RO method.

10.3.3 Independent Development in Industry and Academia

Thus, in the Japanese private sector, textile companies were mainly involved in research on RO membranes. In the public sector, research on seawater desalination began at the GCIRIT of the AIST, and research on the RO method was conducted as part of that research. However, there was not a close collaboration between the private sector and the public sector in that research process.

This is also reflected in the status of joint applications for RO-related patents.[13] Of the 1,256 RO-related patents filed in the 1970s, only 31 (approximately 2.5%) were joint applications between the private and public sectors. In the 1980s, there were only 118 joint patent applications (approximately 6.8%) against a total of 1,730 applications. During this period, private companies and public institutions seemed to be developing independently. So, what kind of contribution did Japanese public institutions make to industrial development? Chiyoshi Kamisawa, who joined the GCIRIT in 1966 and was involved in the research of RO as an early member, wrote the following.

> It is precisely U.S. and Japanese companies that have made significant progress in seawater desalination. I do not see researchers at the GCIRIT and the various universities in Japan as having directly contributed (to the development of membrane modules, etc., although they did contribute to data analysis). Although we were the first in Japan to conduct basic research on RO membranes, I recognize that our contribution is in the following areas: We published papers and other materials to make the development of this method widely known. We held research meetings based on these results and laid the foundation for the establishment of academic societies. And we developed groups related to membranes within our research institute, which has expanded the possibilities of various membrane processes in addition to water separation.[14]

As can be seen from this comment, public institutions and universities have contributed to the dissemination of technology in the early years through activities

[13] The population of the patent data analyzed is all 9,296 RO patents (1971-August 2017) obtained in Japan by any applicant. All published patents with F-term 4D006GA03 (including lower layer) were extracted from the patent database UltraPatent on August 24, 2017. Ultrafiltration membranes (GA06) and microfiltration membranes (GA07) were excluded from the analysis.

[14] Sato and Kamisawa (2016) p. 48.

that spearheaded research, made new technologies known to the public, and established researcher communities, rather than contributing to industrial development by directly transferring research results to companies.

10.3.4 The WRPC as a Place for Learning

It was in the demonstration tests that began in 1975 at the WRPC in Chigasaki that the trends in membrane development conducted by the private and public sectors independently intersected. Toray, Nitto Denko, and Toyobo obtained a great deal of feedback on their technologies through these demonstration tests. There was no place in Japan other than Chigasaki where large-scale long-term performance tests could be conducted.

As described in Chap. 6, membranes made by foreign companies such as UOP, DuPont, and Dow were used in the first phase of the demonstration test. Membranes made by Japanese companies were used in the second phase. Foreign-made membranes were used in the first phase because Japanese membranes were not yet available. Manabu Senoo and Shoji Kimura of the University of Tokyo wrote about this point: "When we started the experiment in 1975, there were no domestically produced modules (not even imitations of foreign ones), so we had no choice but to use foreign modules that claimed to be able to desalinate seawater in a single-stage system.[15]

Although the membranes themselves were made overseas, the equipment for the demonstration test was developed by a Japanese company. Kurita equipped the RO modules by ROGA of UOP because the company was the Japanese licensee of AJAX, the distributor of ROGA, and supported UOP's market development in Japan. In addition, Sasakura Machinery Works (now Sasakura) developed equipment for DuPont's B-10 (polyamide), and JGC developed equipment for Dow's XFS4167.08 (cellulose acetate). In this way, Japanese equipment and plant engineering companies also had the opportunity to learn from the demonstration tests at the WRPC.

By the end of the first period, the membranes of Japanese companies had reached a sufficient level of performance. Toyobo, for example, had succeeded in developing a hollow fiber RO membrane made of cellulose triacetate that would later support its business. Senoo and Kimura said, "What stood out here (in the demonstration test at the WRPC) was the rapid improvement in the performance of Toyobo's hollow fibers". In 1978, we were able to produce a membrane capable of single-stage desalination.[16] However, Toyobo's new membrane was still in the research stage, and the development team was not confident about its practical application. The fact that the HR8650 with the new membrane showed good operational performance in the second phase of the demonstration test indicates that the Chigasaki desalination center was functioning effectively in promoting development toward practical use.

[15] Senoo and Kimura (1983), p. 97.

[16] Senoo and Kimura (1983), p. 98.

Toray also used the plant in Chigasaki as a place to demonstrate the new membrane. The company initially used SC-5200, a cellulose acetate RO membrane element, for the demonstration tests. Later, when it succeeded in developing the new PEC-1000 membrane, it switched to the SP-110 (4-inch diameter) and the SP-120 (8-inch diameter) using this new membrane for the demonstration tests in June 1981. As mentioned in Chap. 6, in a paper published in 1980, Kurihara wrote that he would proceed with field tests of the PEC-1000 to realize desalination using a single-stage method. This was promptly implemented at the WRPC. PEC-1000, which has both high desalination performance and high permeability performance, was able to maintain its performance for more than 3,000 h of operation in this demonstration test (Okazaki & Kimura, 1983; Kurihara, 1983).

While Toyobo and Toray confirmed the effectiveness of their new membranes, Nitto Denko recognized the limitations of its own tubular modules in the demonstration tests conducted at the end of the first phase. Nitto Denko, which did not necessarily consider seawater desalination to be its main application, half passively participated in the demonstration tests after being approached by the Center. However, when the limitations of tubular membranes became clear, Nitto Denko began to explore a variety of technologies. This led to the development of Nitto Denko's membrane business. In this sense, the demonstration test was considered to have had a certain significance. In addition, while Kobe Steel was responsible for the development of equipment for the membrane modules of Toyobo, Kurita was responsible for the development of equipment for Toray. Thus, these plant engineering companies were also able to obtain important feedback.

Haruhiko Ohya, who conducted research activities with Sourirajan at the National Research Council Canada while working for Showa Denko, and later moved to Yokohama National University as a professor and was responsible for performance evaluation of the demonstration tests conducted in Chigasaki, describes the lack of industry-university collaboration in RO membrane research in Japan and the role played by Chigasaki.

> Basically, university professors didn't do joint research (with membrane companies). I don't think they did. Even if we tried, the companies would not get on board. Professors did not have any big secrets and the companies would think they could do it within their own company. Only when they have any problems, they can just come and ask for help. I did an evaluation of membranes in Chigasaki, so I told them exactly what we got there. (The university's role was to) write a book (on RO membranes) and release that information. We did not do anything more than that.[17]

Toray, Nitto Denko, and Toyobo developed technology on their own, and did not obtain technology directly from public institutions. However, the WRPC played an important role in the early development stage of RO membranes in the sense that it provided a place for learning to promptly demonstrate the results of development by private companies. It was a place where research by public institutions and that by private companies merged.

---

[17] From the authors' interview with Haruhiko Ohya on October 15, 2014 at Takadanobaba. The text in parentheses was added by the authors.

10.4 The Development with Support from the U.S. Government

10.4.1 Du Pont and Dow

The U.S. companies that had a great influence on the development activities of Japanese textile companies were traditional large companies such as DuPont and Dow, as well as divisions and subsidiaries established by large military contractors with the aim of acquiring civilian demand. Among them, Permasep B-9, developed by DuPont in 1967, had a strong influence on Toray.

DuPont began by developing hollow fiber RO membranes for gas separation and then expanded their application to desalination. The development of these membranes was done on its own without financial support from the OSW. Figure 10.2 shows the number of contracts and the amount of money spent by the OSW on desalination research for DuPont and Dow.[18] DuPont received little support from the OSW during this period, that implies Permasep B-9 was developed in-house.

After DuPont developed B-9 in 1969 for high-performance brine desalination, the company moved on to develop membranes for more difficult desalination. This led to an increase in contracts between the OSW and DuPont. In 1971, DuPont was awarded a contract to conduct research on "Development of sea water desalination membranes". The OSW invested $585,300 to support this project. The following year, 1972, $888,454 was awarded for the same project. Then, in 1973, the OSW awarded $983,000 for "Non-cellulosic reverse osmosis system for seawater desalting". Although the number of contracts was smaller than that of Dow, the amount of support per contract was large. The history that DuPont, which had previously developed its products in-house, began to receive massive public support in the 1970s suggests the magnitude of the policy's benefits.

The main research topics for which Dow was contracted were factory wastewater treatment and seawater desalination. Throughout the early 1970s, Dow received support for more than 10 projects each year, with annual contract amounts exceeding $1 million. This shows that Dow, like DuPont, was receiving massive government support for its development.

The results of research and development under the contract with the OSW were to belong to the U.S. government, and the details of the development were also obligated to report to the public. The fact that even Dow and DuPont with a strong incentive to keep their development results secret relied on the support from the OSW suggests that the financial benefits were very significant.

[18] Since the commissioned amounts in the Saline Water Conversion Report are only available for 1965, the 1964 information is presented based on Legislative History (Vol. 5, No. 2).

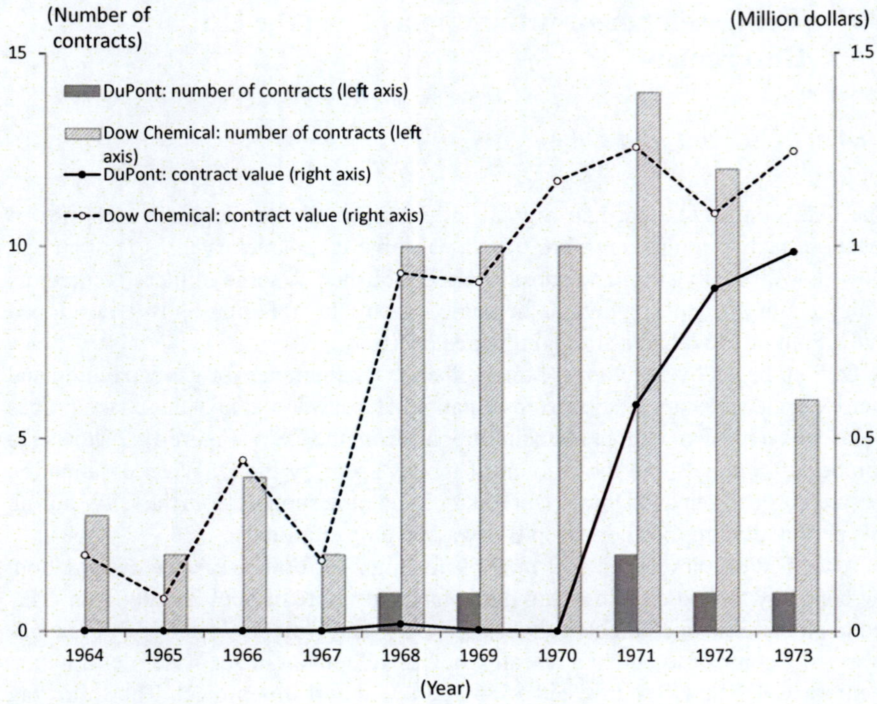

Fig. 10.2 Contracts with DuPont and Dow by the OSW. *Source* Prepared by the authors based on Legislative History (Vol. 5, No. 2) and the Saline Water Conversion Report

10.4.2 The Conversion of Military Enterprises to Civilian Demand

In addition to textile and chemical companies such as Dow and DuPont, a variety of other companies received support from the OSW to develop RO membranes. Among them, military contractors and their subsidiaries received particularly large support. These included Aerojet and its subsidiary Envirogenics, General Dynamics and its GA division, and Aqua Chem.

Figure 10.3 shows the number of cases and the amount of support for these military enterprises. You can see that the number of cases and the amount of support for these enterprises increased steadily throughout the 1960s, and then gradually decreased in the 1970s as the official role of the OSW was waning.

Aerojet received 24 cases of assistance in 1970. In the same year, the OSW's total support was 243 projects, meaning that Aerojet alone accounted for 10% of the total. Envirogenics, a subsidiary of Aerojet, also received substantial support, particularly in 1971 when one of its contracts exceeded $4 million. In 1973, 12 of the OSW's 103 contracts were for Envirogenics. However, despite this extensive support, Aerojet and Envirogenics did not make history in the development of RO membranes.

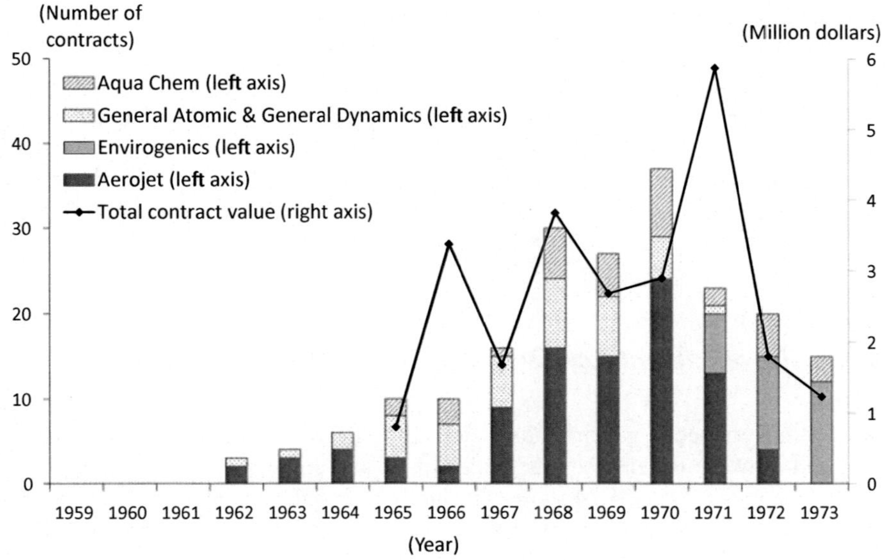

Fig. 10.3 Trends in support for military contractors. *Source* Prepared by the authors based on the Saline Water Conversion Report

The OSW's support for General Dynamics and GA is also notable. The first contract awarded to the company in 1962 was for "Study of the reverse osmosis process," indicating that Riley and his colleagues had intended to use the RO method to desalinate water from the beginning. Aqua Chem, a company that developed and manufactured small desalination equipment for the U.S. military, also received many contracts from the OSW. For example, in 1966, the company's contract reached nearly $2 million, the highest amount of the year.

Most of these munitions companies were located in California. Their conversion to civilian demand was unintentionally supported by the OSW. In the process of civilian conversion, many spin-out companies were born, especially from GA. Public support for desalination research promoted the conversion of military contractors around the military base host of San Diego to private demand and contributed to the formation of industrial clusters through the spin-out of companies. In this regard, Riley reflects as follows.

> Commercialization of reverse osmosis began at General Atomics in the late 1960s. The commercial products were sold under the label of General Atomic Reverse Osmosis (ROGA). In the mid 1970s the company was sold to Universal Oil Products (UOP) and became known as Fluid Systems Division of UOP, Inc. The interest in the reverse osmosis desalination process soon spread worldwide. Since the research and development of reverse osmosis was funded by the U.S. Department of the Interior, the information and rights were in the public domain and available to both domestic and foreign countries through the Freedom of Information Act. For this reason, many Fluid Systems employees left the company to start their own companies and most have been successful. If you look at the industry today, both foreign and domestic, many of the companies can trace their roots to ROGA technology.

The ROGA spiral-wound element design became the accepted commercial configuration. We were the graduate school for the industry. It was all good and consistent with the U.S. Government's original objective.[19]

As mentioned by Riley, the development results of companies that received massive public support became widely available for reference as public goods. This led to the emergence of a wide variety of companies in the United States. In addition, the results of development were also transmitted overseas, encouraging the development of Japanese companies.

10.4.3 Development Activities of U.S. Research Institutes

The U.S. government's generous financial support for desalination research stimulated not only the development activities of private companies, but also those of research institutes. Figure 10.4 shows the number of consignment contracts entered into with the OSW from 1959 to 1973 by type of contractor. We classified contractors as either research institutes or private companies depending on their names, and as "other" when we could not classify them. The line graphs show the changes in the budget of the OSW, indexed with 1959 as 1.

From this figure, we can see that the number of commissioned research projects, which had been about 50 in the late 1950s and early 1960s, increased with the rise of the budget for the OSW. The number of commissioned research projects from 1959 to 1973 reached 2,533. On the other hand, in the 1970s, the budget for desalination research began to decline as the Nixon administration, faced with a trade deficit, adopted spending reduction measures. In 1974, the OSW and the Office of Water Resources Research were merged into the Office of Water Research and Technology that greatly reduced its official role. The sharp decline in the number of commissioned research projects greenlit in 1973 may be due to this organizational integration and budget reduction measures.

The breakdown of the number of contracts shows that in 1959, out of the total 44 contracts, 26 were from research institutes, 17 were from private companies, and 1 was from "other". After 1960, however, contract research by private companies became more active. This is because the focus of development shifted to the downstream operation toward the practical application of desalination.

U.S. government support effectively facilitated development by research institutions and private companies because of the program in which they could be easily financed through contracts with the OSW. Baker of Bend Research, who received support on the private side, recalls the structure and benefits of the program as follows.

Bend research was a contract research company. 'Contract research' in those days, and even today, was something America does. But not normally done in other places. But in America, the government will put out a solicitation. It will be a request for proposals. They call it an

[19] From the authors' interview with Robert (Bob) Riley at Separation Systems Technology, Inc. at Morena Boulevard on March 5, 2015.

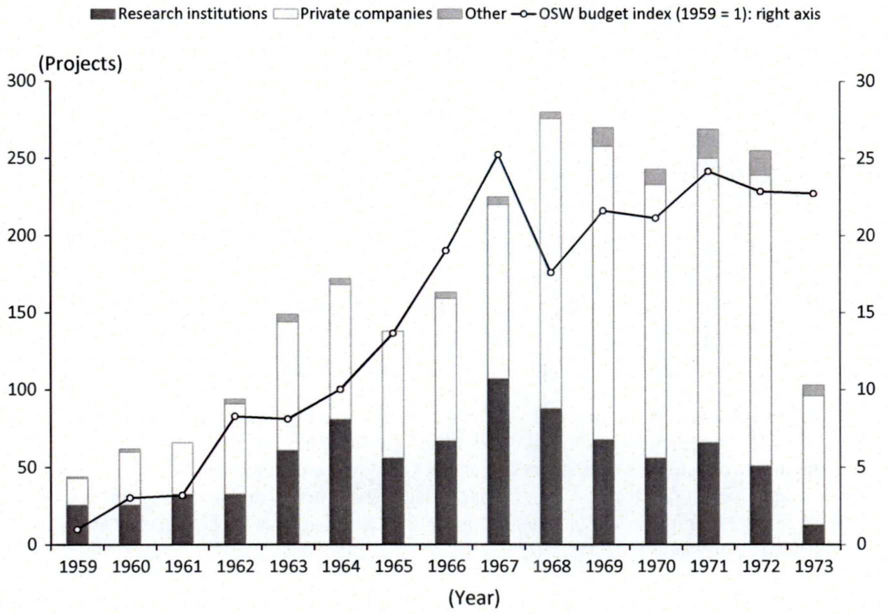

Fig. 10.4 The number of commissioned research projects by institution type. *Source* Prepared by the authors based on Saline Water Conversion Reports

> RFP. They solicit people to do research on some topics. And you could send them a proposal. You have a chance to get a contract. If you get it, it's great. Because you send them a report, they send you a check. Next month, you send them another report, they send you another check. It is a great system.[20]

The U.S. government's active support for the development of this technology attracted attention in Japan from early on. For example, the Science and Technology Agency published the "Report on the Development of Desalination Technology" in 1967 and noted the following.

> The research policy of the OSW is very thorough. All possible methods of brine desalination are considered and studied. Therefore, the scope of their research is extremely wide, ranging from basic issues such as the properties of water, electrolytes, and aqueous solutions, phenomena such as evaporation, condensation, osmosis, and electrolysis, scale formation, and corrosion of equipment, to research on various desalination processes, practical application research using a pilot plant and demonstration plant, research on technical evaluation and cost, and research on future large-scale equipment. In addition, universities, public and private research institutes, private companies, and individuals are involved in these studies and various development projects, all of which are carried out under contract by the OSW. Therefore, a large amount of the national budget has been allocated for this purpose.

Despite the large amount of support, it did not necessarily produce the initially intended results to solve water shortages. The General Accounting Office (GAO)

[20] From the authors' interview with Richard Baker on March 3, 2015 at Membrane Technology and Research, Inc. in Newark.

report submitted to Congress in 1979 stated "Desalting water probably will not solve the nation's water problems, but it can help."

10.4.4 Spillovers to Japan

The U.S. government's support was naturally aimed at U.S. companies. However, contrary to its intention, the development by the U.S. government indirectly benefited Japanese companies across the sea. Of course, there was no financial support from the U.S. government to these Japanese companies. This point can be seen in the following remarks made by Ohya.

> The U.S. government gives money to various companies to commercialize. For example, take research on the spiral method. All the manufacturing equipment of the company that was the first to develop it totally depended on the U.S. government money…After all, if the government doesn't pay for it, it won't work. It's the government's money everywhere. So the U.S. says, "Don't spend it in Japan." I guess they say that because (Japan) has a good understanding of what was going on.[21]

Although the funds from the U.S. government did not flow directly to Japanese companies, most of the research results were disclosed to the public as public property, and the information on the development results of U.S. companies was transmitted to Japan. As already mentioned, Toray's Kurihara, who was studying in the U.S., obtained reports from the OSW and informed his contacts in Japan of the development trends. Information on active development activities and research results in the U.S. was constantly being sent to Japan. This is thought to have contributed to giving legitimacy to development activities within Japanese companies.

Japanese companies also received various inspirations from companies spun out of GA. Kurihara developed a relationship with Riley who was spun off from GA. Nitto Denko received technical advice from Baker of Bend Research. In this way, although Japanese companies did not receive direct support from the U.S. government, they quickly absorbed and learned a multitude of advanced information through various publicly released information and organizations.

10.5 Why Has U.S. Public Support Been So Massive?

10.5.1 Combination of Policies

Why was U.S. public support so large in the first place? The most commonly cited reason is that the U.S. government took the possibility of future water shortages very

[21] From the authors' interview with Haruhiko Ohya on October 15, 2014 at Takadanobaba. The text in parentheses was added by the authors.

seriously. Los Angeles, with its low annual rainfall, was suffering from drought, and population growth was added to its woes. New York City on the East Coast was also experiencing significant population growth, and there was concern about future water shortages. However, from a national perspective, these were regional issues, and it is doubtful that they could justify large-scale national budget allocation in the long term.

Another possible perspective is the policy intention of peaceful use of nuclear energy. In 1953, President Eisenhower delivered a speech on "the peaceful use of nuclear energy" to the international community, which led various military contractors and institutions to actively explore the uses of nuclear energy. Desalination, which consumes large amounts of energy, attracted attention as a promising peaceful application. In addition to the concern about water shortages, a move to explore peaceful uses of nuclear energy was another reason that ultimately resulted in desalination research for seawater desalination gaining stronger legitimacy and stable public support. This is evidenced by the fact that many of the companies that received support from the OSW were military contractors.

A further boost to desalination research came from space exploration that was being promoted at the same time. The Kennedy administration, which took over from the Eisenhower administration in 1961, focused on space development and desalination research with the major political goal of sending a man to the moon and making the desert bloom. NASA's budget, which was about $750 million in 1961, was raised to $1.2 billion in 1962 and to $2.55 billion in 1963.

NASA aimed to establish a complete water circulation system in their spacecrafts and actively supported the development of RO membranes with its large budget. For example, Bend Research was contracted by NASA to develop "Novel Reverse-Osmosis Module for Spacecraft Washwater Recycle" and "Development of a Non-Phase-Change Preprototype Wastewater Reclamation Subsystem".[22] Ohya, who worked with Sourirajan in Canada in the late 1960s on the design of a water recirculation system in a spacecraft, said "NASA had a project like that, so they transferred money to the OSW, and the OSW contacted various places and asked them to develop a membrane that would fit the water recirculation project... (and) North Star Research Institute developed NS-100, the first composite membrane that excelled in the separation of urea. Those were the days. That's why there was a lot of money flowing to various places."[23]

This trend led to an increase in the size of the OSW budget, although it was smaller than that given to NASA. Public Law 87-295, enacted by the Kennedy administration in September 1961, provided for a massive $75 million in funding for desalination research between 1962 and 1967. As a result, the budget for the OSW for the same period was raised from $3.795 million in 1961 to $9.805 million in 1962 and $9.6

---

[22] Based on Lonsdale et al. (1988). NASA contract numbers are, respectively: NAS 9-17306 and NAS 9-17523.

[23] From the authors' interview with Haruhiko Ohya at Takadanobaba on October 15, 2014. The text in parentheses was added by the authors.

million in 1963. Although relatively small in comparison, desalination research benefited from the generous public support for space development as one of the two major goals of the Kennedy administration's aggressive fiscal policy.

10.5.2 How Did the Peaceful Use of Nuclear Energy Combine with Desalination Research?

How did the peaceful use of nuclear energy combine with desalination research? The germ of the project dates back to 1959, when the U.S. Department of the Interior and the Atomic Energy Commission (AEC) planned to build a small-scale evaporative desalination plant with the dual purpose of desalination and nuclear energy utilization. Even though it was a joint project, it was still very small at that time. This plan was ultimately passed over because a suitable site could not be found.

The impetus for the move toward collaboration came in 1962 when Phillip Hammond and others at the AEC's Oak Ridge National Laboratory discussed the economic feasibility of a large-scale dual-purpose plant. This caught the attention of Jerome Wiesner, Director of the Office of Science and Technology, and a group was formed in January 1963 to study the issue. It was Johnson, who succeeded Kennedy as president in November 1963, that elevated this movement to national prominence.

Johnson had been interested in desalination since he was a senator, and decided to promote a joint nuclear power and desalination project. He urged both the Department of the Interior and the AEC to realize large-scale desalination. James T. Ramey, who was chairman of the AEC at the time, described Johnson's interest in the project as follows during a hearing held in August 1964.

> President Johnson is vitally interested in this technology as a means to meet national and worldwide needs for freshwater. As you know, he has requested the Department of the Interior, in close collaboration with the Atomic Energy Commission and in consultation with the Office of Science and Technology, to develop a detailed plan for an aggressive, imaginative program for large-scale water desalting, including using nuclear heat sources.[24]

Johnson himself said, "The time is coming when a single desalting plant powered by nuclear energy will produce hundreds of millions of gallons of fresh water—and large amounts of electricity—everyday."[25] The combination of nuclear power and desalination research was an initiative endorsed by the president. The Joint Committee on Atomic Energy positioned the Use of Nuclear Power for the Production of Fresh Water from Salt Water as one of the issues to be discussed.

It was not until the Johnson administration that the Oak Ridge National Laboratory received its first contract from the OSW in 1964. Although a theme of this contract research was on radioisotopes, not directly related to desalination processes, the

[24] Use of Nuclear Power for the Production of Fresh Water from Salt Water: Hearing Before the Joint Committee on Atomic Energy, Congress of the United States, Eighty-eighth Congress, Second Session. August 18, 1964.

[25] Urrows (1966) p. 13.

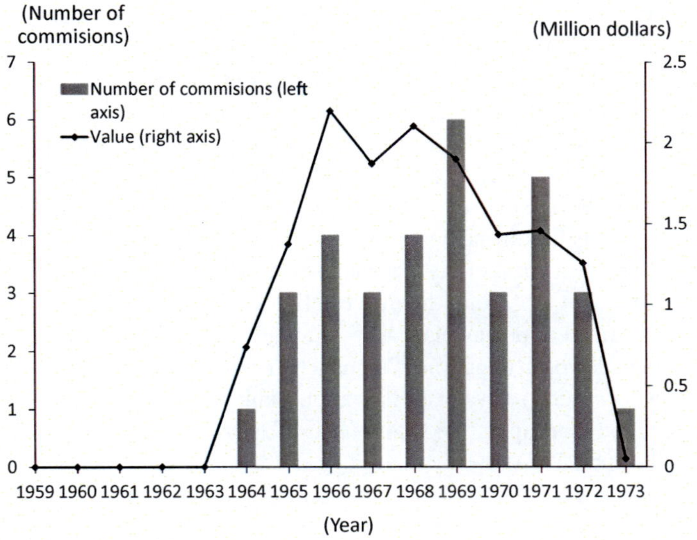

Fig. 10.5 Commissions to the AEC. *Source* Prepared by the authors based on the Saline Water Conversion Reports and Legislative History

amount of support received reached $740,000. This was the largest amount of support received by any research institute that year. This suggests that it was a policy priority to make the peaceful use of nuclear energy compatible with desalination research.

Figure 10.5 shows the number of research projects and the amount of money that the AEC was entrusted with by the OSW. As shown in the figure, the AEC did not receive any support from the OSW between 1959 and 1963. However, from the beginning of 1964, the number and amount of research projects commissioned by the AEC increased steadily. This means that the peaceful use of nuclear energy projects, encouraged by the Johnson administration, entered the budget that was raised to $75 million from 1962 to 1967 by Public Law 87-295, passed by the Kennedy administration. In 1966, the Metropolitan Water District of Southern California passed a bill to build a large-scale nuclear desalination plant (two units of 900,000 KW, three units of 100,000 m^3/day) in Orange County, California.[26]

It is interesting to note that even after 1967, the number of nuclear projects increased in 1968 and 1969. This suggests not only that nuclear projects entered the expanded desalination budget line but also that nuclear projects might have underpinned the budget line. In 1968, the OSW's budget decreased from $29.85 million to $20.8 million, while the AEC's contract research increased in both number and amount.[27] Without the nuclear projects, it is likely that the total budget would have been even lower. The year 1969, when the Johnson administration was out of office,

[26] This plan was not completed as originally envisioned.

[27] The reduction was less than this because $6.42 million carried over from 1967 was added to the 1968 budget.

was the year that OSW's budget was restored to $25.55 million, and at the same time, it was the year that the AEC received the highest number of commissions.[28] In addition, the Saline Water Conversion Act (Public Law 92-60), which was enacted in 1971, specified the AEC as an agency with which to work.

These efforts to combine nuclear power and desalination were not a unique move taken only by the United States, but a global effort. The International Atomic Energy Agency (IAEA) issued a report on desalination using nuclear energy in 1964, and hosted a symposium on nuclear desalination in Spain in 1968. In the city of Shevchenko on the east coast of the Caspian Sea (now Aktau in Kazakhstan), which was Soviet territory at that time, the construction of a seawater desalination plant (BN-350) with a nuclear power plant was also underway. The operation started in 1973 and attracted world attention (Domitriev, 1970; Konishi & Minato, 1999). In Japan, a dual-purpose seawater desalination plant with nuclear power was also studied by the Study Group on the Multipurpose Use of Nuclear Power Generation of the MITI (Tamura, 1974).

Thus, nuclear desalination was a new system that attracted the interest of many countries, including the Soviet Union that was competing with the United States in technological development. Collaborating with nuclear energy was the cutting-edge trend in the world. This may be another reason supporting a large budget for the OSW.

GA was a typical military contractor that enjoyed public support at the civilian level, which was expanded on the back of such a policy partnership. The company, whose name contains the word Atomic, conducted research on the desalination process by the RO method while receiving policy support for the possible peaceful use of nuclear energy. In fact, ROGA continued its development with support from the Department of Energy (established in 1977) that was the successor of the AEC (abolished in 1974) while the support from the OSW diminished. This is another historical fact that supports the policy collaboration.

Originally, the policy integration between the peaceful use of nuclear energy and desalination research was supposed to have importance as long as the evaporation method is used for desalination, which consumes a large amount of energy. On the other hand, the RO method developed by ROGA is a relatively energy-saving technology, so it is not necessarily a good match for the peaceful use of nuclear energy. In this sense, the development of RO membranes may have proceeded with the unintended byproduct of policy collaboration.

10.6 Conclusion

This chapter attempts to answer the question of why the development of RO membranes began in Japan, a country rich in water resources, from the perspective

[28] Note that the administration ended on January 20, 1969, so it did not exercise direct influence for the entire year.

of policy spillover. We clarified that at the private level, Japanese textile companies started to develop RO membranes as one of the competitive imitation behaviors, referring to the preceding U.S. companies. On the other hand, Japanese public research institutes focused on desalination technology, which was being developed on a large scale in the United States, as an ideal destination for salt production technology development projects.

Certainly, Japan's water resources have not always been plentiful. In fact, when Ishizaka and his colleagues applied for desalination research, they cited the Ministry of Construction's concern about future water shortages. Therefore, it is necessary to examine the extent to which concerns about drought in Japan actually affected the development of desalination technology.

In the 1960s, when desalination research began, Japan faced a serious drought problem in 1964, the year the Tokyo Olympics were held. That year, Tokyo's annual precipitation fell to nearly 30% of the normal level, and water usage restrictions were implemented for a long period of time during the summer and early fall. However, it is unlikely that this water shortage, later referred to as the "Tokyo Olympic drought", was the main cause for the strong push for desalination research. Since Japanese companies did not officially start developing membranes until the late 1960s at the earliest, it is unlikely that the Tokyo Olympic drought was a direct trigger. It is more plausible that Japanese companies started to develop membranes not because of the social demand of water shortage concern, but because they followed the development results of American companies as a "leading example".[29]

In terms of public organizations, the GCIRIT began research on seawater desalination before 1964. Although the Science and Technology Agency issued a report on the development of seawater desalination technology in 1967, it was in 1963, before the Tokyo Olympics, that the Subcommittee on Seawater Utilization, Marine Resources Division for Natural Resources Research that issued the report was established.[30] It was in 1964 that the UJNR was established and desalination research was taken as a theme, but it was before Tokyo faced a drought. In summary, desalination research in Japan was not driven by an urgent social demand.

Based on the above, it is reasonable to think that it was the large financial support from the U.S. government that indirectly induced and stimulated R&D activities in Japan. Even DuPont, which had promoted the early membrane development by internal development, came to depend on public support from the U.S. government. Public support from the U.S. government drove research institutes and companies to desalination research more strongly. Because this was a public project, the results of the development became public and spilled over to Japan. This triggered and stimulated research and development by Japanese companies. The overall flow in this chapter is summarized in Fig. 10.6.

[29] There is no evidence that the Japanese companies considered elaborate scenarios for building competitive advantage at the time of entry. The development efforts by Japanese firms seem to have been dominated by the logic of reference group neglect (Camerer & Lavallo, 1999; Moore & Cain, 2007), which is used to explain over- or under-entry behavior in the market (Moore & Cain, 2007).

[30] Matsuda (1970).

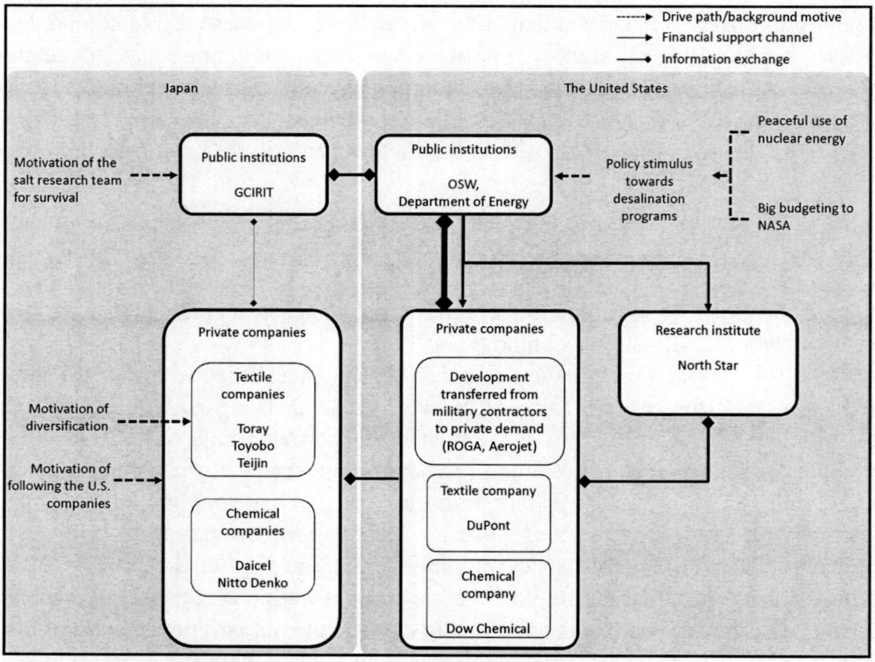

Fig. 10.6 Summary of this chapter. *Source* Prepared by the authors

References

Camerer, C. F., & Lovallo, D. (1999). Overconfidence and excess entry: An experimental approach. *American Economic Review, 89*, 306–318.

Domitriev, I. D. (1970). Kōen: 2 jūmokuteki genshiryoku hatsudensho no gijutsuteki/keizaiteki hyōka" (Lecture: Evaluation of technic and economic in dual purpose atomic generation and desalination). *Journal of the Atomic Energy Society of Japan, 12*, 38–40.

Guan, J., & Chen, K. (2012). Modeling the relative efficiency of national innovation systems. *Research Policy, 41*, 102–115.

Honma, H. (1968). Kaisuiyōsui hatsudensho o riyōsuru gyakushintō hō kaisuitansuika ni tsuite" (On reverse osmosis method seawater desalination using seawater pumping power plant). In *Denryoku Chūō kenkyūjo Otemachi Kenkyūjo "Kenkyū Hōkoku" Dai III shū (Central Research Institute of Electric Power Industry, Otemachi Research Laboratory, Research Report. III)*, 169–184.

Ishizaka, S. (2009). *Kagaku kataribe dai 8 kai: Ishizaka Seiichi sensei interview (Chemical storytelling part 8: Interview with Dr. Seiichi Ishizaka)*. Nihon Kagakukai Kagakuisan Iinkai (Chemical Heritage Committee, Chemical Society of Japan).

Ishizaka, S. (1968). Desalination by reverse osmosis process. *Kobunshi (polymers), 17*, 300–305.

Jaffe, A. B., Newell, R. G., & Stavins, R. N. (2005). A tale of two market failures: Technology and environmental policy. *Ecological Economics, 54*, 164–174.

Kensetsu Shō, Kasen Kyoku (Ministry of Construction, River Bureau). (1968). Zenkoku mizu jukyū no tembo ni tsuite (Outlook for National Water Supply and Demand). *Suiri Kagaku, 12*, 117–124.

Kitanihon Shimbun. (2010). Ishizaka Seiichi. In *Waga hansei no ki: Etchūjin no keifu dai 13 kan (The book of my half-life: Genealogy of people from Etchu)* (Vol. 13, pp. 221–251).

Konishi, T., & Minato, A. (1999). Nuclear energy application for seawater desalination and IAEA's activities. *Journal of the Atomic Energy Society of Japan, 41*, 15–20.

Kurihara, M. (1983). Proceedings: concentration and recovery of dilute aqueous solutions containing valuable materials by synthetic composite reverse osmosis membrane. *Membrane, 8*, 97–112.

Kōgyō Gijutsuin (Agency of Industrial Science and Technology), Ministry of International Trade and Industry. (1962). *Showa 37 nendo shiken kenkyūjo kenkyū keikaku (FY1962 research plan for the experiment laboratories).*

Lonsdale, H., Friesen, D. T., & Ray, R. J. (1988). Novel hollow-fiber membranes for reverse osmosis and coupled transport. *Sen'i Gakkaishi (Journal of Fiber Science and Technology), 44*, 27–35.

Matsuda, T. (1970). Sekai kakkoku ni okeru kaisuitansuika no genkyō (Current status of seawater desalination in various countries in the world). ***Mizushori Gijutsu (water Purification and Liquid*** *Wastes Treatment), 11*, 7–19.

Ministry of Land, Infrastructure, Transport and Tourism. Water Resources in Japan. (2004). *Nihon no mizushigen, heisei 16-nemban (Water resources in Japan, 2004).*

Moore, D. A., & Cain, D. M. (2007). Overconfidence and underconfidence: When and why people underestimate (and overestimate) the competition. *Organizational Behavior and Human Decision Processes, 103*, 197–213.

National Institute for Land and Infrastructure Management. (2006). *Kokudo gijutsu seisaku sōgō kenkyūjo project kenkyū hōkoku dai 4 go (National institute for land and infrastructure management project research report, No. 4).* Accessed April 9, 2018, from http://www.nilim.go.jp/lab/bcg/siryou/kpr/prn0004.htm.

Okazaki, M., & Kimura, S. (1983). Recent advances of reverse osmosis technology for seawater desalination. *Nihon Kaisui Gakkaishi (Bulletin of the Society of Sea Water Science, Japan), 37*, 149–158.

Sato, M., & Kamisawa, C. (2016). Kaisui tansuika gijutsu no kaihatsu (Development of seawater desalination technology). In *Kenkyu Hiwa Website (AIST Research Secrets website).* Accessed January 13, 2018, from https://sankoukai.org/secure/wp-content/uploads/untold_stories/masahito-sato&chiyoshi-kamizawa_final.pdf.

Senoo, M., & Kimura, S. (1983). *Shinkinō zairyō "Maku": Seitai maku no kinō o mezashite (New functional material "membrane": Toward the function of biomembranes).* Kogyo Chosakai.

Tamura, T. (1974). Genshiro ni yoru kaisui datsuen (Nuclear desalination). *Journal of the Atomic Energy Society of Japan, 16*, 165–172.

Tokyo Kōgyō Shikenjo (Government Chemical Industrial Research Institute, Tokyo). (1960). *Tokyo kogyo shikenjo 60-nen shi (60-year history of Government Chemical Industrial Research Institute, Tokyo).*

Urrows, G. M. (1966). *Nuclear Energy for Desalting.* U.S. Atomic Energy Commission/Division of Technical Information.

Chapter 11
Exploration of Initial Markets: Application of Unbalanced Performance Bundles

Abstract In this chapter, we focus on the role of the “initial markets”. The emergence of commercial markets played a role in supporting business continuity by raising expectations of future market expansion. Of particular importance was the emergence of ultra-pure water production applications for semiconductors. This market cropped up like an amazing miracle in the late 1970s, when the miniaturization of semiconductors was advancing and high cleanliness was required in the manufacturing process. The market made it possible for the companies to continue development and led to the establishment of technology that could be applied to seawater desalination.

11.1 Difficulties in Continuing the Development of New Technologies and the Role of Markets

11.1.1 The Importance of the Initial Market

It is not easy for private companies to develop new technologies with high technological uncertainty from the beginning. In the case of technological development aimed at solving social issues, such as seawater desalination, market uncertainty is high because the expected customers do not have sufficient ability to pay. Technology development with such a high level of uncertainty is difficult to start only by the voluntary action of private companies. That is why government support played a major role in the early development of the RO membrane industry. However, government support is not permanent. In fact, the amount of U.S. government support declined dramatically in 1974.

To develop an industry, the private sector should conduct continuous research and development and investment, breaking away from government support. Even if a certain technology is established with government support, there are still many hurdles to overcome before the technology is commercialized and the products are accepted by the market. In the case of seawater desalination, it is necessary to achieve

M. Fujiwara and Y. Aoshima, *Mechanisms for Long-Term Innovation*,
Advances in Japanese Business and Economics 31,
https://doi.org/10.1007/978-981-19-4896-1_11

desalination performance that can withstand the necessities of producing drinking water and actualize water at a low enough cost to be used as water for daily life. The questions are, is it possible to improve such performance? Even if the performance is improved, will the market accept it? In the face of these uncertainties, the private sector must continue to make risky investments. How will this be possible?

It is true that companies may engage in this type of development from a CSR perspective. However, such a position alone is unlikely to be sufficient to support the development of businesses that require large-scale investment. It is difficult to continue to invest for decades in a business that is not expected to be profitable, even if the company is committed to contributing to society. There could be cases in which technological development proceeds successfully, the initially intended market successfully crystallizes, and becomes profitable at an early stage. However, such an ideal situation is rather rare. There are countless cases in which a company withdraws from a new business because the market does not emerge as expected.

Thus, it is important to find application markets that are different from the original target market but can justify the continuation of the business for the time being (Kemp et al., 1998; Schot & Geels, 2008). For example, the development of the carbon fiber industry, which gained a large market in aerospace applications, was partly supported by the early flourishing sports markets for fishing rods, golf shafts, and tennis rackets. The developers did not envision these application markets from the beginning. However, this was an important factor that enabled Toray to continuously invest in the carbon fiber business, which took more than 40 years to become a stable and profitable business (Aoshima & Kasai, 2005).

When a company develops a new technology and aims to commercialize it, it is natural for it to target areas where a large market can be expected in the future. In particular, if a new technology can replace a large market that already exists, the company will be able to anticipate a reliable business scale. A recent example of this is the development of technologies aimed at replacing LCDs with OLEDs in the display industry. However, technological development that aims to replace the market on a large scale is often accompanied by difficulties. This is because the new technology must be superior in every way and also be cost-competitive to completely replace existing technology. This requires a great deal of time and resources.

In the development of RO membranes, many companies were aiming for drinking applications from the beginning, for which there was already a great obvious need. However, drinking water is generally taken from rivers and filtered, which has a very low cost. In the Middle East, where there is no rainfall and river water and groundwater are scarce, desalination by the evaporation method has already been used for a long time. The RO method must at least surpass the desalination performance and cost performance of the evaporation method to win applications for drinking water. To achieve this, various performance requirements must be met at a high level and at a low cost, which takes a long time. In fact, the use of RO membranes for seawater desalination did not accelerate until the 2000s, more than 30 years after Japanese companies began development.

Even if there is a strong promoter, it is extremely difficult for a company to continue development for 30 years without commercialization. However, even if the market is not as large as expected, development may be continued if a market is found for the application of the new technology and the business becomes viable, even if small. As long as development can be continued, performance will improve even if progress is slow, and eventually it will be able to meet the performance and cost requirements of the large market that was initially targeted. Finally, the original objective will be fulfilled.

In the history of the development of the RO membrane business, the emergence of various application markets such as food, wastewater, and industrial pure water made it possible to continue technological development over the years. Continuous performance improvement eventually led to the actualization of the seawater desalination market. In fact, in the domestic market at the end of the 1970s, 68% of the application market for RO membranes was for the desalination of brine, mainly for boiler water, and 24% was for ultrapure water production for the electronics industry (Bairinji, 1984). The emergence of these early markets allowed for continued development of the technology over the years and, through sustained performance improvements, eventually led to capture the desalination market.

Table 11.1 shows the breakdown of the sales volume of the RO method (83,415 m^3/day) by Japanese manufacturers at the end of the 1970s. For raw water, the desalination of brine was by far the largest volume. This was followed by wastewater and industrial water, indicating that there was a certain scale of demand for factory-related water treatment. In terms of produced water, the demand for boiler water, industrial water, and ultrapure water was high, indicating the large scale of applications related to factories and power plants. The demand for ultrapure water is at the top of the list because the purity of the water required for cleaning electronic components increases with their performance and miniaturization.

While exploring these various applications, companies continued to develop and improve the performance of their technologies and eventually succeeded in capturing

Table 11.1 RO capacity sold by Japanese manufacturers at the end of the 1970s (by application)

Raw water	m^3/day	Produced water	m^3/day
Brine	51,479	Boiler water	37,010
Wastewater	14,103	Industrial water	22,680
Industrial water	9,356	Ultrapure water	13,802
Clean water	6,156	Drinking water	3,471
Deionized water	1,783	Recycled water	3,120
Other	394	Effluent	1,084
Pure water	144	Reclaimed water	760
		Purified water for pharmaceutics production	554
		Plating wastewater	540
		Medical water	274
		Bacteria-free water	120

Source Sōgō Hōsō Publication (1980)

the seawater desalination market, which was the original target. In this way, the RO membrane industry has grown to its current scale. Therefore, in order to understand the development of the RO membrane industry, it is necessary to understand the process of creating an application market prior to the expansion of the seawater desalination market. This is the purpose of this chapter.

So how did the application markets emerge? Companies did not always anticipate the markets that would emerge when they began developing RO membranes. Nor were all of the application markets discovered as a result of deliberate exploratory efforts by RO companies. But, this does not mean that they were completely exogenous and coincidental to the companies.

As discussed below, the potential for the emergence of various applications is already contained in the process of developing technologies to realize their ultimate goal. For a technology to be put to practical use, it is necessary to achieve a high level of compatibility between various performances that have a trade-off relationship with each other. In the case of RO membranes, there are various performance requirements, such as desalination performance, water permeability, pressure resistance, chlorine resistance, and heat resistance, and only when all of them are satisfied can the technology be put to practical use. A technology is a "bundle of performances." Different combinations of the levels of each performance in the bundle correspond to different application markets. In general, a "difficult" practical technology is one that requires all of the multiple performance requirements to be met at an extremely high level. The RO membrane for seawater desalination was such a technology. On the other hand, if all of the performance requirements are not met at the same time at a high level, this does not mean that the technology is completely useless. Different applications have different performance requirements. Even if some of the performances do not reach their original targets, it is possible that the combination of the performances at that time matches a specific application. This is the process of industrial development, i.e., the application market that is created during the transition period.

Technology has a wide range of performance and can be evaluated from multiple perspectives. It is not uncommon for a technology to flourish in an unexpectedly high evaluation of a specific performance, leading to a different application than expected. Such application markets emerge through the deliberate search activities of companies and the knowledge of customers, but it is important to note that even if technology development is not pursued with a specific application market in mind from the start, the possibility of diverse application markets can be uncovered as a byproduct of technology development activities aimed at markets that require difficult technologies.

Therefore, to understand the emergence of the application markets that supported the development of the RO membrane industry, it is necessary to understand the evolutionary process of the technology that brought about the new application markets and to clarify the search process by each company that manifested the potential markets. Therefore, we will first show that various potential application markets have emerged on the way to the final goal of technological development aiming at seawater desalination, after clarifying the combination of performance requirements for each application market. Second, we clarify the process by whom and in which the potential application markets indicated by the technology were manifested.

11.1.2 Potential Application Markets Indicated by Technology as a Bundle of Performances

The seawater desalination market did not expand until the 2000s. It had already been more than 30 years since many companies had begun development. Even in the 1980s, when DuPont began to develop the desalination market ahead of other companies, it took nearly 20 years to bear fruit. The desalination market remained in its infancy for longer than expected.

The reason it took so many years for the desalination market to grow to its full potential is that the performance requirements for seawater desalination were so diverse and contradictory that they all had to be met at a high level simultaneously. Takashi Nakane, who worked at the GCIRIT of the AIST, wrote the following in 1974.

> Desalination of seawater by the RO method is difficult because the salt concentration of seawater is about 35,000 ppm and the osmotic pressure is nearly 25 kg/cm^2 at room temperature. In order to obtain potable water with a salt concentration of 500 ppm or less in a single stage, a highly semipermeable membrane with a salt separation rate of 99% or higher is required. Furthermore, to obtain a practical permeate volume, the system must be operated at a high pressure of 80 kg/cm^2 or higher. The membrane must be able to withstand the deterioration of membrane performance due to consolidation at such a high pressure.[1]

When desalinating seawater with a high salt concentration, a strong operating pressure must be applied during desalination to deal with the high osmotic pressure. Therefore, it is necessary to develop a membrane with high pressure resistance. However, if the membrane is made thicker to increase the pressure resistance, the permeability of the membrane will decrease. The pressure pump needs more electric energy to increase the operating pressure, which will raise the cost of desalination.

There are other difficulties unique to seawater desalination. Since a variety of organic and inorganic substances are dissolved in seawater, it is necessary to satisfy not only salinity but also various other removal performances. However, the permeability decreases when various removal performances are improved. Since dissolved substances contaminate the membrane, the requirements for disinfection and cleaning are also severe. Low-cost disinfection requires highly chlorine-resistant membranes, but it is not easy to achieve both chlorine resistance and water permeability. To reduce cleaning costs, chemical resistance over a wide pH range is needed, but this is also difficult to balance with desalination performance and water permeability. To develop the RO method to obtain drinking water, companies had to solve the

[1] Nakane (1974a, p. 112).

extremely difficult problem of balancing multiple performance requirements, which often contradict each other, within cost constraints.

Figure 11.1 shows the comparison of the importance of the performance requirements for RO membranes by application. Since the performance required for desalination membranes varies depending on the raw water environment and the purpose of use, it is difficult to specify the performance required for each application in a general sense, and it is not possible to make an exact comparison. With this in mind, the relative importance of the required performance is presented here as a rough average figure, with the cooperation of the developers in Japanese companies.[2]

As shown in Fig. 11.1, RO membranes for potable seawater desalination must achieve the most diverse performance at a high level. Among the main performance characteristics of RO membranes, which are separation properties such as desalination and TOC, and related properties such as water permeability and pressure resistance, seawater desalination is at the top of the list in terms of all kinds of performance. Since seawater with high salinity is handled, pressure resistance is also the most stringent requirement. The requirement for water permeability performance is second to that for ultrapure water production for semiconductors, and the demand for heat resistance is second to that for food concentration. In terms of operation-related properties, the requirements for contamination resistance and chlorine resistance are the most stringent for seawater desalination applications. Thus, seawater desalination applications need to realize the most diverse performances at a high level.

On the other hand, for applications other than seawater desalination, not all the performances need to be at the same high level. For example, in the case of food applications such as fruit juice concentration, which is one of the early applications, sanitation and contamination resistance are important because the membranes deal with something that enters the human body. On the other hand, high desalination performance is not necessarily required. Therefore, before the establishment of flat membrane technology, tubular RO membranes as well as UF membranes sometimes were used for this application. In addition, since food products such as fruit juices have higher added value than drinking water, they were not subject to the same stringent cost requirements as seawater desalination applications. This was an advantage in the initial stage of high membrane cost. Even if the separation performance and cost advantage were not fully established, food applications could be opened if sanitation, resistance to contamination, heat resistance, and low elution were satisfied.

As in the case of food, resistance to the contamination of membranes is also important for wastewater treatment applications that were pioneered in the early stages.[3] In

[2] In addition to the interviews, a questionnaire was sent to three developers who played a central role in the development of RO membranes at the three Japanese companies. In the questionnaire, we asked the developers to evaluate the severity of the performance requirements for each application on a relative basis, and the authors converted the score into 100 points. The figure shows the average of the responses from two respondents. We would like to thank the three people who cooperated in this very difficult evaluation.

[3] A typical example of a membrane suitable for this application is the low fouling composite membrane developed by Nitto Denko. This membrane, which is characterized by its resistance to fouling, was used in the NEWater project in Singapore, which aimed to obtain potable water from

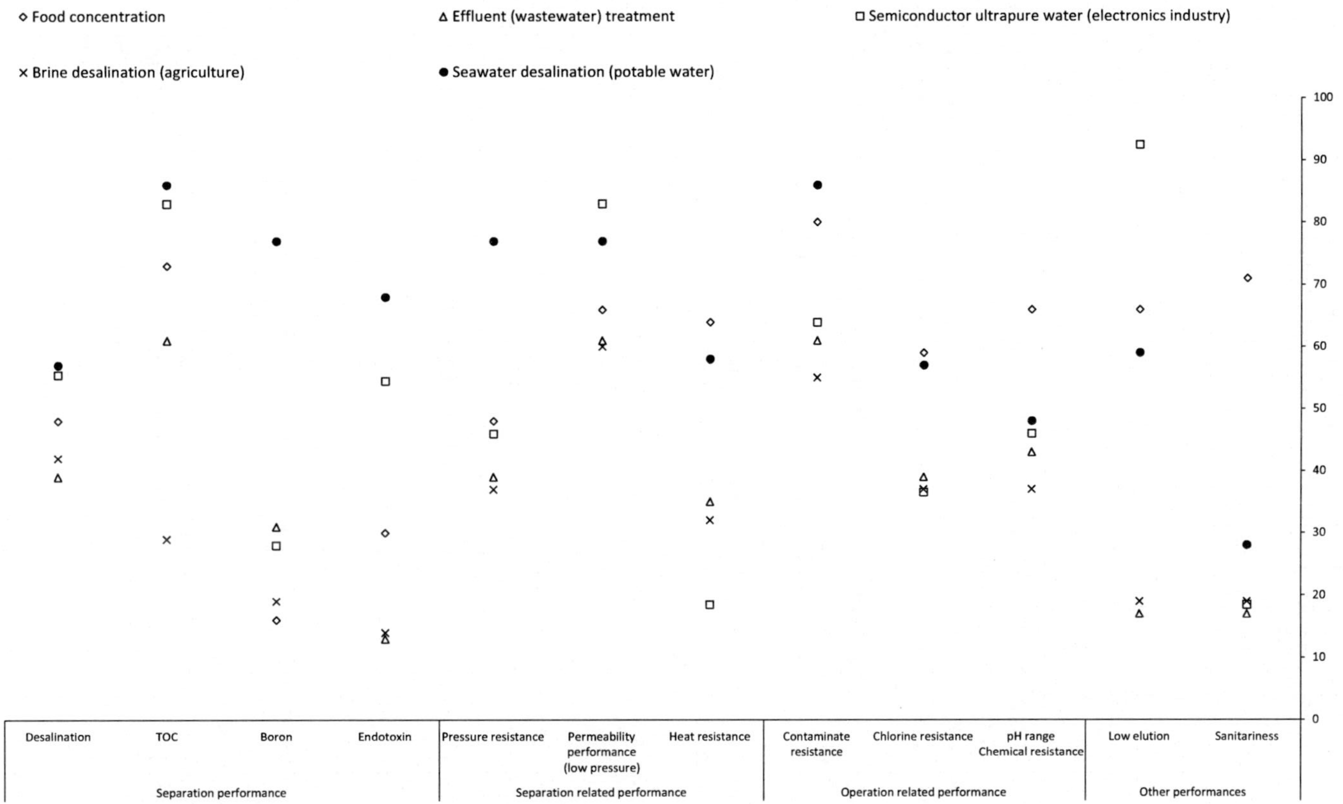

Fig. 11.1 Relative importance of performance items by application. *Source* Prepared by the authors based on research. *Note* TOC relates to low molecular weight organic solutes (including advanced purification treatment targets)

the case of wastewater, TOC removal is also important, but high desalination performance is not required because the wastewater is not for drinking. Therefore, even the initial membranes with low desalination performance could be used.

For the desalination of brine water with low salinity, high separation performance, including desalination performance, was not required as in seawater desalination. In addition, brine has a lower salt content than seawater and can be handled at low pressure, so high-pressure resistance is not required. Furthermore, since the membranes are less likely to be contaminated than those for seawater, the requirement for chlorine resistance is not high. Therefore, large-scale plants using the RO method started with brine desalination.

Among the various application markets, the electronics industry, especially the production of ultrapure water for semiconductors, has benefited the RO membrane companies greatly and supported the development of the industry. With the rapid miniaturization of semiconductors since the late 1970s, the effects of impurities in the cleaning water used in the manufacturing process have become increasingly serious, requiring ultrapure water, purer than pure water. As conventional ion-exchange membranes are no longer sufficient to produce high-purity water, RO membranes have attracted considerable attention.

As the semiconductor industry grew, so did the market for RO membranes. It was a much smaller market than today's desalination market, but it was a market that justified the development of RO membranes, which were struggling to find a viable business. Without this market, RO membranes might not be what they are today.

As the history of Nitto Denko's entry into the market for semiconductor manufacturing shows (see Chap. 9), the companies did not start developing membranes with a focus on semiconductor manufacturing. The engineers of each company were struggling to develop membranes that could achieve both desalination and water permeability at a high level. Some of the membranes developed in the process did not achieve the target desalination performance, but their TOC removal performance and elution that was lower than that of ion exchange resins made them ideal for use in the production of ultrapure water for semiconductors. Because of the low osmotic pressure, membrane pressure resistance was not such a high requirement. Instead, the permeability performance at low pressure was important, as shown in Fig. 11.1. The importance of water permeability at low pressure is highest for ultrapure water applications, and the requirements for TOC removal performance are as strict as those for seawater desalination. In contrast, the requirements for pressure resistance, heat resistance, and boron removal performance, which are required for seawater desalination, were not high.

Therefore, even though the membranes were at an insufficient stage for seawater desalination, they were sufficient for ultrapure water applications. In the case of Nitto Denko, the discovery of this promising application was the result of a proposal from Kurita, an equipment manufacturer. Although the engineers had not developed the

sewage, and later became a foothold for Nitto Denko to expand into special wastewater treatment applications.

membrane for use in ultrapure water, this was an example of a customer discovering the potential for a new application for an imperfect technology.

In this way, the development of RO membranes, which started with the goal of actualizing seawater desalination, overcame a difficult period by finding unexpected application markets. These markets were compatible with the incomplete state of the technology that was biased toward a specific performance. We can say that the niche application market was discovered because the combination of performances was in a sense inconsistent and a specific performance was prominent.

11.1.3 Search Activities in Two Layers

Even if the potential application market expands in the process of pursuing the extreme performance required by the final target market, it is the people and companies who actually search for, discover, and develop the potential market.

So how were the application markets for RO membranes explored? Even though RO membrane companies search for applications, in many cases, their direct customers are water treatment equipment companies, and they do not often have direct contact with their end customers.

Then how could the RO membrane companies find applications? As it turns out, the water treatment equipment and plant engineering companies as well as the membrane companies played a major role in the market search. RO membrane companies are constantly striving to improve their basic performances to achieve their ultimate goal of desalination. Once the performance of the membranes is good enough for desalination, other markets will naturally open up. However, at the developing stage, companies had difficulties in finding business potential. Although membrane companies try to search for markets by contacting end customers in such a situation, the water treatment equipment and plant engineering companies have wider contacts with end customers. Therefore, new markets were found by combining the information on end customers held by the engineering companies with the technical information held by the membrane companies.

For example, the development of ultrapure water applications for semiconductor manufacturing, which saved the RO membrane companies, was certainly a tailwind that came from outside the company. However, it was not a bolt from the blue or an unexpected exogenous variable. As will be explained later, this market was opened up by the efforts of customer companies.

11.2 Unbalanced Development of Performance Bundles and Application Expansion

11.2.1 Food and Beverage Concentrate Applications: Emphasis on Low Transformation Properties

With the advent of L-S membranes, the first practical application of RO membranes was the desalination of brine in Coalinga. Although research activities had been conducted from the beginning with the aim of desalination, the performance requirements were too stringent to be easily met. Therefore, activities to search for other raw water sources in addition to seawater were promoted.

In the 1960s, the concentration of foods and beverages was considered as an application of the RO method mainly at the Western Regional Research Center (WRRC) of the U.S. Department of Agriculture. Watanabe (1985) of the National Food Research Institute of the Ministry of Agriculture, Forestry and Fisheries (MAFF), in his review of the early history of the development of food applications by the RO method, wrote that the WRRC began to publicly discuss the development of food applications of RO membranes in 1965, and in 1969, the institute held a symposium in which 130 food researchers and engineers from the United States and Canada participated. In Japan, a symposium was held in 1970, and the Japanese Society for Food Science and Technology took up the development of food applications by the RO method as one of the themes. The fact that the topic was explored in Japan only one year after the symposium was held in the U.S. indicates that the public efforts had crossed the border from early on.[4]

One of the main reasons why the RO method attracted attention in food applications was its low alteration properties. Since, in the conventional evaporation method, even the aroma is lost during concentration, an alternative method was needed. Concentration by freezing was possible, but it was expensive, leading to consideration of the RO method that does not alter the source solution during concentration. In 1982, the Membrane Research Association in Food Industry (MRAFI) was established under the auspices of the MAFF, and membrane manufacturers and food manufacturers were cooperating to develop its applications. It was in this association that Nitto Denko, mentioned in Chap. 9, participated.

The concentration of food and beverages by the RO method also had problems. One of them was the problem that some aroma components were permeated through the cellulose acetate membrane and did not remain in the concentrated liquid. Therefore, the cellulose acetate membrane was not expected to be used for the concentration of coffee whose product value is determined by the combination of various aroma components. This problem was solved when it was discovered that polyamide materials had sufficient blocking performance. Food concentration using RO membranes

[4] Food enrichment applications were a focus for Sourirajan as well, and his first research topic after moving to the National Research Council Canada was maple syrup enrichment (based on the authors' interview with Haruhiko Ohya on October 15, 2014).

then began to be considered in earnest. Based on the application-specific evaluation dimension of the rejection performance of aroma components, polyamide was superior to cellulose acetate.

In addition to the inhibition of aroma components, other issues for food applications were low fouling, water permeability, separation performance, heat resistance, and chemical resistance (Watanabe, 1985). These qualifications were not significantly different from those required for seawater desalination. However, a high separation performance was not required, as in seawater desalination. In addition, the cost requirement for membranes was not relatively high because the target was food products that had a higher added value than drinking water. Therefore, even the membranes inadequate for seawater desalination could be used for food applications.

On the other hand, the membranes were easily contaminated during the concentration of food and beverages, and because they were to be used for items that enter the human body, it was important that they have high chlorine resistance and can be chlorine-sterilized relatively easily. The polyamide composite membrane NTR-7450 developed by Nitto Denko is a good example of a membrane that has been evaluated in this respect. Although the separation performance of this membrane is not as high, it has excellent water permeability and high chlorine resistance that was highly evaluated by the MRAFI.

11.2.2 Industrial Applications: Emphasis on TOC Rejection Performance

In the 1970s, industrial water use was pioneered in the Kashima Coastal Industrial Zone. In Kashima, which was designated as a special industrial development zone in 1963, the construction of power generation facilities was underway to prepare for the expansion of factories, and the demand for pure water for power generation was increasing. Initially, the demand for pure water for power generation was sufficiently met by taking water from Kitaura and using ion exchange resins. However, when seawater flowed in around September 1970 and the salinity increased, it became necessary to prepare a new pretreatment process for raw water to achieve a higher level of desalination.

The RO method was used for this pretreatment process. The RO desalination plant, built in 1971 and put into operation the following year, had a desalination capacity of 3,000 m^3/day, the largest in the world at the time.

It is interesting to note that the RO desalination plant not only continued to be used but also expanded even after the inflow of seawater subsided and the salinity decreased. The desalination capacity of the plant increased to 5,000 m^3/day during 1972 when the plant started operation and to 13,400 m^3/day by 1975. The main reason for this was that the RO method was highly effective in removing silica and other small amounts of dissolved organic matter and had excellent performance in lowering Chemical Oxygen Demand (COD) that indicates water contamination.[5]

Because there was a large amount of algae in Kitaura, the removal of organic matter by the RO method in the pretreatment process was very effective. It lightened the load on the ion exchange resin used in the water purification system in the subsequent process and saved the amount of chemicals needed. Since the amount of chemicals used in the ion exchange method was originally much larger than that required in the RO method, the total amount of chemicals added greatly decreased, and the running cost was lowered (Taniguchi, 1981; Taniguchi & Kremen, 1973).

The high removal performance of dissolved organic matter, even though the desalination performance is low, led to the establishment of such industrial applications. However, it is unlikely that the developers focused on these characteristics in advance when developing the product for industrial use. Although they may have been aware that RO membranes could remove dissolved organic matter, the significance of the performance of the membranes was intensely recognized only after they were actually applied and tested in the field. In fact, Kurita Kogyo Kabushiki Gaisha (2000) reported that the results of the test operation showed that the system could remove not only salts but also organic matter, algae, and silica. This is how the RO method came to be widely used for industrial water applications.

11.2.3 Expansion of Semiconductor Applications: Emphasis on Low-Pressure Permeability

Since the 1970s, as the miniaturization of semiconductors has progressed, the purity required for cleaning water has increased dramatically, and the importance of cleaning has grown significantly.[6] Under these circumstances, conventional ion exchange resins could not meet the required purity levels because of the dissolution of organic matter, and the RO method came to be the focus of attention. What was expected of the RO method was its ability to remove low-molecular-weight organics.

The processes that require ultrapure water in manufacturing semiconductors correspond to the colored areas in Fig. 11.2. According to Ohya (1988), approximately 10% of the ultrapure water required in the entire manufacturing process is used in the first cutting process and 80% in the last three processes. In the early 1980s, the cleaning process was so important that it was said to account for approximately 30% of the entire process (Yamano, 1987).

The first RO desalination equipment (10,000 gallons per day) for the production of ultrapure water for semiconductors was delivered in 1969 to the Dallas plant of

[5] COD is used as an indicator of water contamination.

[6] One example of a quality problem caused by inadequate cleaning is that if the cut wafers are not cleaned properly, the organic matter remaining on the wafers decomposes and mixes with water vapor in the next oxidation process. The water vapor then partially accelerates the oxidation of the wafer, causing the oxidation layer to thicken.

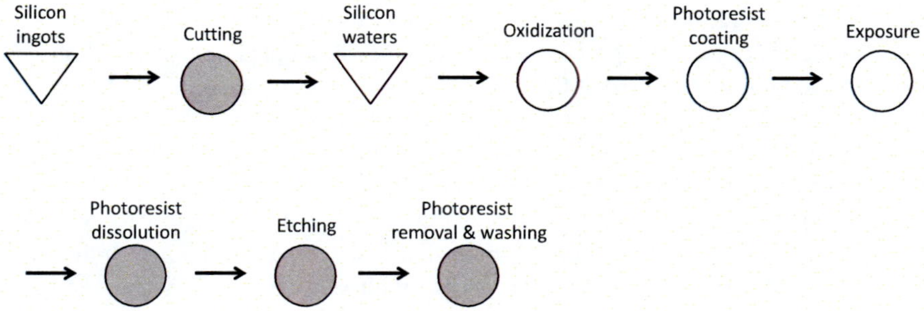

Fig. 11.2 Flowchart of the semiconductor manufacturing process. *Source* Prepared by the authors based on Ohya (1988)

Texas Instruments (TI), the leading company in the semiconductor market at that time (Takashima, 1985).[7]

Initially, the RO desalination system was introduced in the pre- or post-process of the ion exchange system. It was expected to play a role in removing organic matter from raw water in the previous process and in removing TOC from the ion exchange system in the post-process.[8] The equipment using the RO method was extremely expensive. Hisao Sato, Director and General Manager of the Technology Department at Nomura Micro Science, which was engaged in the ultrapure water production equipment business competing with Kurita and Organo, pointed out that "The equipment cost of RO accounts for about 30–40% of the cost of a new ultrapure water production system, so RO is considered expensive and it should be debated if introducing RO is needed or not."[9]

The reason semiconductor companies adopted the RO equipment despite high cost was that Japanese semiconductor companies were becoming more competitive in the market, earning large profits, and having sufficient budget. In addition, even though the equipment was expensive, the total running cost could be lower for the RO method. For example, if the RO method is introduced in the pre-treatment process, the load on the ion exchange resin that follows is reduced and the running cost is significantly lowered. Sato (1976) mentions that the introduction of the RO method stabilizes the water quality and reduces the load on the ion exchange resin to one-tenth, thus increasing the life of the resin tenfold and lowering the total running cost by 30–50% compared to the conventional method.

For the production of ultrapure water for semiconductors, the source of water intake is relatively clear lake or river water, which is less polluted, and the need for sterilization and cleaning is not as high as for seawater desalination. However, since chlorine sterilization is applied in the pretreatment process, a certain level of chlorine resistance is needed. However, for the post-process after the ion exchange system,

[7] 10,000 gallons converted to cubic meters is approximately 37.85 m^3.

[8] Koike (1989).

[9] Sato (1976, p. 43).

chlorine is not used, so at that stage chlorine resistance is hardly a problem. Moreover, since the desalination process has already been completed, high desalination performance was not required at this stage.

NTR-7250 developed by Nitto Denko in 1981 was well suited to the special circumstances of this post-process. This new membrane was inferior in desalination performance, but superior in organic removal performance. Thus, although it was not suitable for pretreatment, it was ideal for post-processes that were mainly responsible for TOC removal. It was not Nitto Denko that discovered this compatibility, but Kurita that had a direct business relationship with semiconductor manufacturers. Furthermore, this membrane had the advantage of saving electricity costs because it could achieve high water permeability even under low-pressure operation. Thus, semiconductor companies began to accept and integrate ultrapure water production equipment using the RO method before and after ion exchange equipment.

Throughout the 1980s, as the performance of semiconductors became more sophisticated and the industry expanded, the amount of fresh water used to process and clean products continued to rise steadily. At the time, 1.2 tons of ultrapure water was needed to manufacture a single 6-inch wafer, so the demand was quite large. Figure 11.3 shows the number of establishments responsible for manufacturing integrated circuits and the amount of freshwater used for product processing and cleaning at each establishment. The number shows an increasing trend until 2000. The amount of freshwater used per plant has also continued to increase over the long term, despite the waves of the silicon cycle. In line with this increase, ultrapure water production equipment using the RO method also became widespread.

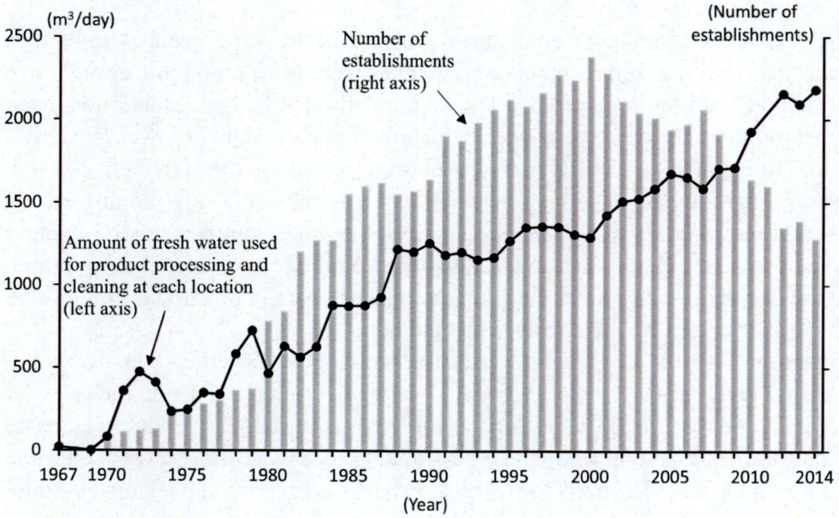

Fig. 11.3 Amount of freshwater used for product processing and cleaning at each integrated circuit manufacturing establishment. *Source* Partially calculated by the authors based on yearly "Industry Statistics Tables" from the MITI. *Note* Data for 1968 and 2011 were not available

Who exactly developed these applications and how were they done? As already mentioned, not only the membrane companies but also the equipment companies, which were their direct customers, played an important role in the application search. To clarify this point, in the next section, we first review the descriptions in Part 3 and summarize the application development activities of the three Japanese companies. Then, in Sect. 11.4, we will clarify the explorative activities of applications by the equipment companies.

11.3 Search for Applications by Three Japanese RO Membrane Companies

11.3.1 Changes in the Use of Raw Water and Produced Water

Having various application markets other than the seawater desalination market in the early stage played a very significant role in the continuation of development by companies. In this section, we capture what kind of applications Toray, Nitto Denko, and Toyobo were conscious of in the development by analyzing patents with F-terms related to each application of raw water and produced water. Figure 11.4a–c show the trends in raw water usage by application for each company up to 2010.[10] Although the ratios are slightly different, it can be seen that all three companies have shifted from "liquid waste and wastewater" to "seawater" as a major trend. This is fairly consistent overall with the descriptions of each company in Chaps. 7–9.

As for Toray, the majority of membranes were used for "liquid waste and wastewater" before 1975, followed by "seawater", with the peak for "seawater" occurring around 2000. A trend in the early 1970s indicates that cellulose acetate membranes, which had been developed earlier, were deployed for "liquid waste and wastewater". After that, the number of patent applications for "seawater" increased in the latter half of the 1970s, when the development of the PEC-1000 progressed, and increased sharply in the 1990s. UTC-80 was successfully developed and delivered as SU-800 to Kita-Daito Island and Minami-Daito Island in Okinawa in 1991, which is consistent with the fact that the company started to develop its business for seawater desalination around that time.

For Toyobo, although the number of cases is small and it needs to be carefully interpreted, there was also a heavy emphasis on "liquid waste and wastewater" in

[10] The number of patents with the F-term group of PB01 (fluid to be treated), which corresponds to raw water, is 7,057, and the number of patents with the F-term group of PC00 (field of use, application), which corresponds to produced water, is 5,181. For raw water, the measurement covers the "treated fluid" (PB01) and all its subterms. For produced water, since there are a wide variety of subterms, the measurement was made by aggregating each upper term according to the procedure described in the Appendix. In this book, only the applications with the largest number of patents are shown, because the visibility of the graph would be significantly reduced if all the raw water and applications were shown.

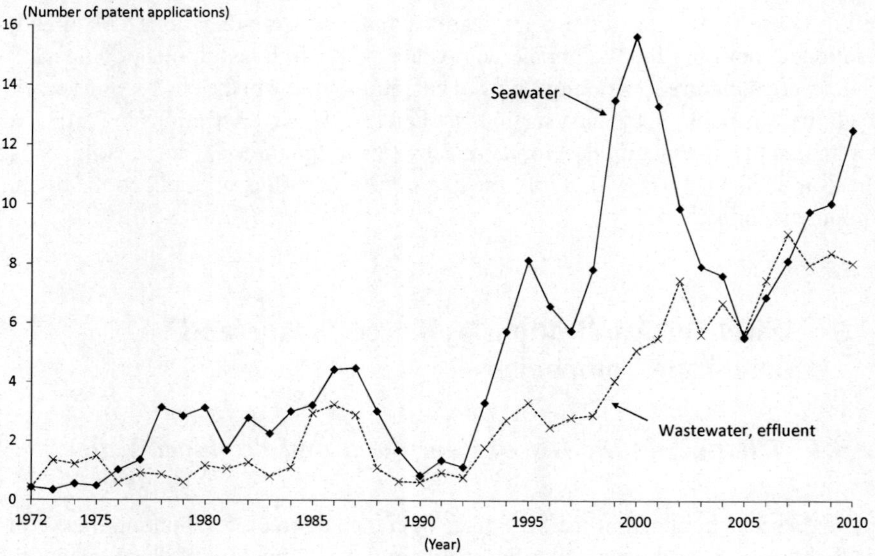

Fig. 11.4a Applications of raw water (Toray). *Source* Prepared by the authors based on data from UltraPatent. *Note* Three-year moving averages

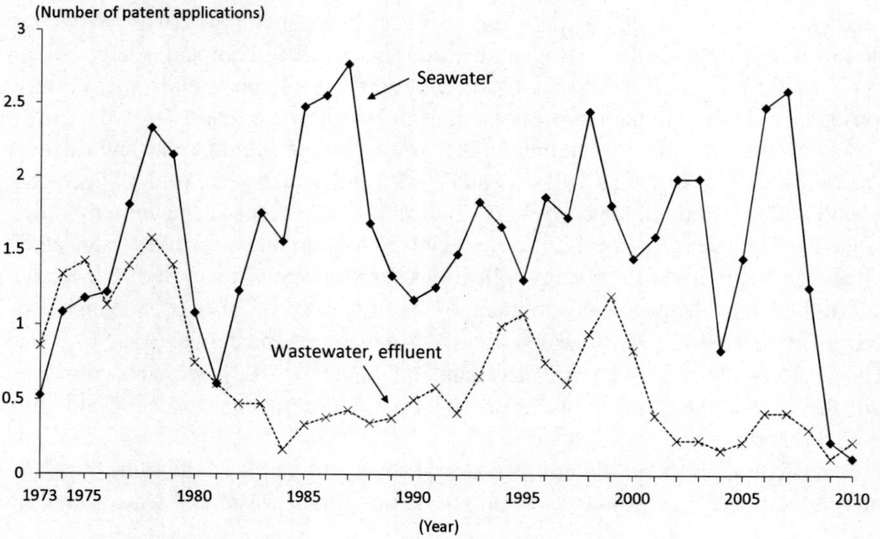

Fig. 11.4b Applications of raw water (Toyobo). *Source* Prepared by the authors based on data from UltraPatent. *Note* Three-year moving averages

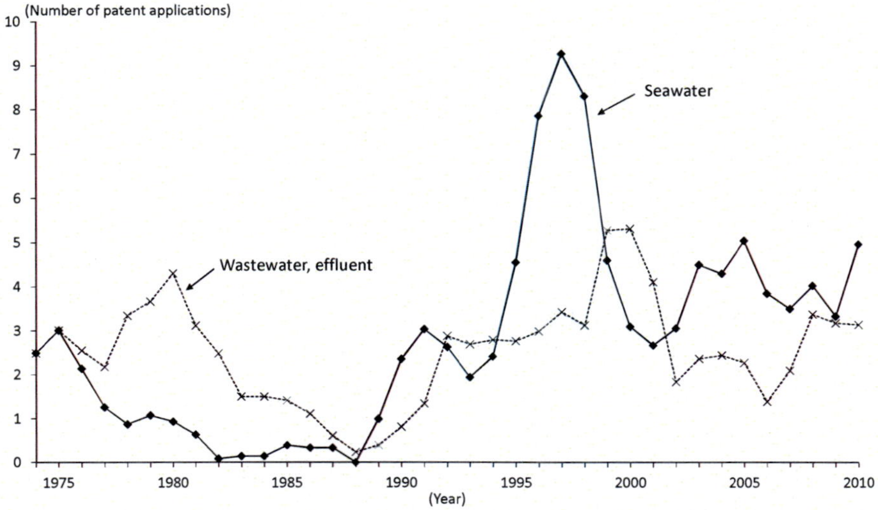

Fig. 11.4c Applications of raw water (Nitto Denko). *Source* Prepared by the authors based on data from UltraPatent. *Note* Three-year moving averages

the early years, as with Toray. This was followed by a large increase in "seawater" in the mid-1980s. Compared to Toray and Nitto Denko, Toyobo's development has been biased toward "seawater".

Although Nitto Denko was aware of "seawater" in the first half of the 1970s, the number of "wastewater and effluent" soon increased. The period when the number of such applications was large continued for a long time. As already mentioned in Chap. 9, Nitto Denko was actively developing for applications other than seawater desalination. The number of applications related to "seawater" increased rapidly around 1987, when Nitto Denko acquired Hydranautics, and it can be confirmed that the inclination toward seawater progressed rapidly.

Figure 11.5a–c show the transition of produced water by application for each company. As a major trend, these figures indicate that the use of produced water for food and medical purposes was relatively high for each company in the early 1970s, and that in the 1980s, the use for semiconductors increased. Only for Toray, we added "Other fields of use and applications" to the graph because it was one of their top uses. It should be noted that the F-term classification does not give a specific F-term to the use of drinking water. Therefore, it is not possible to grasp the trend of desalination with drinking water in mind from this graph, and it is not possible to make a strict comparison with other uses. However, we assumed that desalination for drinking water use is included in this "Other" category.

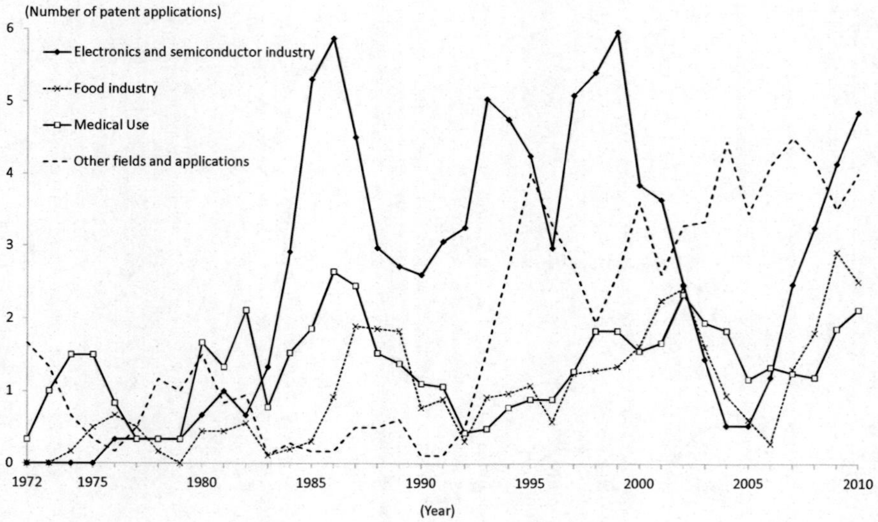

Fig. 11.5a Applications of raw water (Toray). *Source* Prepared by the authors based on data from UltraPatent. *Note* The number of patents with the relevant F-term is calculated as a three-year moving average. Note that, unlike the other two companies, Toray's top three applications in terms of the number of patents include "Other fields and applications". However, to compare the three companies in the same application, "Food industry" is also shown in the figure

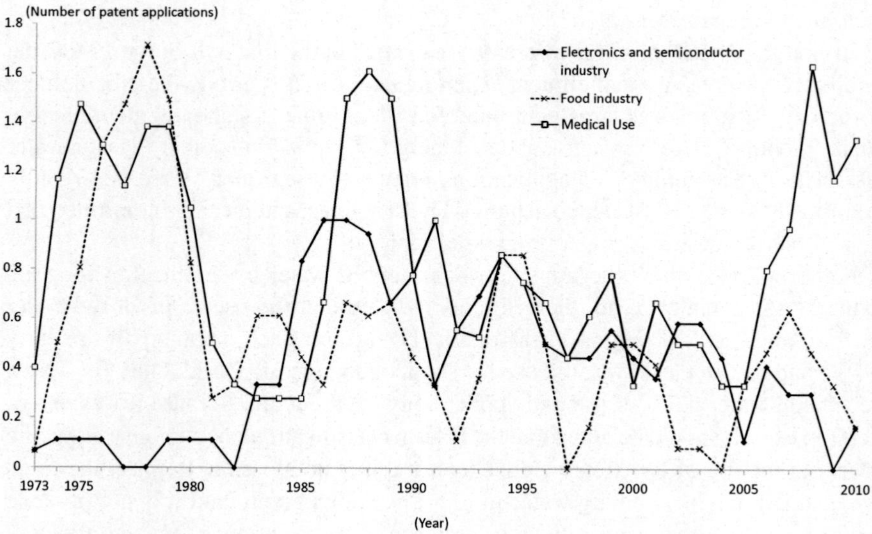

Fig. 11.5b Applications of raw water (Toyobo). *Source* Prepared by the authors based on data from UltraPatent. *Note* Three-year moving average

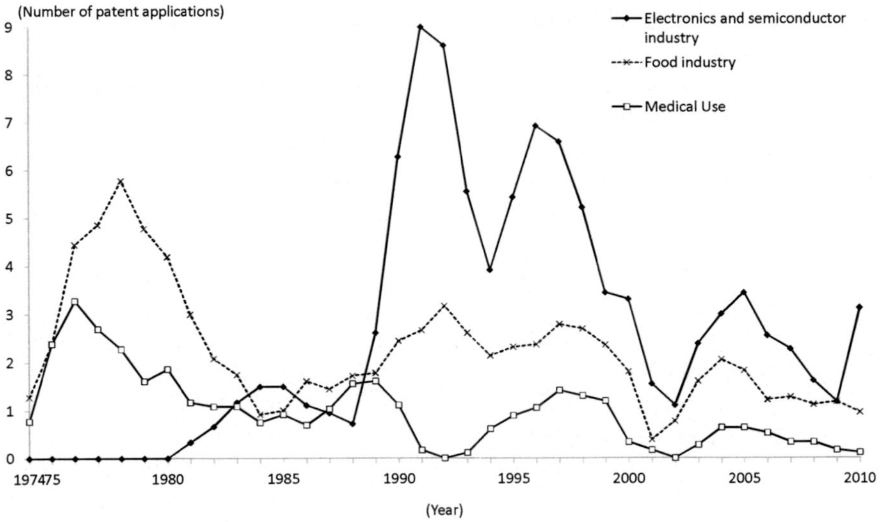

Fig. 11.5c Applications of raw water (Nitto Denko). *Source* Prepared by the authors based on data from UltraPatent. *Note* Three-year moving average

11.3.2 Initial Search Activities by Each Company

From the start of the development process, all companies agreed that seawater desalination was the ultimate application to be achieved. In the early stages of the project, when sufficient performance and cost did not line up, each company was searching for applications other than seawater desalination to continue its business and development.

When Toray began developing cellulose acetate materials in 1968, it was thought that it would be difficult to develop a seawater desalination market with these materials. Toray had not developed the materials in-house, and it was not possible to achieve both desalination and permeability performance at a high level. In the 1970s, Toray's cellulose acetate RO membranes were indeed used in the demonstration plant of the WRPC, but the development group for cellulose acetate membranes did not recognize that seawater desalination would be its main application. In fact, Toray's first delivery to IBM's Yasu Plant, for which it received an order at the end of 1976, was for wastewater treatment.

This is consistent with Kawabata's comment, "You could desalinate seawater with CA, but the pressure resistance was a problem. (So) you can use it for slightly salty brine, or for reclaimed water after secondary or tertiary treatment of wastewater that has undergone primary treatment. And then, if you use ordinary water, you get very clean water, which is ultra-pure water."[11] At that time, the development

[11] See Chap. 7.

team for cellulose acetate materials was aiming at wastewater treatment and brine desalination.

Although Toyobo had the recovery of valuable resources in mind when it first began development, it soon came to consider seawater desalination as its main target. However, it was expected that it would take time to achieve large-scale desalination for domestic water. Therefore, Toyobo first decided to develop a small desalination system for ships. Their hollow fiber RO membrane modules had the advantage of being compact because a large number of membranes can be packed into a module. Although the market was small, it was an important market for the company's business continuity.

Of the three Japanese companies, Nitto Denko was the most active in developing a variety of applications. Nitto Denko began the development of RO membranes as part of its plan to establish a membrane business and did not have a specific application in mind. Rather, they were groping in the dark. Yamamoto, who led the project, says, "No one was studying it with any solid certainty about what direction they should advance in... No one knew how big the market would become. No one knew what applications would grow in the future. It turned out to be a project characterized only by the name 'membrane', with almost no future plans or dreams in sight."[12] Of course, the company knew that the desalination market would open up as the development progressed, but the company, which was under strong pressure to earn profits, could not agree to target seawater desalination, which would not lead to immediate profits. To continue development, the company needed to aggressively search for applications that could generate immediate revenue. Figure 11.5c also shows that Nitto Denko focused its development activities on food and medical applications.

In fact, Nitto Denko's first commercialized product was a cellulose acetate tubular RO membrane module produced with technology from Kobe Steel, which was delivered in 1976 for use in treating starch waste liquids in Hokkaido. After that, the company participated in a seawater desalination demonstration at the WRPC but learned that the technology would not be competitive. In 1982, the company participated in the Food Industry Membrane Utilization Technology Research Association and jointly developed with Ajinomoto a membrane for use in decolorizing flavored liquids. The company also sought to develop other applications, such as the concentration of grapes for wine and the concentration of fruit juice flavors. In other applications, the company supplied tubular UF membrane modules to a major paper manufacturer in Shikoku for the treatment of pulp waste and entered the field of water reuse (grey water) for large buildings and factories. In addition, the company pioneered other fields, such as electrodeposition coating, oil-water separation of wastewater, and manure treatment.

As Nitto Denko was smaller than Toray and Toyobo in sales, it was difficult to continue uncertain technological development aiming at a market that might open up in the distant future. Therefore, more than Toray or Toyobo, Nitto Denko was actively searching for applications to survive.

---

[12] See Chap. 9.

11.3.3 Search for Applications from the 1980s Onward

While the seawater desalination market did not expand as expected, companies continued to search for applications from the 1980s onward. In the process, the market for the production of ultrapure water for semiconductors emerged. Toray and Nitto Denko both enjoyed the demand from Japanese semiconductor companies, which gave them internal recognition for their RO membrane business and justified continued development to further improve performance and reduce costs.

Toray's development group had been consistently developing technologies with seawater desalination in mind but faced a major obstacle in commercialization. PEC-1000, a polyamide composite membrane developed by Toray, was susceptible to dissolved oxygen in seawater, and the company gave up on commercialization. PEC-2000, which had been ready for mass production, was discontinued just before it was put on the market because of its poor chlorine resistance. It was the business of ultrapure water production for semiconductors that supported the continuation of business and technological development despite the setback in the seawater desalination business. As Uemura recalls, "It was an amazing gift," and this application had an extremely important meaning in the history of Toray's RO membrane development.

Toray first entered this market with cellulose acetate-based materials. It was judged that it would be difficult to immediately replace the desalination process, which had been operated using the ion exchange method, with the RO method. Thus, Toray initially entered the market by adding the membrane before and after the desalination process. In 1987, an RO membrane element made of polyamide material was supplied to a newly built DRAM plant of Toshiba, which was at the forefront of the semiconductor industry.

Toray developers knew that cellulose acetate membranes could not capture the desalination market. The key was to develop a polyamide-based membrane. The cellulose acetate RO membranes sold to the semiconductor industry filled the time required to successfully develop such membranes.

Nitto Denko also benefited significantly from the semiconductor industry, and this was one of the factors that enabled it to continue its business and development. Nitto Denko began to be aware of the electronics industry later than Toray in the 1980s. The discovery by Kurita of the value of NTR-7250, which had a low desalination rate but was excellent at removing TOC, was the catalyst for the expansion of business for semiconductors. NTR-7250 was a failure from the developers' point of view because they were trying to improve desalination performance. However, a new opportunity opened up when Kurita discovered the functions required in the semiconductor cleaning process.

From Fig. 11.5c implies that Nitto Denko's business for semiconductors increased rapidly from the latter half of the 1980s, and thereafter, it consistently became the most important area for produced water applications. Around this time, Nitto Denko created its own category of membranes called "loose RO" that collectively referred to membranes that did not have the same separation performance as RO membranes but were superior in other performance dimensions. The company sought to develop products in this category.

Among the three companies, Toyobo was the only one that failed to capture the market for semiconductors. This was because the amount of water permeability was important for semiconductor applications, the chlorine resistance characteristic of the company's cellulose acetate membranes was not so important. The company could not find any advantage over polyamide membranes. In response, Toyobo also developed polyamide hollow fiber membranes and attempted to enter this industry but was unable to compete in terms of performance and cost.[13] This led Toyobo to focus its efforts on seawater desalination, particularly in Saudi Arabia.

Therefore, Toyobo needed to justify the continuation of the RO membrane business for reasons other than the acquisition of an applied market. Particularly in the 1990s, when the entire company was performing poorly, it was not surprising that the unprofitable RO membrane business was targeted for withdrawal. One of the reasons this did not happen was that the RO membrane business was placed under the same umbrella as the highly profitable artificial kidney business. This was particularly effective under Tsumura leadership, which at the time strictly selected profitable businesses by stating that "profits are good and losses are bad."[14] This process will be described in detail in Chap. 13.

11.3.4 Move Toward Seawater Desalination

While continuing to develop membranes for a variety of applications, from the 1990s onward, companies began to expand into the seawater desalination market, the main market for RO membranes.

Toray's full-scale development of seawater desalination began in 1991, when it commercialized SU-800. After SU-800 was delivered to Kita-Daito Island and Minami-Daito Island in Okinawa, it was commercialized as SU-820 and installed at the desalination center in Chatan in 1996. However, Maeda was not satisfied with the scale of the business, and in 1994, he gave strict orders to the developers to build an efficient seawater desalination system. In response to this order, the developers developed highly pressure-resistant membranes and created a highly efficient two-stage desalination system, which increased the water recovery rate from seawater to 60% and reduced desalination costs by approximately 20%. This system was adopted in Maspalomas, Spain, Curacao in the Caribbean Sea, and Trinidad and Tobago in 1999.

---

[13] Considering the fact that DuPont was unable to develop ultrapure water production applications for semiconductors although it was using polyamide materials, the major problem might be less in materials and more in the hollow fiber shape of the membranes perhaps being less suitable for ultrapure water production applications.

[14] For more information on the company-wide structural reform process, see Fujiwara and Aoshima (2016).

It was Nitto Denko that most notably promoted the shift to seawater desalination. An important turning point was the acquisition of Hydranautics in 1987. The management of Nitto Denko, realizing that it could not expect to expand its business in Japan, decided to make this major acquisition even though the RO membrane business was in a slump. The impact of obtaining a large amount of customer information from Hydranautics, which was already doing business with customers around the world, was significant. Kamiyama, who was involved in development at the time, said, "… the number of membrane modules per case is very large. In Japan, even if you build an ultrapure water facility, there are only a few hundred modules at most. It's not even a thousand. On the other hand, Hydranautics' facilities (at that time) already had a thousand units. It's a whole new level."[15] With this acquisition, Nitto Denko was able to develop its business on a completely different scale than before.

However, as described in Chap. 9, Nitto Denko was sued by Dow as a result of this acquisition. It was claimed that Hydranautics was infringing on a patent invented by Cadotte of FilmTec that had been acquired by Dow. Ultimately, Nitto Denko won the case, and the U.S. government was found to own Cadotte's '344 patent. As a result, companies were able to use the technology, which had a major impact on the subsequent expansion of the seawater desalination market. The convergence of technological approaches to polyamide flat membrane technology intensified competition, improved the price-performance ratio, and greatly expanded the potential for seawater desalination applications. This will be discussed in more detail in the next chapter.

Toyobo constructed a competitive advantage by quickly resolving the problems and gaining credibility when its RO membranes supplied to a seawater desalination plant in Saudi Arabia experienced performance problems in the 1980s. However, sporadic large orders meant that the desalination business was not immediately secure, as plant utilization rates fluctuated wildly. At the time, the Saudi Arabian government was experiencing financial difficulties, and the company was having difficulty seizing growth opportunities.

It was not until the 2000s that they expanded their business for seawater desalination. Toyobo developed the 10-inch-diameter HB10255FI and delivered it to a seawater desalination center in Fukuoka. HB10255FI became the company's main product. Furthermore, DuPont's withdrawal from the market provided Toyobo with a commercial opportunity to enter the area where DuPont had retreated. Many of DuPont's products were being used in desalination plants in the Middle East, so Toyobo expanded its business by introducing products with the same shape as DuPont's, making it easier to capture replacement demand.

[15] See Chap. 9.

11.4 Value Exploration by Client Companies

11.4.1 Clients' Wider Contacts

As we have discussed, while a large market such as seawater desalination has been difficult to establish, RO membrane manufacturers have continued to develop membranes by exploring and developing various applications. However, there is a limit to what can be explored by RO membrane companies alone. Membrane companies do not always interact directly with their end customers, such as water treatment companies and plant owners. Typically, they interact with their direct customers, water treatment equipment and engineering companies.

It is water treatment equipment and engineering companies, rather than RO membrane companies, that have operational information on how their membranes will be used in various environments. Therefore, information on the compatibility of functions provided by RO membranes with various applications often comes from equipment and engineering companies. Thus, the expansion of applications for RO membranes depends not only on the exploration of the RO membrane manufacturers themselves but also on the extent to which the water treatment equipment and engineering companies, which are their direct customers, conduct a broad search.

In fact, for example, the developers at Nitto Denko were not initially developing membranes aimed at food and beverage concentration applications. The instructions given to Ikeda, who was in charge of development, were "Just make a strong membrane." NTR-7250 was not developed with the semiconductor industry in mind, and while developing membranes with excellent desalination performance, the company accidentally created membranes with inferior desalination performance but excellent TOC removal performance and water permeability performance, which were highly evaluated by the customer Kurita. The customer network developed by Hydranautics, which was also engaged in plant engineering for seawater desalination, became important for the international development of Nitto Denko's membrane business.

In this section, we describe the application search activities of water treatment equipment and engineering companies in the development process of RO membranes by focusing on the ultrapure water production application for semiconductors, which has become a major application. Three companies, Kurita (established in 1949), Organo (established in 1946), and Nomura Micro Science (established in 1969), are prominent players in the water treatment equipment and engineering industry in Japan. Of these, Kurita and Organo have the longest histories and the largest market shares. Thus, we review how these two companies have expanded their RO water treatment business.

11.4.2 Kurita's Exploration and Growth

Kurita, which was founded in 1949 and diversified into the water treatment equipment business in the 1950s, showed interest in seawater desalination soon after the start of the 1960s. When it was reported in Japan that the U.S. State Department put an evaporation process demonstration plant into operation in 1961, the company quickly responded and began research and studies on seawater desalination in 1962. In 1963, the company opened the Yokosuka Coastal Research Institute where it conducted a series of desalination studies and demonstrations.

However, in the 1960s, Kurita was facing financial difficulties. In 1965, Kurita entered into a business tie-up with ITOCHU to receive financial assistance and management guidance. This business tie-up had a very significant meaning because information from around the world began to flow into Kurita through ITOCHU's overseas offices.[16] This included information on the RO method. Kurita began exploratory research on RO membranes in 1965, the very year it formed a business alliance with ITOCHU (Ishizaka, 1970). In 1967, Kurita, which pioneered the practical application of seawater desalination using the evaporation method, released a small desalination device for fishing boats, which made it easy to secure drinking water on board and was well received.[17]

In November 1967, Kurita decided to introduce the manufacturing technology of RO membranes from UCLA. This was just around the time that Toray began researching RO membranes, indicating that Kurita's decision was quite early. The vice president of ITOCHU Shinzo Kaiishi replaced Haruo Kurita as president of Kurita in December 1967, and the company started joint research with Daicel in 1968. Furthermore, Kaiishi instructed his staff to negotiate with GA, which had a partnership with UCLA, for the purchase of membranes.[18] After negotiations with GA and AJAX, its general sales agent, a sales agent agreement was signed in March 1971, making Kurita the licensee in Japan. Thus, the business activities of ROGA, which were being promoted with the support of the U.S. government, influenced Kurita's decision across the sea.[19]

Kurita was interested in all applications and technologies related to water treatment and proceeded with their development. It also focused on the electrodialysis method to develop the desalination of seawater and brine, and in 1968, it received an order from the Japan Meteorological Agency for a desalination plant using this method. Rather than thinking of Kurita as trying to cope with future water shortages, it is more appropriate to think of it as seeking various desalination methods in order

[16] In February 1966, Haruo Kurita himself made a 40-day trip overseas to absorb as much information as possible.

[17] By 1979, the company had achieved great success, selling its products to some 670 fishing vessels.

[18] This order was issued at the board meeting in September 1968. The name of GA at that time was Gulf General Atomic (GGA), but for the sake of readability, we use the name GA here. After this, the company name of GGA was changed to UOP.

[19] This is also a kind of spillover effect.

to develop its water treatment equipment business, and as one of these methods it began research into the RO method.

However, the performance of the water treatment equipment division was still in the red. The key issue for the division was to win orders for water treatment equipment at any cost and quickly return to make a profit.

What came at that time were two demands that accompanied the creation of the Kashima Waterfront Industrial Zone. The first demand was producing pure water for power generation. In 1968, the Kurita equipment division received an inquiry from Mitsubishi Heavy Industries (MHI) for pure water equipment and responded by developing the "High Series" of ion exchange resin-based pure water equipment. This device was such a large hit that it almost monopolized the supply of deionized water equipment not only from MHI but also from other companies in the same industrial zone.[20]

The second demand was desalination, which came when seawater flowed into Kitaura in September 1970. Kurita received an order from Sumitomo Metal Industries' Kashima Works in June 1971 for a desalination plant using the RO method. Since the RO method had never been used for desalination before, Yoshikazu Hirose, who was the general manager of the plant division at Kurita, wrote that he was told by Horio, the manager of the planning section[21] at Sumitomo Metal Industries, "Mr. Hirose! We have crossed the Rubicon River together, haven't we? There's no going back now, is there? Neither you nor me!" Regarding this, Hirose recounts, "I couldn't stop the nervousness and excitement in my heart."[22] As a result of the development and delivery of equipment using a variety of desalination methods, Kurita's equipment division returned to make a profit during the fiscal year that ended in September 1971. Its profit exceeded that of the mainstay chemicals division.

In December of the same year, three desalination units equipped with ROGA's RO membrane elements 4100STD, 4160HR, and 4160STD were installed at Kashima Works by Kurita and started operation in February 1972. Since each unit showed good operational performance, additional units were quickly installed. By 1975, a large desalination system with a total of 10 units and a capacity of 13,400 m^3/day was established (Taniguchi, 1981). In 1979, Kurita changed some of the RO membrane elements for the ninth unit and adopted Toray's SC-3100.[23]

For ROGA and AJAX, the Japanese licensee Kurita was a major sales partner. According to an estimation in 1980, 65% of ROGA's sales were to Kurita.[24] The remaining 35% was accounted for by Nomura Micro Science (15%), Organo (13%), and Abco Japan (7%), indicating a high degree of dependence on Kurita.

At the same time, that Kurita was responding to demand from the Kashima Coastal Industrial Zone, it was also considering the development of ultrapure water production equipment for semiconductors. After word got out that TI had introduced an RO

[20] By 1996, the company had recorded 122 deliveries (Kurita Kogyo Kabushiki Gaisha, 2000).

[21] Kashima Steel Works Construction Headquarters.

[22] Kurita Kogyo Kabushiki Gaisha (2000, p. 77).

[23] Bairinji (1984, p. 272).

[24] Sōgō Hōsō Publication (1980, p. 98).

system in 1969, a report on the "Current Status and Future Prospects of RO Systems" was presented at a board meeting held in Kurita in March 1971. In this report, it was confirmed that approximately 60% of AJAX's sales were to the electronics industry. Kurita launched sales activities to capture this trend, and in 1973, it entered this market by delivering the first RO ultrapure water production system to Nagano Electronics Industrial. This equipment was introduced as a pre-treatment for the deionized water production process using ion exchange resins and contributed to reducing the burden on the ion exchange resins and raising the yield rate of semiconductors from approximately 50% to 95%.

In 1975, Kurita set up a Membrane Sales Promotion Department to aggressively collect information on semiconductor production lines. While semiconductor companies, which were competing at the cutting edge of the industry, tended to keep information secret, Kurita's proactive activities were important for sensing the problems common to each company and plant and for delivering individually optimized ultrapure water production equipment.[25] In the course of these activities, NTR-7250 developed by Nitto Denko in 1981 emerged. Kurita immediately recognized the value of NTR-7250. It succeeded in signing an exclusive supply contract with Nitto Denko and began delivering ultrapure water production equipment using the membrane in 1983. Kurita's share of the ultrapure water production equipment market immediately increased rapidly, and within the same year, it overtook Organo that had previously held the top position.[26]

In April 1983, Kurita established the Ultrapure Water Service Section within the Water Treatment Sales Department to strengthen its maintenance operations for ultrapure water production equipment and to accumulate experience in long-term stable operation. This function was subsequently transferred to and integrated with Kurita Maintenance in April 1987, which changed its name to Kurita Technical Service in October 1987. In 1991, Kurita Technical Service established an ultrapure water training center and focused on improving its maintenance system by, for example, conducting practical training using ultrapure water production equipment that could handle 16 Mb DRAM, the most advanced DRAM at the time. Kurita Technical

[25] In fact, obtaining information from semiconductor manufacturers was important for the membrane manufacturers and equipment manufacturers located upstream. Akihisa Inoue of Kurashiki Boseki Technical Research Institute said in a roundtable discussion held in 1986, "There were many times when we were troubled because we did not have as much information as we do now about how semiconductors are used in factories. It took us a long time to learn the decision-making standards of semiconductor manufacturers. Even now, I can't say it's enough…" (Yoshida, 1986). In response, Kamiyama of Nitto Denko, who was present at the round-table discussion, said, "That probably means that there was a wall between membrane manufacturers, engineering manufacturers, and users. I think it was difficult for you to get information from them though you wanted to learn about applications. There might be a gap. Also, as there are different criteria for judging among users, I think it's difficult to know what is an accurate absolute evaluation." Nitto Denko has been able to exchange and share information with its customers relatively smoothly, for example, by holding technical exchange meetings with Kurita. This may be the reason for such a response.

[26] Nikkei Sangyo Shimbun (July 24, 1984, p. 8).

Service grew to become a core business entity within the group and was merged into Kurita in 1997.

Since maintenance cannot be performed without operational information on the user side, the growth of technical services suggests that the Kurita Group has become an accumulator of important on-site information and an important nexus between the needs of semiconductor manufacturers and the business ideas of membrane manufacturers.[27] Yoshio Taniguchi, who was involved in the development of membrane applications at Kurita and served as the general manager of the company's research center, describes the role his company played.

> We in engineering had much more information about such needs (ultrapure water) than the membrane manufacturers … (For example) we probably provided information such as "For **hospitals, we need high flux at as low a pressure as possible," to the membrane manufacturers.** (Membrane manufacturers) have the seeds for business, and what kind of needs they can meet with those seeds will grow through the combination of both, won't they? So you can't say "Membranes are good for ultrapure water" from the beginning. On the engineering side, **we know early on that there is a need for ultrapure water in this field. I think the reason** why we were able to combine needs and seeds was due to our interaction (with membrane manufacturers).[28]

Expansion into seawater desalination was also explored. Kurita delivered equipment equipped with ROGA modules for the demonstration tests at the WRPC that began in 1975 and also conducted operational tests using various RO membranes at its own research laboratory. In 1978, the company obtained the result that PA-300 was the best. Meanwhile, in 1979, the company installed Toray's RO membrane and proceeded with the demonstration at the WRPC. While experimenting with such a wide variety of membranes, Kurita sought to expand its business into desalination applications, including receiving orders for equipment for the Middle East. However, the focus of the Middle East market became large-scale desalination plants, an area in which companies specializing in larger-scale plant engineering would play a central role.

[27] Taniguchi, who served as the deputy director of Kurita's research institute, wrote in 1981, "To maintain stable operation and management of the desalination plant and to fully actualize the intended performance of the stand-alone equipment, it is necessary to be familiar with the prescribed operation and management standards, such as scale prevention, fouling prevention, and corrosion prevention, as well as the establishment of peripheral technologies, so that the performance of the entire system can be maintained in a stable manner over a long period of time." As of 1984, Yoshito Motomura, who was in the Engineering Department of the Water Treatment Equipment Division, also stated, "With the development of the semiconductor industry, the quality of ultrapure water is also progressing at a rapid pace toward higher purity. To achieve this, not only the improvement of the ultrapure water production system but also the research and development of peripheral technologies, such as water quality analysis, with a thorough understanding of the overall operation and management, are greatly expected" (Motomura, 1984; Taniguchi, 1981).

[28] From the authors' interview with Yoshio Taniguchi at the WRPC on March 16, 2015. The text in parentheses was added by the authors.

11.4.3 *Exploration of Organo*

It was in 1957 that Organo, which had been involved in the development of pure water systems, developed an ion exchange pure water system and launched it for use in the electronics industry (Shimizu, 1972). As the level of requirements in electronics industrial applications increased, Organo recognized the need for new developments. In 1967, the company began research on the application of UF membrane and RO methods[29] and became a distributor of Havens equipment.[30]

In October 1972, the company delivered its first RO system to NEC's Tamagawa Plant, and in March 1973, it delivered a full-scale ultrapure water production system to a semiconductor plant.[31] Unlike Kurita, which mainly adopted ROGA's modules, the RO membrane that Organo mainly used at this time was DuPont's B-9.

Organo continued to strengthen its expansion into electronics industry applications, and in 1977, the company conducted a pilot test of RO water purification at Fujitsu's Kawasaki plant with positive results (Organo Corporation, 1981). Based on the results of this test, the company conducted further sales activities. In this way, Organo was the first to market its pure water production equipment to semiconductor companies, including major companies such as NEC, Fujitsu, and Hitachi, and opened up the electronics industry to new applications.[32] However, in the 1980s, as mentioned earlier, the company was overtaken by Kurita.

In June 1978, Organo exported its first portable RO seawater desalination plant to Saudi Arabia. However, as seawater desalination plants grew in scale, the company soon began to specialize in the production of ultrapure water for industrial applications.

As indicated in the above description, the widespread acceptance of the RO method for the ultrapure water production for semiconductors was not only because of the exogenously expanded cleaning demand along with the miniaturization of semiconductors, but also the endogenous market creation through the efforts of equipment and engineering companies such as Organo and Kurita.

11.5 Conclusion

This chapter analyzes the process by which the development of RO membranes, which started with governmental policy support, progressed to a stage where the business could be sustained through the expansion of ultrapure water production

[29] Nakane (1974b, p. 50).

[30] The fact that Organo was the sales agent for Havens Industries is mentioned in Ishizaka (1970).

[31] With hospital and laboratory applications in mind, Organo also developed a compact sterile water purification system using the RO method, which it commercialized as Osmoclear in April 1973. The company also established an aseptic room in its Research and Development Department and proceeded with the development and verification of various peripheral technologies.

[32] Nikkei Sangyo Shimbun (November 12, 1983, p. 6).

applications for semiconductors, which was encountered in the course of activities to develop various applications, including food and industrial applications.

To actualize seawater desalination, all the various performance requirements for factors such as desalination, permeability, pressure resistance, durability, and washability must be satisfied. However, since in ultrapure water applications, for example, desalination is carried out using raw water with low salinity, it is not necessary to meet all the performance requirements at a level as high as that required for seawater desalination. Even a developing membrane technology was able to meet the performance requirements of this market. For RO membrane companies that had been struggling to develop profitable applications, the ultrapure water application for semiconductors was a great boon. The existence of this initial market played an important role in justifying the continuation of development by each company.

The results of the analysis in this chapter suggest the importance of companies upstream in the value chain to collaborate with their customers in the search for strategic niche markets. In the case of RO membranes, the performance requirements of the membranes vary considerably depending on the environment and purpose of use. This also means that field operations will vary. Therefore, it is necessary for membrane companies to develop application markets by determining the interrelationship between the characteristics of the membranes they develop and the conditions at the site of use. In this context, the equipment manufacturers that developed maintenance activities for semiconductors, for which long-term stable operation was a top priority, accumulated a wealth of on-site information. In doing so, equipment manufacturers played a valuable role as a nodal point connecting the seeds on the part of membrane manufacturers to the needs.

As described in this chapter, RO companies have developed various initial markets to continue and develop their business. However, they still face performance and price challenges in capturing a large market. For the large-scale desalination of seawater for beverages suitable for human consumption, a new membrane was needed that could be manufactured at low cost with at least high levels of both desalination and water permeability. This could not be achieved simply by developing a variety of applications.

To develop the industry that would lead to the expansion of the desalination market, a breakthrough technology that would converge the activities of the various players was necessary. In the next chapter, we will analyze the emergence and industry impact of the new composite membrane invented by Cadotte as such a breakthrough. We will argue that Cadotte's invention was so revolutionary that it attracted the attention of many players and caused the convergence of various technological approaches into membrane development. The next chapter argues that the convergence of the various technological approaches to membrane development, and that the resulting fierce competition in a limited technological field have led to higher performance and lower cost, contributing to the growth of the industry.

References

Aoshima, Y., & Kasai, S. (2005). Toray: Tanso sen-i no gijutsu kaihatsu to jigyō senryaku (Toray: Technology development and business strategy of carbon fiber). *Hitotsubashi Business Review, 52*, 120–145.

Bairinji, R. (1984). Characteristics and operating history of cellulose acetate reverse osmosis membrane. *Kamipa Gikyōshi (japan TAPPI Journal), 38*, 265–275.

Fujiwara, M., & Aoshima, Y. (2016). Toyobo: Promoting fundamental corporate reforms. *Hitotsubashi Business Review, 64*, 124–141.

Ishizaka, S. (1970). "Gyakushintōatsu hō no kaihatsu to ōyō" (Development and application of reverse osmosis method). *Kagaku Kōgaku* (*Chemical Engineering*), *34*, 28–32.

Kemp, R., Schot, J., & Hoogma, R. (1998). Regime shifts to sustainability through processes of niche formation: The approach of strategic niche management. *Technology Analysis and Strategic Management, 10*, 175–196.

Koike, K. (1989). Production of ultrapure water for semiconductor. *Ōyō Butsuri (applied Physics), 58*, 945–946.

Kurita Kogyo Kabushiki Gaisha (Kurita Water Industries Ltd.). (2000). *Mizu o kiwamete 50-nen: Kurita Kogyo 50-nen shi (50 Years of mastering water: Kurita Water Industries' 50-year history).* Tokyo.

Motomura, Y. (1984). The present situation of ultra pure water treatment. *Suishitu Daku Kenkyū (journal of Water Pollution Research), 7*, 476–479.

Nakane, T. (1974a). Application of reverse osmosis. *Bulletin of the Society of Sea Water Science, Japan, 28*, 110–124.

Nakane, T. (1974b). Nihon ni okeru gyakushintō hō no kenkyū to kaihatsu no genjō (Current status of research and development of reverse osmosis method in Japan). *Bulletin of the Society of Sea Water Science, Japan, 28*, 46–51.

Ohya, H. (1988). A review on ultrapure water. *Bulletin of the Society of Sea Water Science, Japan, 42*, 97–108.

Organo Corporation. (1981). *Organo 35-nen no ayumi (Organo's 35-year history).* Tokyo.

Sato, H. (1976). Chōjunsui/mukinjunsui no seizo (Production of ultrapure water and sterile pure water). *Journal of Environmental Conservation Engineering, 5*, 38–44.

Schot, J., & Geels, F. (2008). Strategic niche management and sustainable innovation journeys: Theory, findings, research agenda, and policy. *Technology Analysis & Strategic Management, 20*, 537–554.

Shimizu, H. (1972). Ion kōkantai no tenbō (Prospects for ion exchangers). *Journal of Synthetic Organic Chemistry, Japan, 30*, 973–977.

Sōgō Hōsō Publication. (1980). Kōbunshi *takōshitsu maku no kaihatsu dōkō to shijō jittai: Gyakushintō maku, gengai roka maku, seimitsu roka maku, ion kōkan maku* (*Development trend and market status of polymer porous membranes: Reverse osmosis, ultrafiltration, microfiltration, and ion exchange membranes).* Okinawa: Sōgō Hōsō Publication.

Takashima, S. (1985). Technology of ultra pure water by water treatment maker. *Journal of Environmental Conservation Engineering, 14*, 58–64.

Taniguchi, Y. (1981). Desalting technique applications in water treatment processes. *Sen'i Gakkaishi (journal of Fiber Science and Technology), 37*, 23–30.

Taniguchi, Y., & Kremen, S. (1973). "Yōsui oyobi haisuishori ni okeru spiral gata gyakushintō sōchi no seinō to jitsuyōrei" (Performance and practical examples of spiral type reverse osmosis equipment in the treatment of water and wastewater). *Bulletin of the Society of Sea Water Science, Japan, 26*, 282–289.

Watanabe, A. (1985). Present situation of membrane technology in food industry. *Journal of Japan Oil Chemists' Society, 34*, 847–851.

Yamano, A. (1987). Cleaning and estimation technology of semiconductor manufacturing process. *Jitsumu Hyōmen Gijutsu (practical Surface Technology), 34*, 44–50

Yoshida, N. (1986). Zadankai "Needs kara mita sen-i kōgyo e no chūmon" handōtai kōgyō to sen-i gijutsu no kongo no kadai" (Roundtable discussion: orders to the textile industry from the perspective of needs: future issues for the semiconductor industry and textile technology). *Sen'i Gakkaishi (Journal of Fiber Science and Technology), 42*, 391–396.

Chapter 12
Emergence of Breakthrough Technologies and Convergence of Technological Approaches

Abstract In this chapter, we focus on the role of "breakthrough technologies". In particular, we shed light on the breakthrough technology invented by John E. Cadotte (1925–2005) of FilmTec. Cadotte's patent for polyamide composite membranes, filed in 1979, is known as the '344 patent, from its patent number 4,277,344. The '344 patent was an innovative technology that achieved a high level of compatibility between various contradictory performances through the use of membrane composition and interfacial polycondensation. This led to the convergence of diversified technological approaches.

12.1 Questions and Perspectives

In the previous chapter, we described how developers, who had been struggling with the lack of profit prospects, were able to continue their technological development aiming at an expected future large-scale market by developing applications other than seawater desalination. In particular, the emergence of ultrapure water applications for semiconductors was a boon for RO membrane companies. The existence of this initial market played an important role in justifying the continuation of development by each company. However, for seawater desalination, a new kind of membrane was needed that could provide both desalination and permeability at a higher level and that could be manufactured at a lower cost.

As will be shown later, the price of membrane modules fell by approximately 70% in the 20 years between the 1980s and the 2000s, when the desalination market grew significantly. One of the factors that contributed to the price decline was the expansion of ultrapure water production applications. However, the ultrapure water market is a very small market in terms of volume compared to the market for seawater desalination, which developed later, and it is not possible to attribute all of the price decline from 1980 onward to its mass production effect. In addition, application markets such as those for semiconductors were developed as an outlet for incomplete technology in the process of pursuing superior technology to actualize desalination, and

M. Fujiwara and Y. Aoshima, *Mechanisms for Long-Term Innovation*,
Advances in Japanese Business and Economics 31,
https://doi.org/10.1007/978-981-19-4896-1_12

it cannot be said that the development of these application markets itself encouraged technological progress toward desalination.

Therefore, the convergence of technological approaches is what we focus on in this chapter as another mechanism that has led to improved performance and lower costs for RO membranes. As long as each company develops and manufactures nonstandard membranes by adopting different technological approaches for membrane materials and geometries, industry-wide learning through technology spillover across company boundaries will not progress. As each company's membranes are differentiated from each other, market competition is also eased, making it difficult for prices to fall and for the market to expand. However, if the technologies of each company converge on a single approach, industry-wide learning can be expected, and the price bargaining power of customers with choices will increase, leading to a decline in market prices.

This chapter analyzes the convergence of these technological approaches that have occurred during the development of the RO membrane industry. The question is as follows: How did RO technologies converge, and what impact did this have on business behavior and industrial development? To answer this question, we show that the advent of the breakthrough technology invented by Cadotte, especially because it was so innovative, led to a rapid convergence of technological approaches across industries and had a significant impact on subsequent development behavior.

To clarify this point, Sect. 12.2 shows the convergence of membrane technologies in terms of materials, shapes, and manufacturing methods by analyzing patent applications. Section 12.3 refers to Cadotte's '344 patent that caused this convergence, and analyzes its impact. Finally, in Sect. 12.4, we show how the convergence of technological approaches has led to price declines and the impact on technology development activities.

12.2 The Convergence of Membrane Technologies

12.2.1 The Change and Convergence of Membrane Materials

In this section, we demonstrate how the materials, shapes, and manufacturing methods of membranes have converged by analyzing patent applications. First, the convergence of materials is shown in Fig. 12.1. This graph shows the share of materials used in RO membranes based on the number of patent applications.[1] 3,527 patents related to RO membranes with F-term groups (MB groups) related to materials were extracted from patents in Japan. The shares of the top three materials, polyamide, polysulfone, and cellulose, are shown on the line graph. The

[1] See the appendix for more information on this process.

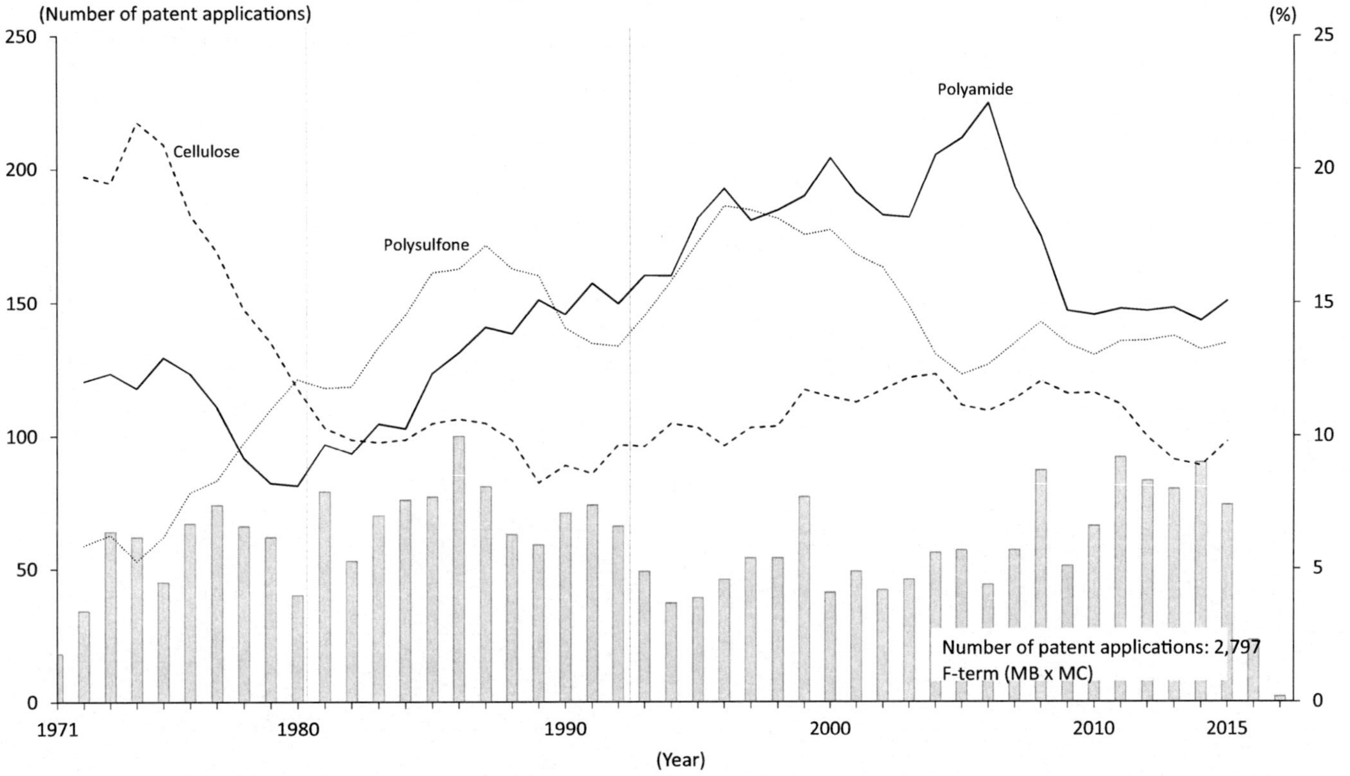

Fig. 12.1 Share of major materials. *Source* Prepared by the authors based on data from UltraPatent. *Note* The number of patent applications (left axis) on the bar graph is the actual yearly number of applications. The ratios expressed by the line graph (right axis) were three-year moving averages

graph is based on three-year moving averages to eliminate the effects of short-term fluctuations. The bar graph shows the number of patent applications in each year.[2]

The graph shows that while the number of patent applications for mainstream cellulose-based materials decreased in the 1970s, the number of patent applications for polyamide-based materials rapidly increased and became mainstream from the 1980s. By the beginning of the 1980s, the share of cellulose-based materials had fallen to 10%. Polysulfone replaced cellulose in the application market. Polysulfone is a typical material that was developed for the support layer of composite membranes. This indicates that the trend toward non-cellulosic materials and the accompanying trend toward composites progressed in the 1970s.[3]

In the 1980s, the application share of polyamide-based materials increased rapidly, and this trend continued until the late 2000s. Looking at the upward trend in detail, it can be seen that the application share first increased rapidly from the beginning to the latter half of the 1980s, then leveled off for a while but then increased again from the beginning to the middle of the 1990s. The timing of this rise is thought to have been influenced by the appearance of Cadotte's '344 patent and the ruling that the patent belonged to the U.S. government, as discussed below.[4]

Figure 12.2 clearly shows the significant change in the diversity of materials in the 1970s and 1980s. This line graph expresses the diversity of materials by using the idea of the reciprocal of the Herfindahl-Hirschman Index (HHI). The value shown in the graph is the reciprocal of the HHI that indicates the concentration of the materials to be developed. It shows how many kinds of materials constitute the entire group of patent applications if the number of patent applications for each material is the same. In other words, it is the number of equally-sized patent classes of materials. The more diverse the materials to be developed are, the larger the value becomes. The narrower the membrane materials that are selected, the smaller the value becomes.

The graph shows that the diversity of materials increased throughout the 1970s, peaked in the 1980s, and then declined significantly. Although the data are a moving average and thus cannot specify data specifically for any one year, it is interesting to note that the diversity peaked in the early 1980s. The '344 patent of Cadotte appeared in 1981, suggesting that the '344 patent may have influenced the convergence of membrane materials.

---

[2] To avoid the overlapping of line graphs and bar graphs, the upper limit of the left axis on which the bar graphs rely was set to 250 patent applications.

[3] Although the share of polyamide-based applications also decreased in the 1970s, many of the patents classified as polysulfone were designed for composite use with polyamide-based materials. Therefore, the decrease in share of polyamide-based separation membranes was not due to a decrease in the number of applications, but rather to an increase in the number of patent applications for support materials due to the composite use of membranes. In fact, 67.2% of polysulfone patents (F-term: MC62 group) in the 1970s had F-terms (MC54 group) for polyamide materials as well. In other words, nearly 70% of the polysulfone development patents had also included the polyamide materials. This ratio dropped slightly to 57.2% in the 1980s, but rose again to 66.3% in the 1990s.

[4] Since the end of the 2000s, the share of polyamide material development has been rapidly declining, and this can be attributed first, to the increase in the application share of polysulfone and second, to the development of polyalkenyl halogen (MC25 group).

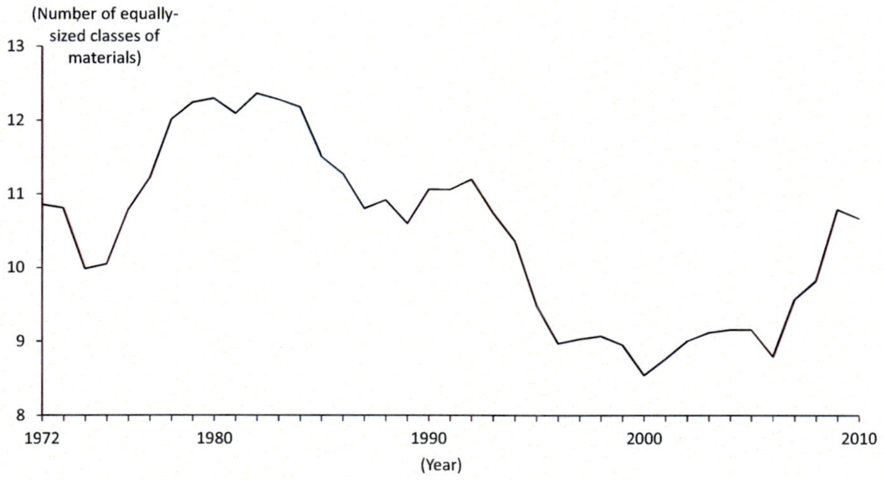

Fig. 12.2 Changes in the diversity of materials. *Source* Calculated by the authors based on UltraPatent. *Note* Three-year moving averages

The graph also shows that diversity ceased to decline in the late 1980s and early 1990s. This was the period when Dow began to assert patent infringement claims against its competitors in connection with the '344 patent. Therefore, there is a possibility that competitors avoided developing similar materials that might infringe upon the '344 patent and sought other materials.

On the other hand, after 1993, when the patent lawsuit between Dow and Nitto Denko was settled, the diversity of materials dropped significantly again, as the ruling that the '344 patent belonged to the government allowed competitors to develop materials similar to the '344 patent, resulting in a convergence of membrane materials. This is consistent with the results of the graph.

Thus, the values in the graph move in a way that implies the impact of the appearance of the '344 patent (1981), the patent litigation by Dow in the late 1980s, and the ruling that the '344 patent belongs to the government (1993).

Apart from these fluctuations caused by patent litigation by Dow, the diversity of material development followed a downward trend after around 1980, which was accompanied by a material shift from cellulose to polyamide.

12.2.2 The Convergence of Membrane Shapes

How did the shapes of the membranes change as the membrane material shifted to polyamide? Fig. 12.3 shows the number of patent applications for polyamide RO membranes classified by the shape of the membrane and the net number of applicants for flat-type polyamide RO membranes.

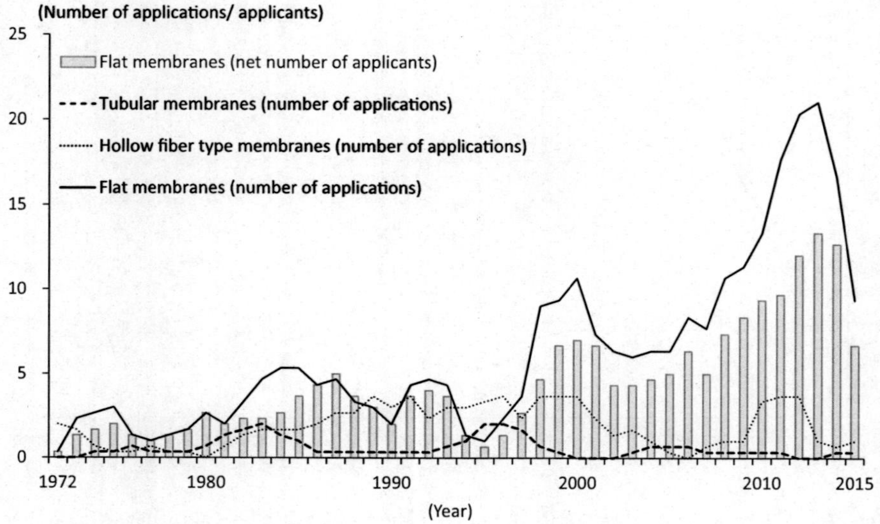

Fig. 12.3 Patent applications and net applicants for polyamide RO membranes by shape. *Source* Prepared by the authors based on UltraPatent. *Note* Exclusively aggregated by shape. Three-year moving average values

The graph shows that the development of flat membranes was dominant for polyamide materials. Although the development of hollow fiber membranes was also carried out in the 1990s, it can be confirmed that the flat membranes have been developed almost consistently from the initial stage. Since it was found that consolidation problems limited the improvement of the desalination rate and water permeability performance of cellulose-based asymmetric membranes, composite membranes have been developed using non-cellulose materials to form separation and support layers separately. The flat membrane type was suitable for forming this composite structure. Although it was not impossible to form composite membranes using tubular or hollow fiber membranes, it was technically difficult, and the membrane formation efficiency was not high enough to make it a practical product.

The bar chart, which is the net number of patent applicants related to polyamide flat membranes, indicates that the development of polyamide flat membranes became more active as the number of new entrants increased.[5] Here, the net number of

[5] In preparing this figure, we extracted patents only with the F-term of polyamide as membrane material, and the F-term of either tubular, hollow fiber, or flat membrane as shape. And we counted the net number of representative names of applicants for each year by area. Representative names of applicants are adopted in order to eliminate as much as possible fluctuations and errors in the notation of applicant names. The representative name does not refer to the first applicant, but to the representative name of each applicant. And for jointly applied patents, the representative names of each side are listed without omission. The reason why the total number of applicants is not counted is to examine whether or not a large number of patents are applied for by a small number of applicants. If a large number of patents are obtained in a specific area by a small number of applicants, there should be a discrepancy between the number of applicants and the number of patents by technical

applicants means the number of applicants excluding the duplication of applicants for each year. The fact that the number of net applicants and the number of patent applications are linked indicates that the increase in the number of patent applications pertaining to flat membranes was accompanied by an increase in the number of applicants, not simply by the fact that conventional companies began to apply for a large number of patents. This indicates that the industry as a whole has come to agree that the flat membrane type is the best choice for handling polyamide materials.

Thus, the development of RO membranes from the 1980s onward has converged on the development of flat composite membranes using polysulfone as the support layer and polyamide as the separation layer. As a result, as of 2018, 90% of the RO membrane market was dominated by polyamide-based flat composite membranes.

12.2.3 The Convergence of Membrane Manufacturing Methods

Finally, we will examine the trends concerning the manufacturing method of RO membranes. Figure 12.4 is a graph depicting the trend of membrane manufacturing patent applications and their share by material.[6] The figure shows that the number of applications for membrane manufacturing patents with polyamide materials has been increasing since the beginning of the 1980s. The application data on a single-year and actual number basis show that the number of applications has been 6 in 1978, 6 in 1979, 5 in 1980, 11 in 1981, 13 in 1982, and 15 in 1983, indicating that the number has been growing since 1981.

RO companies adopted the interfacial polymerization method as the manufacturing method for polyamide composite membranes. For Fig. 12.5, we extracted 612 patents with F-terms (NA01 to NA50 groups) that indicate specific manufacturing methods from 662 polyamide membrane manufacturing patents, and then identified the actual number and ratio of patents with F-terms (NA41 group) related to

area. Here, we attempted to verify the discrepancy by looking at the net number of applicants. In addition, we judged the applicant name of patent application No. 49-109,271 as the same applicant as the patent application No. 50-081,978 (applicant: Montedison S.p.A.) was used.

[6] The NA group (NA00 to NA75) is listed as the F-term related to the membrane forming method. Among the 2,513 patents with F-terms in the NA group, we calculated the ratio of patents with F-terms that refer to cellulosic materials (MC11 group) and the ratio of patents with F-terms that refer to polyamide materials (MC54 group), and made a comparison among materials in the patent applications related to the membrane making processes. To clarify the relative relationship between the two groups, the NA group patents that have both MC11 and MC54 F-terms were excluded from the sample so that patents that have both F-terms were not reflected in the graph in duplicate. As a result, the total number of membrane formation patents with cellulosic material was 370, and that of polyamide material patents was 662. In addition, the data is shown as three-year moving averages to reduce the influence of short-term fluctuations and to make the trend clearer.

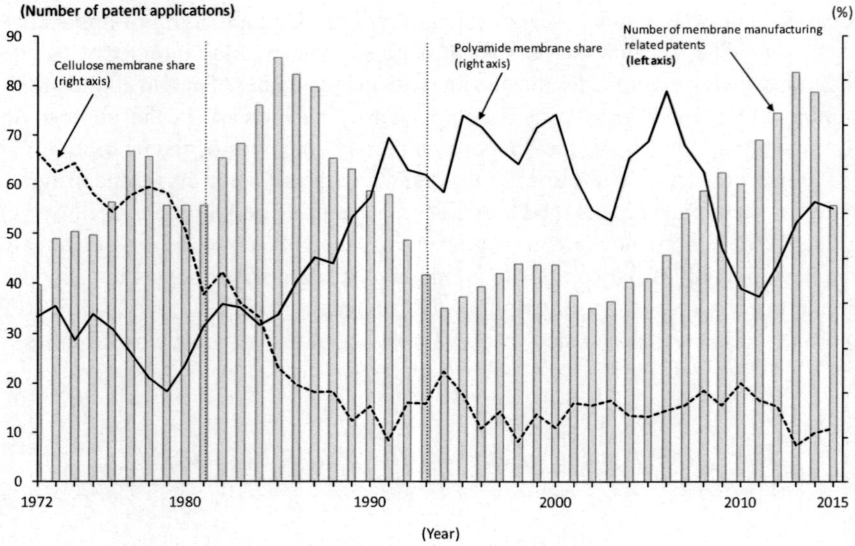

Fig. 12.4 Patents related to membrane manufacturing. *Source* Prepared by the authors based on data from UltraPatent. *Note* Three-year moving average values

"crosslinking and surface polymerization" in these 612 patents.[7] The figure indicates that the number of patents on membrane manufacturing methods related to polyamide materials increased significantly in the 1980s and that the patent applications related to interfacial polymerization supported this increase.[8] The actual number of patents related to the interfacial polymerization method using polyamide materials began to increase significantly, especially in 1985.

The ratio of patent applications related to the interfacial polymerization method using polyamide materials began to increase in the mid-1970s. Although the upward trend calmed down around 1980, it began to increase at a rapid pace again in the mid-1980s, reaching a peak around 1995. Since then, interface polymerization has continued to be dominant, with the ratio remaining at approximately 75%. From these developments, it can be said that the shift to polyamide materials was accompanied not only by a change in shape to flat membranes but also by an inclination toward interfacial polymerization as a membrane-manufacturing method.

[7] Although it is described as on-surface polymerization in the F-term, it is uniformly described as interface polymerization in this book. Additionally, the materials used are calculated in a way that does not allow overlap between cellulosic and polyamide materials.

[8] The patents related to the interfacial polymerization method here refer to the patents with the F-terms of NA41 group.

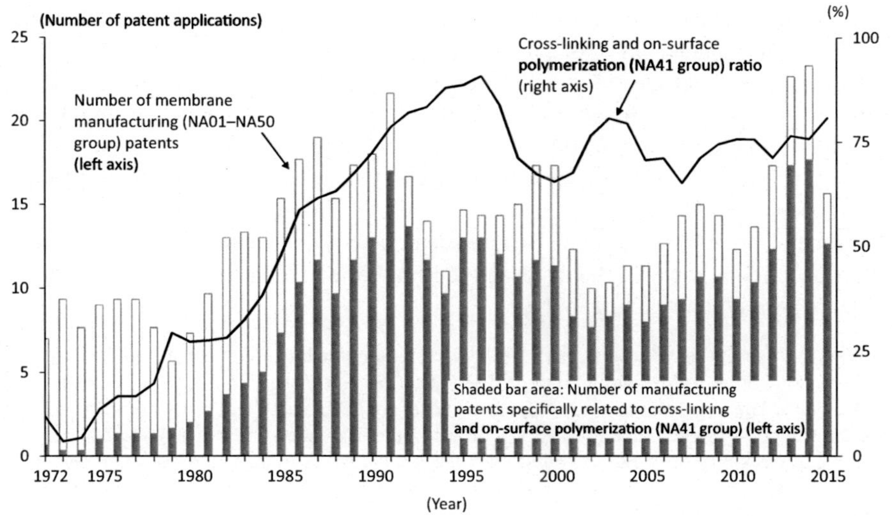

Fig. 12.5 Slant toward interfacial polymerization in the area of polyamide. *Source* Prepared by the authors based on data from UltraPatent. *Note* These patents here are with polyamide material F-term MC54 but without cellulose material F-term MC11. The values are three-year moving averages

12.3 How Technological Approaches Converged: Focusing Through Breakthroughs

12.3.1 Increasing Interest in Cadotte

Cadotte, who had been working with North Star and FilmTec on the development of polyamide flat composite membranes, played a significant role in the convergence of the development targets of the companies to polyamide flat composite membranes by interfacial polymerization.

As mentioned earlier, cellulose acetate materials were mainly used in the early stage of membrane development. The L-S membranes were the first to be developed. However, the asymmetric L-S membrane could not improve water permeability due to the problem of consolidation. Therefore, the development of composite membranes, in which a support layer is made of a different material and a separation layer is layered on top of the support layer, was aimed at, and various materials for the separation layer and the support layer were explored.

In 1967, while working for North Star, Cadotte succeeded in using polysulfone as a support layer to avoid consolidation problems and increase water permeability by five times. This achievement attracted the attention of the industry, and polysulfone became the material to converge on the support layer of non-cellulosic composite membranes.

Soon after, Cadotte succeeded in making the separation layer non-cellulosic, and in 1970, he introduced NS-100, a new membrane in which both the separation layer and the support layer were non-cellulosic. This opened up the market of composite membranes, and he continued the trend toward non-cellulosic membrane materials with the development of NS-200 in 1972 and NS-300 in 1977.[9]

Because North Star's development was under contract to the OSW, its results were reported and made public, and Cadotte's technical achievements attracted researchers' attention. Baker at Bend Research was one of the first to note and report on Cadotte's work. Baker recalls the first time he demonstrated a prototype membrane based on NS-100 approach.

> With CA you got maybe, on a good day, 98% rejection of salt from sea water. Sometimes 99.1, many times less. It was hard to get 99% rejection. It was very important to get better than 99%. The target was always 99.3, almost never could do with CA. The first time I tried John Cadotte's method, it was so cool … I test it, the first time I did it, I thought "this is going to work, hey boy this is so cool." Whatever. 99.5. It was tremendous. I did that once and I thought "this is something, this is really a breakthrough." And it was. So Cadotte's invention changed everything.[10]

However, the performance of these membranes was not yet high enough to meet the market requirements, and the search for suitable materials for the separation layer continued in the 1970s.

It was Cadotte that converged the search for materials for the separation layer. Cadotte developed a new cross-linked aromatic polyamide composite membrane by interfacial polycondensation and presented his findings at an international conference in October 1979, which attracted considerable attention. Cadotte, who had already moved to FilmTec at this time, applied for a patent for this development, which was granted in 1981. It was Patent 4,277,344.

12.3.2 *Innovativeness of the '344 Patent*

This patent, called the '344 patent following the last three digits, was a truly ground-breaking breakthrough. FT-30, which was developed based on this breakthrough, not only achieved a high level of compatibility between desalination performance and water permeability that had been a hurdle for everyone, but also demonstrated outstanding performance in many other areas, including durability and membrane formability. FT-30 demonstrated high performance in field tests and paved the way for full-scale single-stage desalination in seawater desalination that had previously required two-stage desalination.[11]

---

[9] Cadotte (1985a, 1985b), Tomaschke (2000).

[10] From the authors' interview with Richard Baker on March 3, 2015 at Membrane Technology and Research, Inc. in Newark.

[11] Larson et al. (1981).

Table 12.1 shows the overall performance of FT-30. In this table, various RO membranes published at that time are compared along five performance dimensions: desalination performance, permeability performance, pH range, maximum operating temperature, and chlorine resistance. While some membranes perform better than FT-30 in certain dimensions, it is clear that FT-30 is the only membrane that performs well in all five dimensions. This new membrane was an outstanding achievement in the ongoing competition for membrane performance.

The features of FT-30 were highly evaluated by the Japanese side from early on. For example, Kamisawa at the GCIRIT wrote about the excellent features of FT-30 as follows in 1980.

> At an operating pressure of 56 kg/cm^2, the water flux is 0.95 m^3/m^2/day, and the salt rejection rate is 99.5%, indicating that this is an extremely excellent membrane … The water flux increases linearly with increasing temperature up to 60°C, indicating that the membrane has excellent heat resistance. For pH, the membrane can be used in the range of 3 to 11. For chlorine resistance, the membrane was immersed in a 100-ppm sodium hypochlorite solution for 72 hours, and the results showed no abnormalities. It is stated that the membrane withstood continuous operation for 2000 hours against residual chlorine in drinking water…

Table 12.1 Performance of FT-30 and other RO membranes (as of 1985)

Membrane material (Product name)	Company name	Performance at salt concentration of 3.5%		pH range	Maximum operating temperature (°C)	Chlorine resistance (ppm)
		Desalination rate (%)	Permeability performance (l, m^{-2}/d^{-1}/atm^{-1})			
FT-30	FilmTec	99.6	17.9	3–11	60	100
NS-200	North star	>99	10.8–12.0	Good	Good	Bad
NS-200	Envirogenics	99.6	11.1			
PA-300	UOP	98.5–99.4	12–15	3–12	60	?
APA (Permasep B-10)	DuPont	98.5	0.71	5–9	35	0.1
PEC-1000	Toray	99.8	5.36	1–13	55	–
PBIL	Teijin	99–99.7	4.5–6	1–12	70	100
CTA/CN-CA	UOP	99.5	3.59	–	–	–
CTA (XFS 416712)	Dow chemical	98.7	0.51	4–7.5	35	1
CTA (ROGA 2B-TFC)	UOP	98.5	5.56	4–7.5	35	1
CTA (HOLLOSEP)	Toyobo	99.7	0.909	3–8	40	1
SOLCON-P	Sumitomo chemical	5000 ppm–NaCl 98	9.6	1–10	45	0.1

Source Prepared by the authors based on Yoshidome (1985), p. 831

> If the membrane has this kind of performance and has excellent heat resistance and chlorine resistance, it will be one of the membranes that will attract attention in the future.[12]

In a literature review on seawater desalination by the RO method published between 1979 and 1982, Motohiro Okazaki and Shoji Kimura of the Institute of Industrial Science, University of Tokyo, describe FT-30 as having "greater pH resistance and chlorine resistance than cellulose acetate membranes and conventional polyamide membranes, and can be used from pH 3 to 11. The characteristic of this membrane is that it is excellent in desalination at low pressure."[13] Researchers in Japan were strongly aware that FT-30 was an outstandingly promising new membrane.

12.3.3 The Impact made by the 344 patent's innovativeness

Cadotte's achievements had a major impact on the private sector because they were not incrementally but dramatically superior to other membranes. Even though the practicality of the technology was still unknown at that time, Cadotte's results became the main reference framework for the industry, and the competition to develop polyamide-based materials became even fiercer. Dow, which had previously developed its business with cellulose acetate hollow fiber membranes, quickly changed direction and obtained the '344 patent by acquiring FilmTec. The '344 patent was so influential that it changed the development strategy of a major company.

In fact, the '344 patent has been cited in a number of U.S. patents since the late 1980s. Figure 12.6 shows the number of U.S. patent applications that cited the '344 patent (bar graph) and the net number of applications excluding self-citations by Dow (line graph).[14] The graph first shows that the number of citations to the '344 patent increased from the late 1980s to 1993. The number of citations reached its peak in 1993, when this patent was attributed to the U.S. government. After that, the number of citations declined for a while in the mid-1990s, but it became stable in the mid-2000s, and the number of citations increased again in the 2010s. Although the recent surge includes many self-citations by Dow, citations in patents from other companies are also increasing, indicating that Cadotte's technology, invented nearly 40 years ago, has not only become a mainstream market technology but is still referred to as a fundamental technology today.

---

[12] Kamisawa (1980), pp. 354–355. Judging from the time of publication, Kamisawa's description is considered to be based on a presentation made by Cadotte at an international conference held in October 1979. The reason why the values in Table 12.1 reported by Yoshidome (1985) are slightly different from those in Kamisawa's description is probably because they are based on subsequent tests.

[13] Okazaki and Kimura (1983).

[14] According to this database, Teijin was the first company that cited the '344 patent. This suggests that Teijin was relatively enthusiastic about the development of RO membranes at the time.

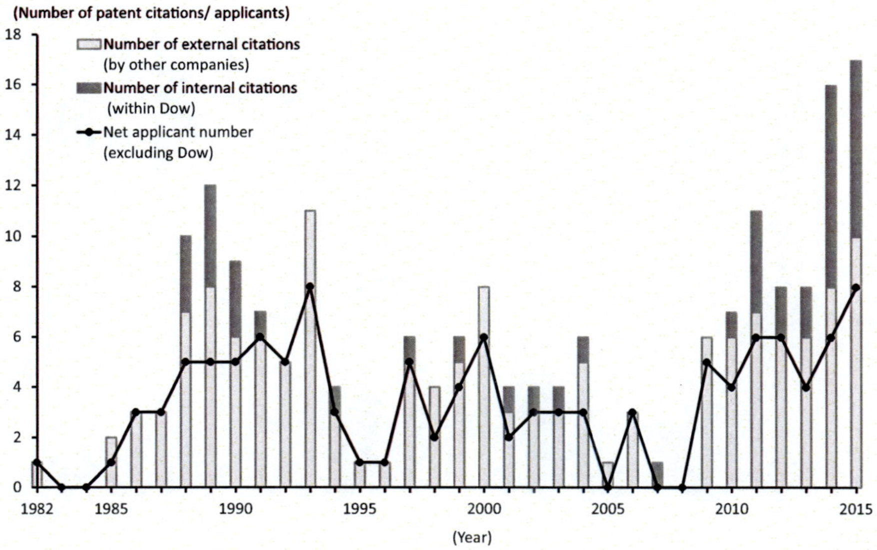

Fig. 12.6 Number of citations of the '344 patent in the U.S. *Source* Prepared by the authors based on data from UltraPatent

Dow, which had obtained the '344 patent, sought to exclude other companies through patent disputes, including UOP's Allied Signal, which had been ordered not to manufacture RO membranes. This fact shows that each company was developing a technology that was quite similar to the content of the '344 patent.

Cadotte's invention also had a great impact on the development activities of Japanese companies. As Fig. 12.3 indicates, companies began to develop polyamide flat membranes after this invention.

The above evidence implies that a consensus was formed throughout the Japanese and U.S. industries, which can be summed up as saying "If polyamide materials are to be handled, it should be in the form of flat membranes."

12.3.4 Reaction of the Three Japanese Companies

The emergence of the '344 patent and the resulting change in the development focus to polyamide materials had a major impact on the development behavior of the three major Japanese companies, Toray, Toyobo, and Nitto Denko.

Figure 12.7 shows the composition ratio of the number of patent applications for the two materials, cellulose and polyamide, divided into six types of membranes crossed by three shapes, i.e., tubular, hollow fiber, and flat membrane, and the number of patent applications with the F-term related to each type totaled for each five-year

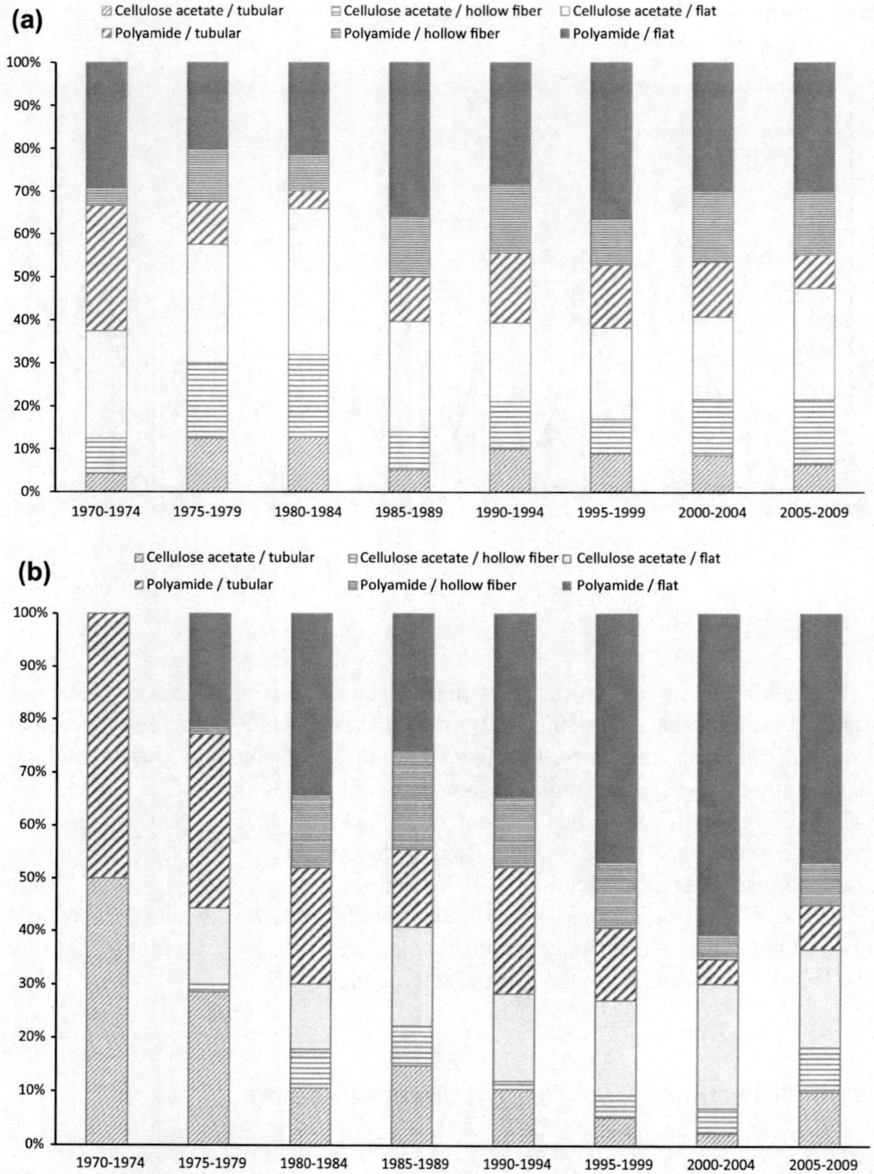

Fig. 12.7 Share of patent applications by technological domain in each Japanese company. **a** Material/shape composition ratio (Toray). **b** Material/shape composition ratio (Nitto Denko). **c** Material/shape composition ratio (Toyobo). *Source* Calculated by the authors based on data from UltraPatent

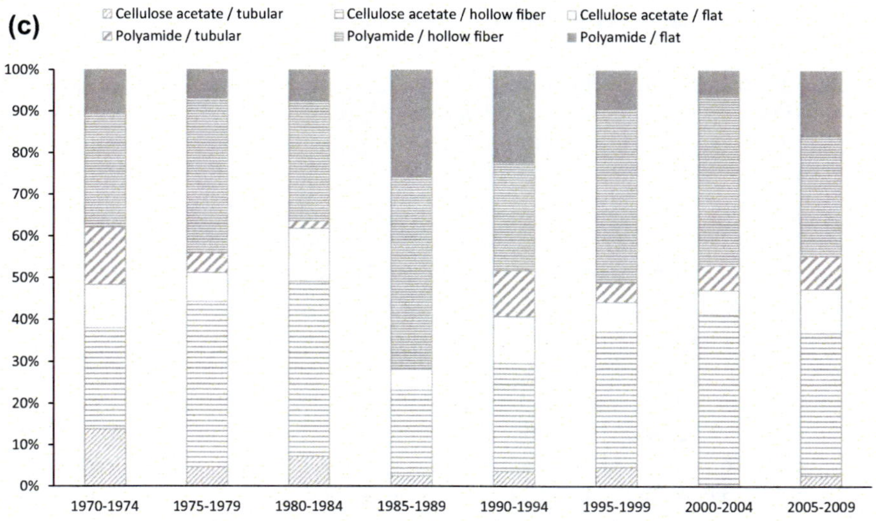

Fig. 12.7 (continued)

period.[15] These figures support the description of the development behavior of each company given in Part 3.

First, Toray's actions show that the development of the cellulosic flat membrane was the mainstream for them in the first half of the 1980s. It was in the latter half of the 1980s that the membrane materials in the applied patents were largely replaced. Patents related to polyamide-based materials were filed simultaneously in 1986 and have continued to increase steadily since then. This is consistent with the description in Chap. 7 that the development of cellulose-based flat membranes was promoted first, and then converted to polyamide-based materials in the mid-1980s.[16]

In fact, Toray had presented its own new membrane, PEC-1000, at the same international conference where Cadotte had presented his development results in 1979, attracting much attention from the audience. Toray was also searching for a non-cellulosic material suitable for the separation layer and was developing its own. However, PEC-1000 could not be put to practical use due to its weakness against dissolved oxygen. Toray then read the results of Cadotte and proceeded to develop a new membrane that would surpass FilmTec in the same material system. As described in Chap. 7, Himeshima, who was in charge of membrane development at the time, said "At that time, Cadotte was a giant. I wanted to catch up with him,

[15] The number of patents is calculated in a way that allows for overlap in the F-term of each area. This is because the number of patents would be extremely small and the influence of short-term fluctuations would be too strong if the number of applications for each single year is shown after eliminating the duplication of F-terms when the number of patents is determined at the level of individual companies. In addition to this, in order to make it easier to compare the three figures, the starting period of the graphs are all unified from 1970 to 1974.

[16] Compared to the other two companies, Toray's ratios in each area are close to each other. This suggests that Toray is operating a fairly broad range of business areas. Hirai of the WRPC commented on this point, "I think each company is trying to figure out where it fits in. Toray is engaged in broad-based development, but other companies focus on their specific areas." (From the authors' interview with Mitsuyoshi Hirai at the WRPC on March 16, 2015).

and I wanted to make something good by noticing even one thing that they (he and other star scientists) hadn't already noticed."[17] While Toray was pursuing its own efforts to make non-cellulosic products, it started to develop membranes following the direction '344 patent indicated and surpass the performance of the technology.

Second, looking at Nitto Denko, it is clear that in the 1970s, the development of tubular membranes, whether cellulose acetate-based or polyamide-based, was the mainstream. This is consistent with the description in Chap. 9 stating that the company's development began with tubular membranes. The development of polyamide flat membranes proceeded in the late 1970s and expanded rapidly in the 1990s, consistent with the company's leap forward in RO membrane development after the acquisition of Hydranautics.

Nitto Denko developed and introduced NTR-759, a membrane with both a high desalination rate and water permeability, based on the technology owned by Hydranautics. However, Dow claimed that this membrane infringed on Cadotte's patent, and Nitto Denko was forced to discontinue the sale of NTR-759. This incident shows that the technology of Hydranautics also had a large overlap with Cadotte's invention.

Last, the Toyobo graph shows that development was intensively promoted in hollow fiber membranes, both cellulosic and polyamide. This is consistent with the description in Chap. 8 that Toyobo promoted the development of RO membranes to utilize its spinning technology.

It is interesting to note that Toyobo, which had consistently commercialized cellulosic hollow fiber membranes, was also enthusiastically developing polyamide hollow fiber membranes. The development results of Cadotte not only shifted the mainstream market to polyamide flat membranes but also attracted development efforts from Toyobo that had been developing its business mainly with cellulosic hollow fiber membranes.

12.4 What Did the Convergence of Technological Approaches Bring About?

12.4.1 Price Decline

Attracted by the promise of Cadotte's achievements, the industry's development activities narrowed down to polyamide flat membranes by interfacial polymerization. This convergence of technological approaches and the ability of many players to develop products based on the '344 patent has had three major effects on the industry.

The first is the price decline of membrane modules due to intensified competition. According to Semiat (2000), for example, the price of membrane modules, which had been declining at a gradual and almost constant pace from 1980 to the beginning of the 1990s, plummeted from 1994 to 1996. Its Price Index, with 1980 as 1, moved

[17] See Chap. 7.

to approximately 0.8 in 1984, 0.6 in 1989, and 0.5 in 1994, then plummeted to 0.35 in 1996 and then to 0.3 in 1999. It can be inferred that this discontinuous price decline between 1994 and 1996 was the result of intensified competition following the decision to vest the '344 patent in the U.S. government. Martin H. Peery, who joined Dow in 1989 and became involved in the development of RO membranes in 1992, recalled as follows.

> The courts ruled that the FilmTec patent resulted from research conducted under the 1971 Saline Water Conversion Act and, therefore, was available for the general public to practice. As a result, competition increased leading to lower market prices for RO modules. This catalyzed efforts to improve cost efficiencies.[18]

Desalination costs have also plummeted in response to the low membrane module prices. Based on Wilf and Bartels (2005), desalination costs, which had been flat in the early 1990s, fell from approximately \$1.4 per m^3 in 1995 to \$1 in 1996 and then plummeted to just over \$0.8 in 1997. By 2000, the cost had fallen to around \$0.5, meaning that the cost of RO desalination had fallen by less than half in just five years.

Thus, Cadotte's groundbreaking invention led to a convergence of technological approaches among companies, which has become a de facto public good through patent disputes, resulting in harsh competition and a significant drop in the price of membrane modules. This, in turn, has led to lower desalination costs, which is considered to be a major step forward in achieving the cost-effectiveness required for seawater desalination applications.

12.4.2 Driving Incremental Innovation

The second effect of the convergence of technological approaches is the stimulation of incremental innovation in membrane development. As the scope of development was narrowed down to polyamide flat membranes by interfacial polymerization and the '344 patent became a de facto public property, companies focused on the development of improvements aimed at gradual performance progress and cost reduction while referring to the development results of other companies. In the midst of intensifying downward pressure on prices due to competition, this was a perfectly legitimate course of action.

[18] From the authors' interview with Martin H. Peery on March 2, 2015 at Dow Water & Process Solutions, Edina.

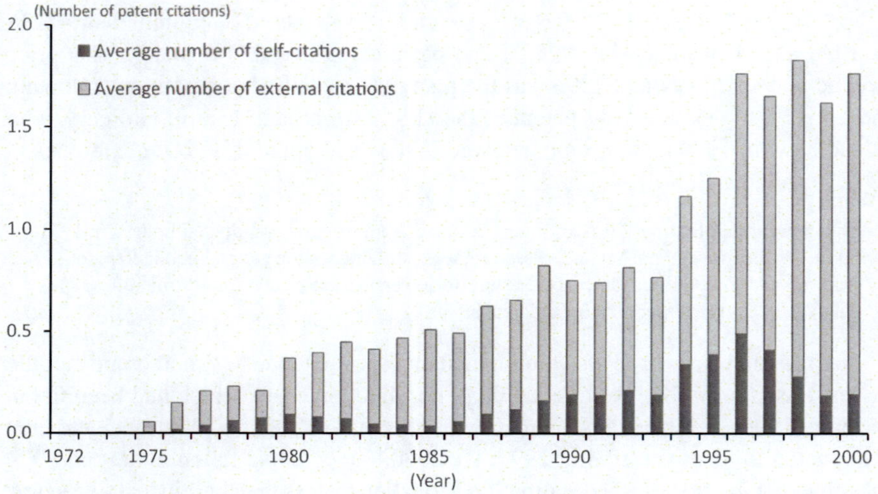

Fig. 12.8 Backward citations of membrane patents. *Source* Calculated by the authors based on data from UltraPatent. *Note* The average number of citations for each year was extracted by first using this formula: [total number of backward citations (self or external) in membrane-related patent applications issued for the year]/[number of membrane-related patent applications for the year]. Then, all values are three-year moving averages based on those calculation results

Figure 12.8 shows the result of the backward citation analysis of applied patents to understand the progress of such incremental innovation by each company.[19] Incremental innovations generally have a cumulative continuity by utilizing an accumulation of previous technology. It is highly likely that the patent examiner will point out and cite the previous patents. Thus, it is thought that an increase in backward citations indirectly indicates an increase in incremental innovation. As a concrete analysis, we first classified the patent applications into three categories, i.e., membranes, membrane modules, and operations, according to the F-term classification and then extracted 3,247 membrane patents from these categories to grasp the transition of the average number of backward citations.

[19] Among the F-term groups for RO membranes (GA03, including sublayers), four F-term groups from MA to NA were defined as membranes, three F-term groups from HA to JB were defined as membrane modules, and six F-term groups from KA to LA were defined as operations. The patents were categorized into membranes, membrane modules, and operations according to which F-term group was most frequently assigned. Since the membrane modules covers only three groups at most, while the operations covers six groups, a simple addition of the number of F-terms may cause a bias in classification. For the detailed procedure, please refer to Appendix 3–3. As a result of the classification of 9,255 patents, excluding 41 patents that do not have any of the above F-term groups, the total number of patents in each area is 3,443 for membranes, 2,864 for membrane modules, and 2,948 for operations. The reason why the number of patents on membranes is reduced to 3,247 instead of 3,443 in Fig. 12.8 is that we excluded the patents for which the highest ratios are the same values and are difficult to be classified into one specific area, and only analyzed the patents that can be clearly classified into the membrane category. In other words, the analysis of this graph does not employ procedure 5) in the Appendix 1.3.3.

The top of the bar graph shows that the average number of backward citations of all membrane patents, which includes both self-citations and citations of other companies, has been rapidly increasing since 1994. It is a natural tendency that the average number of backward citations increases with the accumulation of past patents. However, there is a rapid upward swing after 1994, which is different from an overall trend. Such a rapid change suggests that when the '344 patent became the property of the U.S. government in 1993, many companies started to develop their products under the extension of the '344 patent. The convergence of technology approaches suggests that companies have shifted to incremental development with a high degree of continuity related to previous patents.

The breakdown of the bar graph shows that such continuity is expressed not only in the internal patent self-citations but also in the external patent citations of other companies. The average number of self-citations gradually increased from the middle of the 1980s, rose discontinuously in 1994, and has been on a downward trend since 1997. The average number of citations of other companies, on the other hand, increased continuously until 2000. The fact that the number of citations of other companies is higher than the number of self-citations may suggest that the industry as a whole has become more interdependent.

The result of this convergence of technological approaches is thought to be a normalizing of incremental innovation activities building up the industry, with companies looking not only at their own development information but also that of other companies.

12.4.3 Stimulating Downstream Innovation

The third change resulting from the convergence of technological approaches is the stimulation of innovation in the downstream area. The prospect of establishing technology for membranes has led to more improvements in downstream for widespread use.

It should be noted that membranes are materials that have a close relationship with the operating systems that make them function. For example, Kamiyama of Nitto Denko wrote in 1986, "The future of RO membranes is closely related to how the ultrapure water production system will change. In other words, the emergence of new RO membranes will change the system."[20] In the 1990s, Toray developed membranes with high pressure resistance to withstand the high pressure applied in the second stage, with the aim of building a two-stage desalination system to achieve a high freshwater recovery rate. Toyobo also solved a membrane problem it faced in Saudi Arabia by using a unique sterilization method. In this way, the practical application of membranes requires not only improvements in the membranes themselves but also improvements in the system and in the operation side.

[20] Kamiyama (1986), p. 13.

In the case where the membrane itself and its operation are closely related, the convergence of technological approaches on the membrane side will narrow down the membrane property that should be considered in the system and operation side, which will facilitate innovation activities in the downstream areas of the value chain.

Figure 12.9 shows the changes in the share of patent applications in membranes, membrane modules, and operations. The graph indicates that the share of the membranes declined sharply in the 1990s and that the share of the operations increased instead. This reversal occurred in 1993, and since then, the share of operation patents has been the highest except for 1997, although the shares of membrane patents and membrane module patents have been close to each other. 1993 is the year when Cadotte's invention became the property of the U.S. government.

Advances in the system side as well as in the membranes themselves were very significant for the development of seawater desalination applications. This is because these downstream peripheral technologies were extremely important for cost reduction in desalination applications. For example, the performance of energy recovery equipment evolved significantly in the 1990s, which have contributed to cost reduction.

The importance of anticipating and accumulating peripheral downstream technologies in this way can be clearly seen in the situation when Toray, which had begun development with the aim of desalination, started parallel development. At

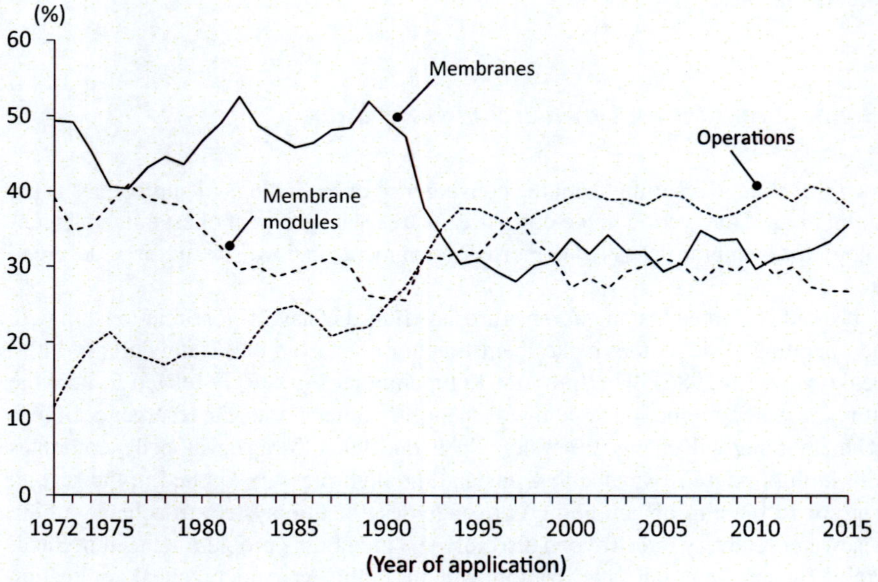

Fig. 12.9 Share of patents by value chain domain. *Source* Calculated by the authors based on data from UltraPatent. *Note* The values are three-year moving averages. The sample size for each domain is 3,443 membrane-related patents, 2,864 membrane module-related patents, and 2,948 operation-related patents

Toray, while waiting for the polyamide materials team to make progress in development, the cellulose acetate materials team was proceeding with development with one mission in mind: to commercialize first and accumulate various complementary technologies needed downstream, such as membrane modularization and operations. As Kawabata said, "in terms of the total water business, we needed a lot of knowhow beyond just membranes, so (CA RO membranes) played the role of being the first to be implemented."[21]

As described above, the convergence of technological approaches in the RO industry has not only stimulated incremental innovation in the membranes themselves but also in their downstream areas. These are likely to have had the effect of encouraging further performance advances as well as creating the scope for further reductions in the cost and price of membrane modules from the mid-1990s onward.

12.5 Conclusion: The Mechanism of Dominant Design

In general, various technological approaches are tried and tested in the early stage of industrial development. In the RO membrane industry, various membrane materials, membrane shapes, membrane manufacturing methods, element structures, etc., were tried. However, as the industry developed and technological development progressed along with market interaction, a variety of technological approaches were gradually eliminated, and eventually a particular product design shared within the industry emerged. This is called the dominant design, and it reflects the technological approach that determines the direction of subsequent development (Abernathy, 1978; Utterback & Abernathy, 1975). Once the dominant design is formed, the diversity of technological approaches is greatly reduced, and the frequency of product innovation declines; instead, process innovation increases. As a result, product costs fall rapidly, and markets expand. What has happened in the RO membrane industry is generally consistent with this pattern.

How, then, is dominant design created? The analysis in this chapter suggests that the emergence of breakthroughs is an important factor in advancing the formation of dominant design. The dominant design in the RO membrane industry is based on three basic principles: polyamide materials, a flat membrane shape, and membrane formation by interfacial polymerization. This basic form is a characteristic of RO membranes that account for 90% of the current market. The '344 patent by Cadotte presented this basic form. The fact that the basic form has not changed means that there have been few breakthrough product innovations since 1980. Instead, development has been focused on improvements and refinements in materials and shapes as well as innovations in membrane manufacturing processes and downstream areas.

Breakthroughs such as the '344 patent serve as a major roadmap for players who are groping in the dark in the midst of technological uncertainty. Even if it is not clear at the preliminary stage how much economic benefit the breakthrough will actually

[21] See Chap. 7.

bring, the innovativeness of the breakthrough will attract many players, and technological development efforts in the industry will converge in a particular direction. As a result of development competition in the same technological field, price decline and performance improvement will both accelerate. Economic efficiency is enhanced as a result.

The discussion in this chapter is summarized in the diagram shown in Fig. 12.10. In the RO membrane industry, the appearance of the '344 patent, a breakthrough technology, led to a rapid convergence of the previously diverse technological approaches to polyamide systems, flat membranes, and interfacial polymerization, and focused the development efforts of each company. If the '344 patent had been owned exclusively by Dow, other companies would not have been able to proceed with development in this area. When the U.S. government decided in 1993 that Cadotte's '344 patent belonged to the U.S. government, the decision prompted Japanese companies that had previously been reluctant to pursue the commercial development of polyamide flat membranes similar to Cadotte's membranes. It can be said that Cadotte's epoch-making invention provided a clear direction to the development activities of the companies, and the elimination of the proprietary nature of the technology through the subsequent patent disputes led to a convergence of technological approaches.

The first result of this convergence was a significant drop in the price of membranes. The price of RO modules plummeted between 1994 and 1996, reflecting the impact of the Cadotte patent becoming essentially a public good in 1993. Progressive innovation and improvements in downstream areas, such as plant operations, led to a steady improvement in basic performance and a significant decline in the cost of desalination using RO membranes. As a result of intensified competition and the concentrated development efforts of various companies, the price of RO membranes fell, and their performance improved. Therein, the expansion of the seawater desalination market, which had been a dream since the initial development stage, finally became a reality.

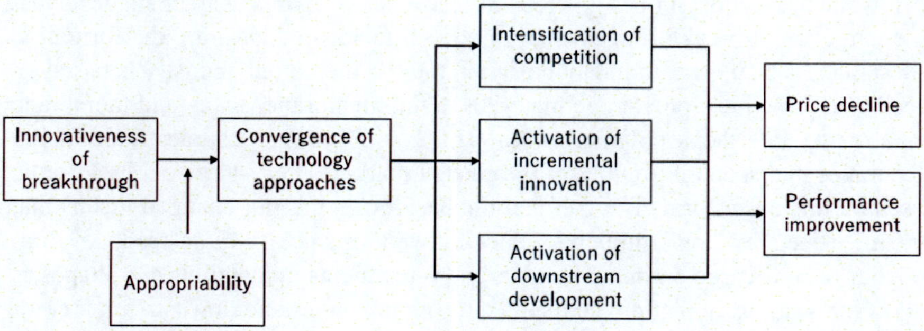

Fig. 12.10 Mechanism of dominant design. *Source* Prepared by the authors. *Note* Although the intensification of competition, the activation of incremental innovation, and the activation of downstream development are thought to affect each other, the rendering of this is omitted to avoid complicating the figure. Column: The standardization of membrane elements

Thus, the convergence of membrane technologies has promoted incremental innovation in the field, and together with the standardization of membrane modules, which was proceeding in parallel, the cost reduction of the RO membrane process has progressed. This has brought about the expansion of the seawater desalination market that has been the longstanding goal of companies. This process has certainly made a significant contribution to industrial development. On the other hand, however, it is also thought to have led to a deterioration in the profitability of each company's business due to intensified competition. We would like to discuss this point again in Chap. 14.

Appendix: The standardization of membrane elements

The de facto standard for membrane element shapes was established at a very early stage, when Fluid (UOP), which acquired ROGA, led the market with 4-inch and 8-inch membrane elements and delivered high-performance pressure vessels to match them, which were highly evaluated. When Fluid's pressure vessels were delivered to many plants, other companies competing with Fluid were asked by their customers to develop elements to fit these pressure vessels. Although some companies tried to differentiate the shape of the element, the shape of the element was gradually narrowed down for competitive reasons. Truby explained the process as follows.

> Standardization part came because the pressure vessel, housing you put the membranes in, has to be somewhat precise. It needs to fit carefully. In the early days, we used steel pipe. The North power station in Kashima, Japan had carbon steel pipes. All the early RO systems had steel pipes.
>
> Steel had many disadvantages one of which was it was not a precise diameter inside. You had to be careful when you selected a pipe material before you made a pressure vessel. We started to make fiber glass. Fiber glass, you wind the fiber glass on a mandril was absolutely precise. You could make pressure vessels very precise. Then you could put the spiral element into it … I think we commercialized the first ones and we sold them to Orange County in maybe 1975, anyway we made the first ones.
>
> Those fiber grass vessels worked so well. Then, everybody who made the spiral wound element was asked to make them fit in these vessels. So that became the standard. Fluid systems created the first membrane and vessel. People who wanted to copy Fluid systems, who was No.1, had to make their membrane that fit in the same vessel … Filmtec made them, so did Hydranautics, everybody made them, they need to fit them into the vessels. Sometimes in the beginning, we had different core tubes and different adaptors. But eventually even those became the same because the customers prefer. So, the standardization came slowly with time. But it really started with the pressure vessels.[22]

The de facto standardization of membrane elements, which started with pressure vessels, progressed rapidly in the 1970s. For example, Toray's PEC-1000, which was developed around 1980, was prototyped in 2-inch and 4-inch sizes (Kurihara

[22] From the authors' interview with Randy Truby at Carlsbad on March 4, 2015. The text in parentheses was added by the authors.

et al., 1980), but the actual elements were SP-110 (4-inch) and SP-120 (8-inch) (Kurihara, 1983). It can be considered that a de facto standard had already been established at this time. This was also because plant engineering companies, which preferred to purchase membrane elements from multiple sources, encouraged their own customers to use the same shape of pressure vessel.

References

Abernathy, W. J. (1978). *The productivity dilemma: Roadblock to innovation in the automobile industry*. Johns Hopkins University Press.

Cadotte, J. E. (1985a). Development of composite reverse osmosis membranes in retrospect. *Membrane, 10*, 117–118.

Cadotte, J. E. (1985b). Evolution of composite reverse osmosis membranes. in D. R. Lloyd (Ed.), *Materials science of synthetic membranes, American Chemical Society* (pp. 273–294).

Kamisawa, C. (1980). Advanced membrane desalination technology and newly developed membrane for desalination. *Membrane, 5*, 348–356.

Kamiyama, Y. (1986). Ultrapure water manufacturing: reverse osmosis membrane. *Sen'i Gakkaishi (journal of Fiber Science and Technology), 42*, 9–13.

Kurihara, M. (1983). Proceedings: Concentration and recovery of dilute aqueous solutions containing valuable materials by synthetic composite reverse osmosis membrane. *Membrane, 8*, 97–112.

Kurihara, M., Kanamaru, N., Harumiya, N., Yoshimura, K., & Hagiwara, S. (1980). Spiral-wound new thin film composite membrane for a single-stage seawater desalination by reverse osmosis. *Desalination, 32*, 13–23.

Larson, R. E., Cadotte, J. E., & Petersen, R. J. (1981). The FT-30 seawater reverse osmosis membrane: Element test results. *Desalination, 38*, 473–483.

Okazaki, M., & Kimura, S. (1983). Recent advances of reverse osmosis technology for seawater desalination. *Nihon Kaisui Gakkaishi (Bulletin of the Society of Sea Water Science, Japan), 37*, 149–158.

Semiat, R. (2000). Desalination: Present and future. *Water International, 25*, 54–65.

Tomaschke, J. E. (2000). Interfacial composite membranes. *Membrane Preparation,* 3319–3331.

Utterback, J. M., & Abernathy, W. J. (1975). A dynamic model of process and product innovation. *Omega, 3*, 639–656.

Wilf, M., & Bartels, C. (2005). Optimization of seawater RO systems design. *Desalination, 173*, 1–12.

Yoshidome, H. (1985). Review of polymer separation membrane. *Journal of the Japan Oil Chemists' Society, 34*, 829–833.

Chapter 13
Company-Specific Rationales

Abstract In this chapter, we focus on the role of "company-specific rationales" and explore the reasons why development has continued to be justified in three major Japanese companies. The chapter reveals that each of the three companies has continued to support development based on their own internal reasons and their own value standards. At Toray, development was carried out in line with the concept of technological superiority, which was a response to the highly challenging application of seawater desalination with advanced technological capabilities. The developers at Toyobo positioned their business within the company-wide policy of diversification, and when faced with a company-wide crisis, they took the path of acting as inconspicuously as possible. At Nitto Denko, the business was justified by the company-wide philosophy of making the membrane business one of the pillars of corporate growth.

13.1 Company-Specific Factors

Thus far, we have explained the rise and development of the RO membrane industry from three aspects: government support, the development of the initial market, and the emergence of a dominant design. The development of RO membranes began in the 1950s in the United States with government support in response to social need. Japanese companies quickly followed this trend. However, the industry did not expand as initially expected. In terms of seawater desalination costs, time was needed to resolve many issues before the RO method surpassed the conventional evaporation method.

The previous chapters described the reasons why the companies were able to continue their technological and business development even in the midst of severe business conditions. Those were the emergence of a new market for ultrapure water for industrial use and the convergence of technological approaches based on Cadotte's invention that opened up the possibility of market expansion through cost reduction. However, these factors alone cannot explain the whole picture of why the four leading companies, Dow, Toray, Nitto Denko, and Toyobo, have been able to continue their membrane-related business activities for a long time.

M. Fujiwara and Y. Aoshima, *Mechanisms for Long-Term Innovation*,
Advances in Japanese Business and Economics 31,
https://doi.org/10.1007/978-981-19-4896-1_13

As mentioned earlier, ROGA, which led the development of the industry in its early days, was later taken over by other companies, including Gulf Oil, UOP, and Koch, in that order. Upon these acquisitions, important engineers who were active in the early days left the company. Even DuPont, which had laid the foundation for technology and business in the industry and served as a role model for later companies, withdrew from the business entirely in 1997, before the desalination market had even begun to take off. In Japan, too, there are companies such as Daicel that developed RO membranes in the early stages but later withdrew from the business. These divestments and withdrawals occurred mainly for business reasons, suggesting that the RO membrane business did not always provide stable and high profits.

The market for the semiconductor industry that developed from the latter half of the 1970s onward did indeed bring in a certain amount of revenue. However, it was not large enough to support the businesses of many companies. Although cost reductions in RO membranes continued, more time was needed to achieve the cost and performance levels required by desalination systems as infrastructure for daily life. Despite these circumstances, why were the four companies able to continue their businesses? Even though this is a business area with strong social demands, it is hardly possible for a private company to continue business activities for half a century on the grounds of social contribution. How is it that three Japanese companies were able to persevere in the development of RO membranes for such a long period of time, even though Japan receives plentiful rainfall and has no social needs or large amounts of government support?

To understand the overall development mechanism of the RO industry, it would finally be necessary to identify the reasons that enabled these particular companies to continue their businesses. To do so, we believe it is necessary to focus not only on institutional and industry-level factors, as we have done earlier, but also on company-specific factors. Therefore, in this chapter, by analyzing the individual cases of three Japanese companies, Toray, Nitto Denko, and Toyobo, we will clarify the reasons why the RO membrane business has been continuously justified within the companies and has survived for nearly half a century. Specifically, the analysis will focus on descriptions in Chaps. 7–9, which indicate the reasons that enable persistent technological development and business operations in three Japanese companies.

If the desalination market were to grow as expected, there would have been no great difficulty in making decisions regarding business continuity. Therefore, the following analysis covers the period before the 2000s, before the desalination market was fully established. We analyze the case of Toray in Sect. 13.2, Toyobo in Sect. 13.3, and Nitto Denko in Sect. 13.4. Then, in Sect. 13.5, we review the case analyses of the three companies and clarify their similarities and differences.

13.2 Toray

13.2.1 The Philosophy of Technological Superiority for Seawater Desalination

The main reason why Toray justified its continuation of its RO membrane business activities was the shared goal of actualizing the world's highest level of technology in desalination products. The actualization of seawater desalination was widely recognized as the goal of RO membrane development, not only by Toray but also by the other Japanese companies. However, Toray particularly focused intensely on development activities targeted at seawater desalination because it was the most technically difficult task.

This Toray philosophy of maintaining technological superiority was not limited to the development of RO membranes but was the basic stance of Toray's overall R&D activities at the time. The goal of developing the world's first or the world's best technology in any specific domain was emphasized in the launch of research projects. Kawabata, who developed RO membranes using cellulose acetate materials, recalls, "What Toray is all about is making things that no one in the world has ever made and making things that have the best performance worldwide."[1]

This way of thinking was evident in the approval process for the development of cellulose acetate-based membranes. As already mentioned, the ultrapure water production application for semiconductors that emerged in the latter half of the 1970s became the savior for Toray's RO membrane business. This market was captured by cellulose acetate membranes that had been commercialized earlier than composite membranes. Although cellulose acetate membranes played a significant role in securing the profitability of the business, Kawabata said developing cellulose acetate membranes must have been unacceptable in the normal process.[2] Cellulose acetate membranes are composed of general-purpose materials available in the market. It is, thus, difficult to differentiate from competitors in material development. This was not in line with Toray's goal of developing the world's best technology. Nevertheless, the development of cellulose acetate membranes was approved because developers persuaded that the development of composite membranes was already in progress and that their practical application would be ready later. The development of the cellulose acetate membrane was approved not only because it would be profitable but also as a "bridge" until the introduction of a new technology that could be "the world's first and highest level".

The development of composite membranes was the most important issue for Toray. Through demonstration tests, it was clear that it would be difficult to capture the seawater desalination market with cellulose acetate flat membranes. For Toray's development team, which considered seawater desalination to be the ultimate goal, the RO membrane business would not be possible without the development of

[1] See Chap. 7.

[2] See Chap. 7.

composite membranes. As Kawabata mentioned, if the development of composite membranes had not proceeded in parallel, the development of RO membranes using cellulose acetate materials would not have been undertaken in the first place. If composite membranes were not being developed in parallel, Toray might not have been able to capture the ultrapure water market for semiconductors at the right time.

13.2.2 Persisting in Technological Perfection and Continuing to Try New Challenges

The idea of technological superiority focused on seawater desalination can be glimpsed in the process seen in the late 1970s and 1980s, when the composite membranes PEC-1000 and PEC-2000 were abandoned by the decision of engineers.

The first composite membrane developed by Toray's development team, which was aiming for the world's best level of technology, was the polyether-based PEC-1000. As shown in Fig. 7.1 in Chap. 7, PEC-1000 was a revolutionary product that greatly surpassed the RO membranes of other companies at the time in terms of desalination rate and water permeability. It was the membrane that embodied Toray's goal of becoming the number one in the world. However, the membrane had two fatal problems when it came to seawater desalination: its weak resistance to chlorine and its vulnerability to dissolved oxygen in the sea. It might have been possible to consider applications other than seawater desalination, but Kurihara, the developer of the membrane, advised the head of research and development to stop mass production, assuming that seawater desalination was the goal. This reflects the mindset of Toray's engineers, who were adamant about the goal of seawater desalination and the pursuit of highly perfected technology. This idea was accepted by the management, and the company abandoned PEC-1000 and proceeded to develop PEC-2000, a polyamide composite membrane.

However, Kurihara advised discontinuing the development of PEC-2000 because, as in the case of PEC-1000, the low chlorine resistance of PEC-2000 was considered to be a problem when desalination was assumed to be the target goal. Its discontinuation was a serious problem because the Ehime Plant had already begun mass production and even established a production department. The technological issues with PEC-2000 were not something that had been pointed out from sources outside the company. Since factory had planned to deal with them by making improvements in the mass production stage, there was a great deal of opposition to the cancellation. Moreover, in 1984, Toray had just announced a plan to increase the sales of the membrane business to 10 billion yen in three years. Despite these circumstances, the developers refused to introduce an incomplete technology to the market, and the decision was approved by the management.

Since commercialization was halted twice with PEC-2000 being halted even at the mass production stage, the continuation of the development and commercialization of desalination technology would have been in jeopardy under normal circumstances.

In fact, Toray also had doubts about the development of desalination technology and decided to focus on the technically easier development of ultrapure water and brine for the time being. However, this did not mean that the company gave up on seawater desalination. Just as cellulose acetate membranes were recognized as a bridge to the actualization of composite membranes, composite membranes for ultrapure water and brine were also a bridge to the actualization of composite membranes for seawater desalination. The developers were working on the development of new membranes in parallel, and finally actualized UTC-70. Again, it is clear that the shared goal of establishing the world's best technology for seawater desalination was the reason why the development was continued despite numerous failures.

13.2.3 Shared Values with Top Management

The goal of capturing the seawater desalination market by developing the world's best technology was shared not only by developers but also by management. This was one of the factors that enabled Toray to persevere in technological development and continue its business activities in that area. The director of R&D from 1969, the early stage of development, was Masatoshi Ito. He was also president from 1981 to 1987, just as the commercialization of PEC-1000 and PEC-2000 was being abandoned. In addition to carbon fiber, which later became Toray's major business, Ito was also interested in the development of RO membranes from early on. The fact that Ito, who had personally embodied Toray's culture of technological pursuit as the head of the Research Center, was the president of the company is thought to have provided a tailwind that enabled the developers to continue the development of RO membranes without it collapsing in the face of numerous difficulties.

Katsunosuke Maeda, who succeeded Ito as president, also contributed to the development of the RO membrane business. Maeda was the Manager of the Ehime Plant when PEC-2000 was almost commercialized. Maeda ordered the transfer of production from Shiga to Ehime and the establishment of a mass production system for PEC-2000. Even though the market launch of PEC-2000 was canceled on the advice of Kurihara, Maeda did not stop the development of RO membranes for seawater desalination. Kurihara stated he has never heard any harsh opinions from the top that would force us to dissolve the development of RO membranes.[3] As these words reflect, management basically supported the development of RO membranes.

After the failure of PEC-2000, the development team temporarily gave up on the development of membranes for seawater desalination and developed new membranes for brine and ultrapure water. This resulted in the successful development of UTC-70 and UTC-80. However, it was Maeda that strongly encouraged the development of these membranes for seawater desalination, who had been president since 1987. In 1994, Maeda ordered developers to build a highly efficient seawater desalination system to further strengthen the desalination application. Maeda's name is on the

[3] See Chap. 7.

patent for this system, which shows the strength of his enthusiasm for desalination applications. In this way, Maeda's obsession with the difficult application of seawater desalination and his commitment to consistently raising the level of technology were shared not only by the developers but also by top management. This shared value of technological superiority was probably the reason Toray was able to continue to develop RO membrane technology and expand its business activities in that area despite the difficult business environment.

13.3 Toyobo

13.3.1 Positioning Within the Company-Wide Strategy

One of the reasons why the RO membrane business has continued at Toyobo for many years is that it was positioned as one of the new business areas expected to bear fruit in the process of changing their business structure. The RO membrane business for Toyobo started as one of the many initiatives for diversification. To cope with the severe postwar textile recession, Toyobo developed and marketed a variety of synthetic fibers from the late 1950s to the 1960s. Although this technological shift to synthetic fibers had a certain effect on the sluggishness of the textile business, it did not provide a fundamental solution to the structural problems of the industry because it soon led to overcapacity and excessive competition. Therefore, Toyobo advocated diversification and began to develop new businesses where it could utilize its own technologies, such as chemical products (Fujiwara & Aoshima, 2016). The RO membrane business also began in the early 1970s as part of this search for new businesses. The company chose the hollow fiber type as its membrane shape because it was in line with the company policy of promoting diversification by utilizing its fiber technology.

Although the developers were aware of seawater desalination applications from the early stages, they were not necessarily focused on that. Initially, they developed a system for the recovery of valuable resources and later developed applications such as seawater desalination equipment for fishing boats and pure water production in hospitals. Although these were not major businesses, they were recognized within the company as new business areas to be developed.

In the 1980s, the company started developing desalination applications in Saudi Arabia following DuPont's strategy. Toyobo conducted caravan demonstrations in 1981. In 1982, the company set the ambitious goal of quadrupling sales to 2 billion yen in three years, and the Iwakuni Plant was expanded to two lines. Toyobo succeeded in winning large orders for Haql and Duba in 1984 and for Jeddah Phase I in 1986. However, the RO membrane business was not all smooth sailing. Once the supply of modules for the large projects came to an end, demand declined, and the utilization rate of the production facilities dropped sharply. Although orders continued to come in, Saudi Arabia's financial difficulties put the company in a difficult business situation in the latter half of the 1990s.

The second half of the 1990s was a time when Toyobo decided on structural reforms that entailed a great deal of pain. As the company subsequently closed its textile plants one after another, the entire company was in a critical situation, with interest-bearing debt exceeding 340 billion yen in FY1999 and a debt-to-equity ratio of 5.39 in FY2000. Even in this difficult situation, the RO membrane business was allowed to continue, even though it did not contribute significantly to earnings. Why is this? According to Ryuzo Sakamoto, who was involved in the structural reforms and later became president, the reason had to do with the urgent need to expand into new fields while promoting structural reforms.[4] RO membranes were one of these new fields. Although there was severe pressure on business profits, it was a necessary investment to develop a company-wide strategy.

For Toyobo, diversification has been a consistent issue since the end of World War II. To achieve this, the company had to not only downsize its textile business but also grow new competitive businesses. In this sense, Toyobo was always eager to find new business areas in which it could take advantage of its strengths. By implementing structural reforms involving the closure of plants from the late 1990s, Toyobo reduced its textile sales from 60% and higher to 30% in about 10 years. Even during this difficult period, Toyobo's management team defended R&D expenses to the utmost extent. This was also based on the idea that if new business areas did not develop, the company would not be able to reform. Toyobo's RO membrane business continued to be positioned as one of the new businesses. Although it did not necessarily achieve high performance, the management did not consider it a target for business liquidation.

13.3.2 Survival by Being Inconspicuous

However, no matter how much a business is positioned in a new business area, it is difficult for management to approve its continuation if a business being in operation for more than 20 years continues to post large losses. This was even more difficult in a situation where the entire company was in a financial crisis.

Another reason why the RO membrane business was allowed to continue despite the severe structural reforms was that the management did not recognize that the business had serious problems. Two reasons the RO membrane business was not recognized as a troubled one were: (1) the resources required were minimized by drastically narrowing the business and R&D activities, and (2) it was included in the same business category as the highly profitable medical-use artificial kidney membrane business.

Although Toyobo's RO membrane business had been developed with cellulose acetate hollow fiber membranes, mainly for desalination in Saudi Arabia, in the

[4] From the authors' interview with Ryuzo Sakamoto at Toyobo's headquarters on April 8, 2016.

1980s, the development team was also searching for and developing other innovative technologies. In particular, they explored polyamide materials and established mass production technology by 1989. In the 1990s, the company developed polyamide-based hollow fiber membranes for the expanding ultrapure water use for semiconductors and even announced their introduction to the market. However, these new membranes were ultimately unsuccessful in the market.

On the other hand, from 1990 onward, Toyobo was in a severe financial situation throughout the company. By 1995, the operating profit margin had fallen to 1.3%. To protect the RO membrane business under these company-wide conditions, Fujiwara, who was in charge of the RO membrane business, took action to shrink the business thoroughly. He decided to stop the development of new membranes and concentrate on cellulose acetate hollow fiber membranes. In terms of the market, the company decided to focus on a large seawater desalination plant in Saudi Arabia. The number of employees in the division was reduced to three. As a result, the business managed to make profits in FY1998. With such a small resource input, the impact on the company performance was small. Since the business division did not stand out within the company, so management did not have to consider it a particularly problematic business to be liquidated. Looking back on that time, Fujiwara says the following.

> We had to narrow it down. We didn't have many people. There's only so much we could do. So, the most efficient thing to do when you are limited in what you can do is to take on large projects and get replacement orders. That's what you have to do. You can't do what you can't do.[5]

Another fortunate thing was that the RO membrane business was considered, as a business related to other functional membranes, to belong to the same business category as Artificial Kidney Hollow Fiber membranes (AKH). While Toray's medical membranes and seawater desalination membranes businesses are separate entities, Toyobo's hollow fiber membranes have been used in both business areas, and the two have historically been part of the same functional membrane group. The plants are also located in the same place in Iwakuni. Therefore, while the earnings of the two business areas are managed separately, the income of the entire functional membrane business is judged as a whole unit. The RO membrane business was sufficiently slimmed down and hidden under the umbrella of the strong medical membrane business. Thus, it was not seen as a particular problem by management. Regarding this point, Sakamoto states the following.

> We didn't make a special distinction between AKH for medical use and RO membranes for seawater desalination use, and we had been developing them together for a long time, historically speaking. We started out with a common technology for membranes called functional membranes, and then we found an application for it with AKH, which went well, so I think there was a sense that we could do various things with membranes. That's why we were able to continue our investment in RO research because AKH took the lead. As a result, although it took a long time, RO has managed to contribute to the gradual expansion of the industry. If there had been no application, (the development of RO membranes) would

[5] From the authors' interview with Nobuya Fujiwara at AJMC on February 2, 2016.

have been given up before it had a chance to come to fruition. I think it was fortunate we had AKH.[6]

In addition, because it shared a location with the medical membranes business, it was possible to flexibly transfer R&D and plant personnel between the business areas. Even in difficult times, the RO membrane business could survive by being recognized as one of the new business areas that would lead to the future and by hiding in the medical membrane business and saving fixed costs. Of course, the manager of RO membrane business took such survival strategy based on the anticipation of future growth. Since replacement demand can be predicted in the seawater desalination business, it is possible to revive the business once it overcomes the difficult period of the production trough. The means to overcome this difficult period was to "keep a low profile" in the case of Toyobo.

13.4 Nitto Denko

13.4.1 The Role of the Management Mission

One of the reasons why the development of RO membranes has continued and the business has developed at Nitto Denko is the role of the mission set by top management. At Nitto Denko, research on RO membranes officially began in 1973. In terms of promoting commercialization, the company-wide policy led by President Hijikata in 1974 had a significant impact. Hijikata clearly stated that the company's three pillars of future growth would be electronics, medical care, and membranes. This policy was followed in subsequent years. The company-wide policy called for a "membrane division" in general and did not specifically mention RO membranes or seawater desalination. Unlike Toray, Nitto Denko did not target seawater desalination that was only one application among many of its membrane. At the time Nitto Denko started its membrane business, neither the application nor the technology was clear. The mission of "membranes" came first. The development of technologies and the search for applications were carried out to actualize that mission. As Yamamoto, who was in charge of the membrane business at the time, recalled "No one knew what applications would grow in the future. It turned out to be a project that simply had the word 'membrane' in the name..."[7]

The management mission set out by Hijikata came into play when Nitto Denko decided to expand overseas in earnest by acquiring Hydranautics. Nitto Denko's decision to acquire Hydranautics came at a time when the newly constructed Shiga Plant was unable to operate due to the semiconductor recession. There were months

[6] From the authors' interview with Ryuzo Sakamoto on June 16, 2016 in Osaka. The text in parentheses was added by the authors.

[7] See Chap. 9.

when sales were only approximately 20 million yen, and the factory staff was doing nothing but prototyping, painting and weeding.

How was this acquisition possible? Yamamoto, who became the general manager of the business division at the time, said "There was a limit to turn membranes into a business as long as we stay in Japan. I thought that it would never take off in the way that people expected. So, I pondered the question of what we should do to actualize membrane business expectations. The acquisition of Hydranautics in 1987 was the answer."[8] Kamai, who became president in 1985, followed the policy of his predecessor Hijikata and was persuaded by Yamamoto's words that if the membrane business was to continue, an acquisition was the only way.[9]

The acquisition of Hydranautics was a major decision that determined the future development of Nitto Denko's RO membrane business. The company-wide policy laid down by Hijikata had an impact on management decision-making 13 years later.

13.4.2 A Thorough Search for Applications

Even though the top management had set the growth of the membrane business as a mission, it would be difficult to continue if it did not make the sales figures. In particular, at that time, Nitto Denko was a medium-sized company, and earnings pressure within the company was so strong that it could not afford to hold on to a loss-making business for a long time. Therefore, those who were entrusted with business development had to search for every possible application. If they found that there was some use for the product and found a customer, they could continue to develop the technology for the time being. As technological development progresses, new applications will come to light. Nitto Denko's RO membrane business in its early days was based on this search for new applications. To find a potential market for anything, developers went to customers to develop applications. In some cases, the membrane engineers even designed the equipment. Even if the characteristics of the membranes created in the development process were not necessarily what they had intended, they enthusiastically pursued market development based on the idea that there might be an application utilizing their ideas.

For example, in the food industry, the company participated in the Food Industry Membrane Utilization Technology Research Association, which was established in 1982, and worked with Ajinomoto to develop new applications. In addition, the company did not limit itself to RO membranes but developed a "department store of membranes" by supplying tubular UF membranes for use in electrical equipment coating, waste liquid treatment, and human waste treatment.

Discussion meetings with customers were also used to develop applications. Yamamoto held technical exchange meetings with Kurita on a regular basis as part of its activities to explore a wide range of applications. The application of NTR-7250,

---

[8] See Chap. 9.

[9] From the authors' interview with Hideki Yamamoto on December 12, 2014 in Osaka.

which had an inferior desalination rate but excellent water permeability and TOC removal performance, to semiconductors was a product of this rich exchange with Kurita.

In this way, the early RO membrane business at Nitto Denko was characterized by the development of a diverse application market without limiting either technology or application. At Nitto Denko, the corporate culture of "Sanshin Katsudo" (Three New Activities) is rooted in the creation of new businesses in the three directions of new product development, new application development, and new demand creation. This corporate culture has encouraged developers to develop various applications. The new applications developed as a result illuminated the future potential of the business and justified the continuation of business activities and technological development, even if the company was not yet generating sufficient profits.

13.5 Unique Rationale of Each Company

13.5.1 Reasons for Business Survival: The Role of a Shared Mission and Shared Values

In the middle stage of new industrial development, both technology and markets are still immature, and it often happens that companies cannot secure sufficient profits from their business activities. Because of the uncertainty of technology and markets, it is often difficult to explain the rationale for continuing to invest in a business on an objective basis. However, even under such circumstances, the development of an industry will not proceed unless companies invest in technological development and market development, researchers and engineers devote their efforts to technological advancement, and business managers are diligent in searching for new markets. This is also the case in the RO membrane industry. The targeted market for seawater desalination did not grow as expected due to the cost inferiority to the evaporation method and the financial difficulties of customers. The market for ultrapure water for semiconductors, which was the only market that brought in revenue, was not a market that could support the scale of business of major companies in the future. Looking back on the unstable history of the RO membrane business, there must have been many occasions when it was difficult to convincingly explain the continuation of business investment to management and investors on objective and economically rational grounds. Nevertheless, the three Japanese companies persevered and continued to invest in technology development and business development, eventually winning the global seawater desalination market. Why was this possible?

The above analysis shows that in the process of development of the RO membrane business, a mechanism peculiar to each of the three companies was working, which supported the business for reasons different from economic rationality in the strict sense.

At Toray, the idea of technological superiority, of actualizing the world's best technology, was shared not only by developers but also by management. This functioned as a criterion for business selection that differed from the simple pursuit of profits. This is why Toray's developers were obsessed with seawater desalination from start to finish.

Toyobo's RO membrane business has been justified in the context of diversification, which has been a longstanding issue since the end of World War II. The company adhered to hollow fiber in terms of technology from the standpoint of creating a new business by utilizing its own technology. The development of RO membranes, which started with the aim of diversification, justified the survival of the endeavor because RO membranes were considered to be on the side of growth rather than being liquidated, even in the midst of severe financial conditions.

At Nitto Denko, the company-wide policy formulated by the president had a major influence in justifying the RO membrane business. Although Nitto Denko did not have any noteworthy proprietary technology or a clear market, the shared policy of making membranes the pillar of the business supported the RO membrane business. This encouraged developers to explore applications based on "Sanshin Katsudo" and led to the bold decision to acquire Hydranautics even under a difficult business situation.

As we can see, the reasons that supported the continuation of the RO membrane business were different for each of the three companies. Each company had its own unique mechanism for justifying the business. However, all three companies are similar in that a common set of values and business policies, which were shared throughout the company, supported their business activities. Toray shared the concept of "the world's number one technology". Toyobo shared the issue of "diversification". Nitto Denko shared the company-wide policy of making "membranes" a pillar of its business. Beyond the criteria of direct economic rationality, there were criteria for selecting projects, which were shared as a superordinate concept.

In the case of new business development with high uncertainty, especially in the case of new business development that requires long-term and patient development such as the RO membrane, judging technology development or business development investment only by objective economic criteria may lead to underinvestment. In the case of new business investment with high uncertainty, it is difficult to show high economic rationality (profitability) in advance, but the rationality may turn out later. If this is the case, the key to business development is how to support new business development by criteria other than economic rationality in the preliminary stage. One such criterion is the company-wide mission and values that have been fostered and shared within the company over its long history. Since company-wide missions and values are unique to each company, it is impossible to propose a general proposition as to what kind of values and mission definitions will promote business development. However, we can say in general terms that each company has its own specific value criteria other than economic rationality and that by appealing to these criteria, continuous business development becomes possible.

13.5.2 The Importance of Justification Efforts by Developers

Even if the shared mission and value standards give legitimacy to new business development, as long as the company is a for-profit company, if it cannot even show the potential as a for-profit business, it will naturally not be able to continue to receive investment in technological and business development, and the business endeavor will disappear. Such efforts were observed in the case studies of the three companies.

Despite repeated failures, Toray's R&D activities of composite membranes have continued because of its value of technological superiority. However, it was not only the company value that contributed to the continuation of the R&D. It was the ultrapure water application for semiconductors that saved Toray's sluggish RO membrane business in the 1980s, and it was the cellulose acetate-based RO membrane that first brought profits to the RO membrane business by capturing the market in a timely manner. The fact that cellulose acetate-based RO membranes took the lead in the ultrapure water market led to the continued development of composite membranes, which had repeatedly experienced failures, and led to their full-scale commercialization. However, to obtain approval for the development of cellulose acetate-based membranes, it was necessary to show that the company had a composite membrane in reserve. The presence of composite membranes embodies the value of technological superiority and the development of cellulose acetate membranes, which was approved in a manner that did not conflict with this value, won demand for semiconductor manufacturing. In doing so, the company demonstrated the potential of the RO membrane business. The combination of these two factors was important for the continuation of the business.

In the case of Toyobo, too, the RO membrane business was not justified by diversification alone. In the face of the need for painful company-wide structural reforms, those in charge of the RO membrane business were determined to show the viability of the business. To achieve this, they focused on specific areas of technology and markets, drastically reduced the number of personnel, and hid under the umbrella of the medical membrane business. By positioning the business in the context of diversification and demonstrating to management that it could be viable as a for-profit business, the business division was able to survive the difficult period leading up to the next leap forward without becoming a target for liquidation. The combination of the mission and the potential as a for-profit business was also important here.

The same was true for Nitto Denko. Although it was important for the continuation of technological and business development that membranes were positioned as a pillar of the company, that alone was not enough for a medium-sized company such as Nitto Denko to continue a non-profitable business. Therefore, in the initial stages, the people in charge of the project conducted a thorough search for applications. Instead of focusing on specific applications, such as seawater desalination, the developers themselves sought out businesses that they thought would be profitable. Early market development played an important role in demonstrating that the technology had the potential to become a profitable business.

Table 13.1 Each company's rationale for continuing the RO membrane business

	Toray	Toyobo	Nitto Denko
Shared values and mission that support the activity	The value of technological superiority	Diversification	Membranes as a pillar of business
Justification efforts by the developers	The combination of old and new technologies	The reduction of input resources	The development of application markets

Source Prepared by the authors

Thus, each of the three companies had its own standards for business judgment that were not based on objective economic rationality alone. At the same time, there was ingenuity and hard work on the part of those in charge of creating a "track record" that would at least indicate that the business had the potential to be viable. Table 13.1 shows the combination of these factors.

Appendix: Differences in Development Areas Between Toray and Nitto Denko

How does the difference in the reasons for justifying development activities influence the organization's innovation activities? As a stepping stone to understanding this, we focus on Toray and Nitto Denko, which were in the same polyamide flat membrane group but had contrasting reasons to support their development.

Toray had the ambition of actualizing seawater desalination with the world's best technology. Therefore, the first priority was to improve the performance of membrane materials to a level acceptable for seawater desalination. Thus, it is estimated that the development of downstream technologies, such as modularization technology and operation technology for RO membranes, tended to lag behind the development of materials.

In contrast, for Nitto Denko, which advocated being a department store for membranes, it was important to establish a membrane business that was not limited to seawater desalination. It is thus estimated that development in downstream areas was promoted at an earlier stage than for Toray.

These differences are confirmed by the patent data in Figs. A.1 and A.2. In this graph, the patents acquired by Toray and Nitto Denko are classified into three categories, namely, membranes, modules, and operations, in the same manner as in Chap. 12, and the trends in these categories are compared.

Looking at Toray, the ratio of membrane development has exceeded 60% since the 1970s and further increased to nearly 80% in the 1980s. Since the 1990s, when the practical application of membranes for seawater desalination started progressing, the ratio of membrane-related patents has been decreasing. However, the fact that the percentage of patents related to membranes still exceeds 50% is noteworthy. This

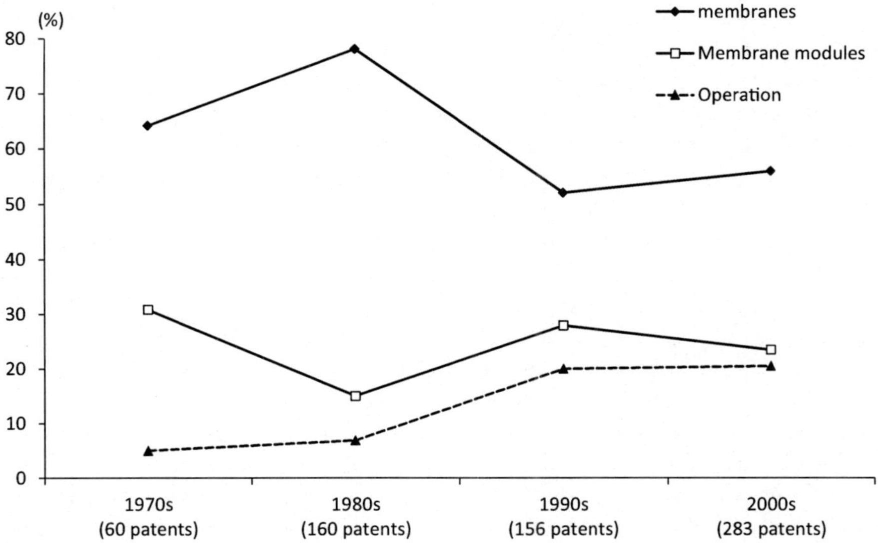

Fig. A.1 Share of patent applications by technological domain (Toray). *Source* Calculated by the authors based on data from UltraPatent

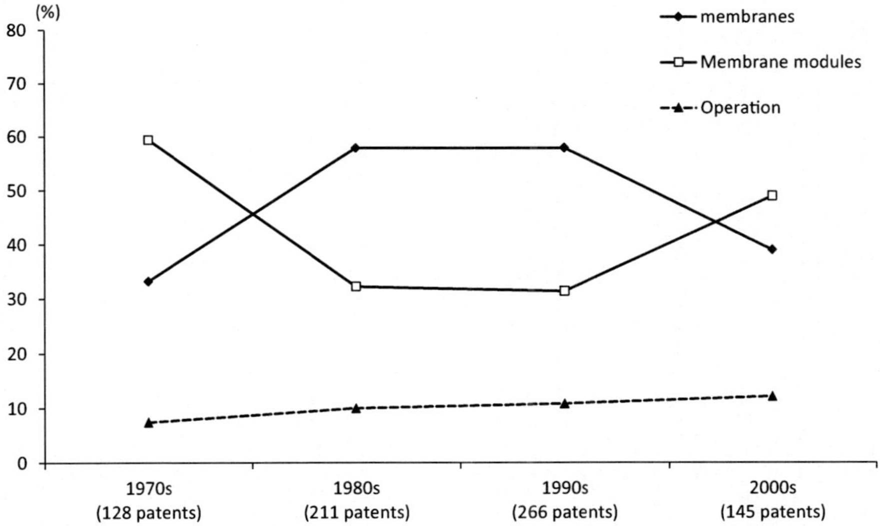

Fig. A.2 Share of patent applications by technological domain (Nitto Denko). *Source* Calculated by the authors based on data from UltraPatent

suggests that Toray is continuing to focus on material development. This is consistent with the company's stance of focusing on seawater desalination and pursuing the world's best technology as a shortcut to justify development.

Nitto Denko shows quite a contrasting trend with Toray. The graph shows that at **Nitto Denko, development in the membrane module and operation fields proceeds at** a high rate as well as the development of the membrane itself. The ratio of patents in the area of membrane modules exceeded that in the area of membranes in the 1970s, and the ratio in the area of membrane modules remained the highest in the 2000s.[10]

Although the above is only the result of an experimental survey, it suggests that there was a gap in the areas of subsequent development between Toray, which started **membrane development with the aim of achieving the highest level of technical diffi-**culty, and Nitto Denko, which began development with the intention of creating a new membrane business. This implies that the way in which the reasons for development are created within the company will determine the subsequent development activities and direction, and eventually the characteristics of the products introduced by each company and the markets to which they are introduced will also change. **This suggests that there might be a path in which the reason for development defines** the direction of development and the strategy that lies ahead.

Reference

Fujiwara, M., & Aoshima, Y. (2016). Toyobo: Promoting fundamental corporate reforms. *Hitotsubashi Business Review, 64*, 124–141.

[10] This transition is a unique feature of Nitto Denko, not seen in Toray or even in Toyobo.

Chapter 14
Long-Term Development Mechanisms Under Uncertainty

Abstract In this chapter, based on the results of the analysis in the previous chapters, we present a descriptive model to conceptually understand the process of long-term continuation of new technology and new business development under four uncertainties. The chapter develops an explanation that the four factors, government policy support, emergence of breakthrough technologies, initial markets, and company-specific rationales, acted in a complementary manner to each other to enable the continuation of the development of new technologies and new business fields that could not be actualized with a simple profit motive. We develop an explanation of how this has been achieved.

14.1 An Integrated Model for Continuous Technology Development

14.1.1 Factors that Enabled the Continuation of Development

Why are firms able to continue long-term technology and business development in the face of high uncertainty? In the previous chapters, we have explored the reasons why firms have been able to sustain their development activities for a long period of time by focusing on four aspects of this question: policy support (Chap. 10), initial market development (Chap. 11), emergence of breakthrough technologies and convergence of technological approaches (Chap. 12), and company-specific rationales (Chap. 13). In this chapter, following those specific analyses, we present a model that integrally describes the mechanisms that make long-term development possible.

As discussed in the introduction, sustaining new technology and new business development activities are difficult because they are subject to a high degree of uncertainty with respect to the technology and the market. Uncertainty refers to a state in which the probability of an event resulting from a decision is unknown or the event itself is unknown (Knight, 1921). In the case of technological development, it is a situation in which it is difficult to estimate in advance the probability of achieving the

M. Fujiwara and Y. Aoshima, *Mechanisms for Long-Term Innovation*,
Advances in Japanese Business and Economics 31,
https://doi.org/10.1007/978-981-19-4896-1_14

target performance requirements, or in which it is impossible to predict technological issues that will arise in the future. In the case of business development, it is difficult to estimate in advance how large the target market will be or to know where the market will be.

When a company develops a new technology or a new market that has never existed before, it is often accompanied by these uncertainties that become barriers to continue its innovation activities. If the probability of success, whether in technology development or business development, is known in advance, even if it is low, companies can make a decision on the appropriate investment of resources in innovation activities after considering the risks. However, in the case of a completely new technology or new business area, where past experience cannot be used as a reference, it is often impossible to estimate even the probability of success. As it is known that people tend to avoid events where the probability is unknown,[1] uncertainty-averse firms may hesitate to invest management resources and development efforts in innovation activities (Kellogg, 2014; Leahy & Whited, 1996). As a result, there is a risk of underinvestment in innovation activities.

Under these circumstances, how can firms mobilize resources for innovation activities? One of the things that became clear from the analysis in the previous chapters was the importance of "expectations" formed within industries and companies, such as "this technology seems promising," "this seems to have marketability," and "this seems to be profitable," even if the probability of success is not known. Expectations here refer to subjective estimates of the likelihood of achieving an objective, formed under uncertainty where the probability of an event occurring is not known. These expectations may be qualitative and ambiguous, but they have provided guidance to companies in making investment decisions for new technology and new business development.

For example, Cadotte's '344 patent overcame multiple technical challenges simultaneously and provided a direction for solving the challenges of composite membrane technology, which led to the convergence of various technical approaches in one direction. The "solution" that the '344 patent indicated not only decreased technological uncertainty, but also have dramatically increased the developers' subjective estimates for the feasibility of their development target because of its novelty.

The unintended emergence of semiconductor manufacturing applications has increased the estimated market potential of RO membranes. Not only has semiconductor application itself become a profitable business in the immediate future, but it has also raised expectations for the future expansion of the potential market. In addition, policy support in the early stages of the industry increased the profitability of the development activities and raised expectations of the future potential of the desalination technology and market with the endorsement of the government. All of these factors may have increased the subjective profit expectations of the companies and enabled them to continue their technological and business development.

[1] This is famously known as Ellsberg's (1961).

Independent of these factors affecting earnings expectations, there were other factors that supported innovation activities. One was the company-specific rationale and reasons that justified innovation activities. For Toray, it was "pursuit of the world's best technology". For Toyobo, it was "the structural reform of diversification". For Nitto Denko, it was "establishment of a membrane business". These were not necessarily directly linked to immediate earnings expectations, but they provided a rationale that supported the continuation of technology and business development from within.

Another factor was the social label that the RO membrane business carried. In other words, the possibility of solving the important social issue of water shortage alone was part of the reason why companies continued to develop RO membranes. Of course, a company seeking to make a profit cannot continue to operate at a loss endlessly. However, fulfilling social responsibility is also a necessary activity for a company to survive. From this point of view, the social nature of RO membranes provided a reason for companies to continue their business, even if only partially.

14.1.2 The Basic Model

A rational company would judge the amount of resources to be invested in technological development by the magnitude of the expectation of profit to be generated by it. The relationship between resource input and profit expectations can be further broken down into three relationships: (1) resource input and technology level, (2) technology level and created value, and (3) created value and profit level, as shown below. These relationships are illustrated in Fig. 14.1.

According to Fig. 14.1, companies first estimate the extent to which technological problems will be overcome by investing resources in technological development and the extent to which the level of technology will improve as a result. The company then predicts the size of the market that will be explored by the technological solution. In other words, the company estimates how much additional value it can provide to its customers by improving the level of technology and how much of a customer base it can acquire as a result. Third, the company estimates how much profit it can earn from the value created. This estimate will vary depending on the competitive conditions in the industry and the competitive advantage of the company. Finally, the company will determine the amount of resources to be invested in technology development according to the magnitude of the estimated expectations. Most of the profits obtained in the early stage of development may be negative. Companies will invest resources in development even if they continue to be in the red for a certain period of time because of the expectation of future profits.

This relationship between resource input and profit expectations is shown in Fig. 14.2.

The horizontal axis of the graph from the origin to the right shows the "cumulative resource input". Normally, the cumulative resource input increases as the company

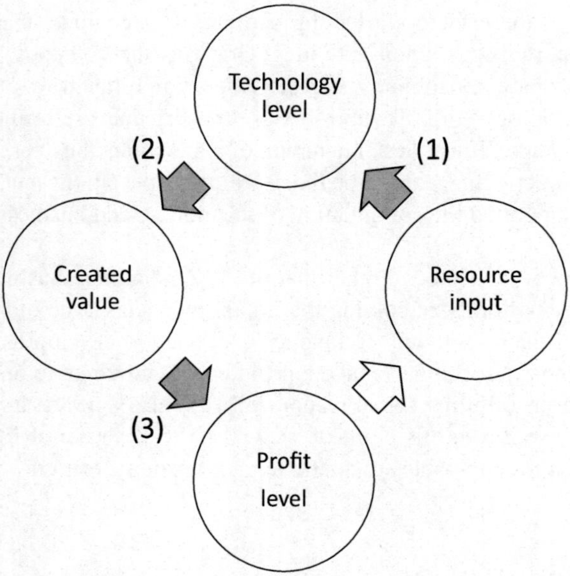

Fig. 14.1 Relationships between resource input into technology development and profit expectations. *Source* Prepared by the authors

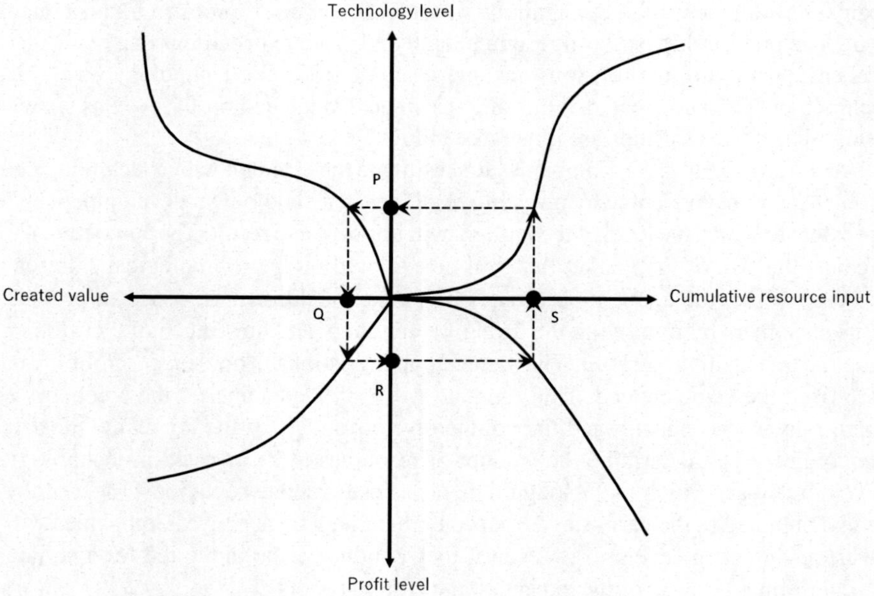

Fig. 14.2 Basic model. *Source* Prepared by the authors

progresses from the R&D stage to the commercialization stage. To advance development activities, actualize innovation, and achieve targeted business expansion, continuous resource mobilization must be enabled to move to the right along this horizontal axis. Second, the vertical axis moving upward from the origin is the "technology level". Normally, the technological level should be expected to improve as the cumulative resource input increases through developer efforts. Third, the horizontal axis moving from the origin to the left shows the magnitude of the "value created". Here, the created value indicates the magnitude of the value that the technology creates for the customer. This can be rephrased as the sum of the customer's willingness to pay for the technology. Under normal circumstances, the larger the market to which the technology is applied, the larger the value created. Fourth, the vertical axis moving downward from the origin is the "profit level" of the company. This can be expressed as the possibility of the value created by the technology being allocated to the company. The lower the axis, the higher the expected profit.

Each quadrant in Fig. 14.2 has a graph line drawn on it, which represents a hypothetical relationship between the corresponding vertical and horizontal axes. The line in the first quadrant indicates that the more the cumulative amount of resources is invested and the more the development progresses, the higher the technological level becomes. Here, an S-shape is drawn in the first quadrant in line with Foster's model, which holds that the relationship between the amount of cumulative resources and the level of technology is an S-shape (Foster, 1986). According to Foster, in the initial stage of development where the direction of technological development is not determined and trial and error are continued, the technological level does not improve easily despite the resources invested. However, once the period of trial and error is over and the direction of problem solving is agreed upon, the efforts of the developers are concentrated in a specific direction, and the technological level improves rapidly as the development efficiency increases. In the latter half of technological development, however, the technology approaches its inherent physical limit, and the rate of improvement of the technological level relative to the input of development resources diminishes. The existence of such S-curve of technology has been clarified in several industries. The shape of the S-curve is drawn a posteriori in the entire industry as a result of the development efforts of the firms, and it cannot be precisely identified in advance by individual firms (Christensen, 1992). However, the path of technological development indicated by the S-curve should be an important clue in estimating the degree to which development efforts lead to the improvement of technological levels.

The line in the second quadrant is also drawn as an S-shape from the level of technology on the vertical axis to the value created on the horizontal axis. This S-shape reflects the results of a series of studies that have clarified the emergence of innovation from the demand side (Adner, 2002, 2011; Adner and Leventhal, 2001). First, no matter how much the technological level improves, value cannot be created for customers while the technological level is low and even the minimum performance required by customers is not achieved. This minimum level of performance required by the customer is called the functional threshold (Adner, 2002). In the case of RO membranes for seawater desalination, if the desalination rate is only 50 or 70%

and the membrane cannot be used in practical applications, the functional threshold has not been reached. However, once the technology level exceeds the functional threshold, it begins to provide value to customers. Then the value created increases rapidly as the technology level improves, and the market enters a period of expansion.

However, as technological development progresses further, customers' willingness to pay for the improvement in technological level gradually diminishes (Meyer and Jonson, 1995). The creation of value is no longer proportional to the improvement in the technological level. When the technology level exceeds the level required by the average customer, which is to say, when what is known as performance overshoot occurs, the customer is not willing to pay additional money for further improvement of the technology level. This situation is generally referred to as commoditization. The S-shape in the second quadrant represents such a change in the demand-side evaluation of technology.

The line on the graph in the third quadrant shows that the greater the value created, the greater the profit expected by the company will be. However, the relationship is not linear. The graph shows that returns gradually diminish as the value created increases. This reflects the fact that competition intensifies as industries develop. If the market expands and the value created becomes larger, the possibility for firms to earn profits also increases. However, the more attractive the market becomes, the more new firms are expected to enter the market. As a result, profits will not be proportional to the expansion of the market. It has often been observed that as industries mature and technologies become more generalized, firms are exposed to fierce market competition and, as a result, become less profitable.

Finally, the line in the fourth quadrant shows a relationship in which the higher the expected level of profit, the more additional resources are mobilized and the higher the cumulative resource input will be. However, the relationship is not linear. As the graph shows, it is assumed that the amount of additional resources mobilized gradually diminishes in relation to the profit expectation. In the early stage of development, it is not unusual for a business to remain in the red because the market does not blossom. Even in such a situation, if the deficit is within a certain range, the company may be willing to invest in development in anticipation of the future. However, such a situation will not last forever. Eventually, a reasonable level of profit will be required to justify the mobilization of additional resources, and the required level of profit will increase. As industrial development progresses and businesses mature, even if they are generating profits as "cash cow", they will not be able to demonstrate further growth potential, and investment in the development of technologies and businesses will not be easily approved. The curve in the graph in the fourth quadrant shows such no-linear relationship between profit levels and resource mobilization.

A situation where there is no uncertainty refers to a situation where the decision maker knows completely in advance the relationship between the vertical and horizontal axes, as indicated by the curves of the graph in each quadrant. For example, as shown in Fig. 14.2, it is a situation where the decision maker knows that if the cumulative resource input reaches S, the technology level will be P, the value created will be Q, and the profit level will be R. As long as profit level R is consistent with the corporate goal, the company will invest resources in technological development up to the level of S.

14.1.3 Four Uncertainties

However, in the early stages of technological development, it is extremely difficult to know in advance how an increased development effort will lead to profitability. In particular, in the development of completely new technologies, it is often impossible to grasp the relationship between the vertical and horizontal axes, even in terms of probability, because there is no precedent in the past. Development activities are exposed to a high degree of uncertainty.

Uncertainties faced by innovation activities can be divided into several types, depending on their sources. We categorize them into four types: (1) technological uncertainty, (2) customer uncertainty, (3) competition uncertainty, and (4) social and organizational uncertainty.

According to Jalonen (2012), who analyzed 101 previous papers that dealt with uncertainty in the innovation process, the most common uncertainties addressed by existing studies were technological uncertainty and market uncertainty. Similarly, McGrath and MacMillan (2000) also identified technological and market uncertainty as uncertainties that new business projects should address. Existing research tends to treat technological uncertainty as a lack of knowledge about the details of a new technology. This implies a lack of understanding of natural laws. Some studies also refer to technological uncertainty as uncertainty resulting from a lack of knowledge about the use of the new technology.

On the other hand, market uncertainty has been treated as uncertainty caused by the inability to foresee changes in the relationship with customers or the inability to foresee the nature of relationships with competitors. The former may be called customer uncertainty, and the latter is called competition uncertainty.

If we incorporate these three types of uncertainty—technological uncertainty, customer uncertainty, and competition uncertainty—into the basic model shown in Fig. 14.2, we can express them as graphs that show the relationship between the vertical and horizontal axes in the first, second, and third quadrants, respectively. This is shown in Fig. 14.3.

The first uncertainty, "technological uncertainty", indicates that the relationship between the cumulative resource input and the technological level in the first quadrant is unknown. It refers to a situation in which it is not known whether the technological problems can be solved as planned or a situation in which unexpected technological issues may arise, when development resources are invested.

The second uncertainty, "customer uncertainty", means that the relationship between the technological level and the value created in the second quadrant is unknown. It refers to a situation where it is not known whether customers will accept the technology even if the technological problem is solved. As already mentioned, in the initial stage of an industry, even if the technological level is improved, it may

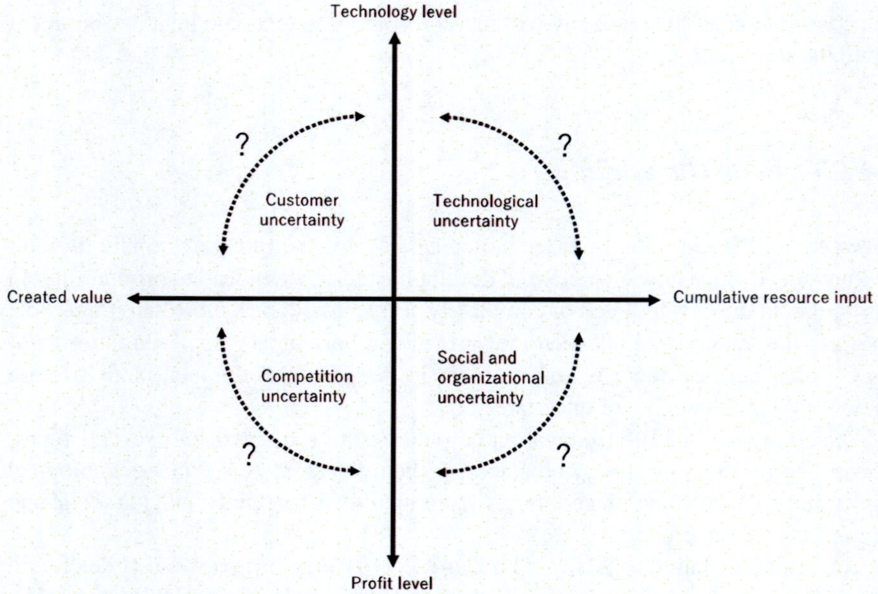

Fig. 14.3 Four uncertainties. *Source* Prepared by the authors

not reach the functional threshold required by the customer, and the value may not be recognized at all. On the other hand, when the technological level transcends the customer's requirements in the mature stage, further performance improvement may not bring value to the customer (Adner, 2011; Adner and Leventhal, 2001; Christensen, 1997). One aspect of market uncertainty is the difficulty in predicting the timing of such changes in the relationship between technology and value.

The third uncertainty, "competition uncertainty", stems from a lack of prior knowledge about the relationship between the value created and the level of profit. Even if the technology and the products and services are highly evaluated by customers, the company may not be able to make profits because of the market price decline under fierce competition. On the other hand, even if the value created is small, the company can secure its own profit if it can maintain a monopolistic position. Uncertainty of competition refers to a situation in which the future competitive environment cannot be grasped in advance and the possibility of capturing the value as profits cannot be seen.[2]

[2] Companies looking to generate revenue from material innovations such as RO membranes need to understand that the competitive structure differs depending on the application. For example, both Nitto Denko and Toray have been able to secure high profitability in the semiconductor industry despite the small market, but it is presumed that they are struggling to gain revenue in the seawater desalination market. The reason for the difference in profitability is the difference in the competitive structure of the two markets. Specifically, there are differences in the customer's industry in terms of (1) the rivalry between companies, (2) the scale of purchasing volume, (3) the performance contribution to customer value, and (4) the capacity to pay. Since municipalities, which are customers in the seawater desalination market, may not compete fiercely with each other, it is unlikely that RO

In addition to these three types of uncertainties, we also focus on the fourth type of uncertainty: social and organizational uncertainty. This is the uncertainty about the relationship between the level of profits and the cumulative amount of resources mobilized, as shown in the fourth quadrant of Fig. 14.3. There are various institutional environments and social relations inside and outside the corporate organization. They affect the resource input to innovation activities for reasons different from pure economic rationality. Therefore, the degree to which the level of profit influences the amount of resources invested in technological development will differ depending on the institutions and social relations surrounding the firm.

For example, if a company is under strong pressure from the capital market and has a high hurdle to profitability, it cannot continue to invest in new technology unless it can expect a reasonably high profit. In contrast, a company that is less dependent on the capital market and can tolerate low profitability may be able to invest more resources for its profit expectation. Some companies are daring to refrain from going public to preserve their investment freedom.

Moreover, firms do not always make economically rational decisions. There may be new projects that, due to political conflicts within the organization, are not allowed to invest resources even with high profit expectations (Pettigrew, 1973; Thomas, 1994). Conversely, there are projects that are continuously developed even if high profitability is not expected due to unique corporate philosophy and social responsibility (Takeishi, et al., 2012).

Not all investment actions can be accepted by society just because profit expectations are high. For example, some technologies, such as genetic modification and cloning technologies, face headwinds from an ethical standpoint. Conversely, there are cases where technological development is promoted by social tailwinds even if it is not highly profitable. As seen in the recent CSR and CSV movements, companies are expected not only to conduct efficient economic activities but also to solve social issues. A number of recent studies have shown that socially responsible corporate behavior leads to higher corporate value in capital markets (Becchetti et al., 2012; Malik, 2015; Peloza & Shang, 2011; Van Beurden & Gössling, 2008), and addressing social issues is becoming increasingly important from an economic perspective. Thus, the extent to which profit levels lead to resource input depends on the institutional and social environment inside and outside the organization. Uncertainty increases to the extent that it cannot be grasped or predicted in advance. Such uncertainty is called social and organizational uncertainty here.

---

membrane companies can take advantage of competition among customers in price negotiations. In addition, since the scale of purchase volume is much larger for seawater desalination plants, this tends to increase the pressure for price discounting. In terms of performance contribution, ultrapure water is an important input that affects semiconductor defects, so it will be difficult for customers to exert pressure for discounts. Furthermore, in the 1980s, Japanese semiconductor companies were making large profits on the back of industry growth and had plenty of money to spend. In contrast, local governments are in dire financial straits, and their ability to pay is not high. Although some countries are blessed with oil resources and are financially well-off, the demand for water is generally high in developing countries where the ability to pay is low. Thus, the price pressure of the same RO membrane differs depending on its application, and companies will need to take this into account when developing their business.

Of the above four types of uncertainty, technology, customers, and competition can be regarded as uncertainties related to the relationship between input resources and profit levels. If a company is unable to develop a practical technology, if the expected customers do not exist, or if unexpectedly fierce competition arises, the company will not be able to earn sufficient profits. Without clarifying these situations, companies cannot make appropriate decisions about investing additional resources. These three uncertainties are based on the profit motive.

In contrast, social and organizational uncertainty is uncertainty about factors beyond the profit motive (what factors other than the profit motive justify resource input, and how) as well as uncertainty about the profit motive itself (how much profit makes additional resource input acceptable).

If the organization can be regarded as a monolithic, unified decision-making entity with no internal contradictions, there is no uncertainty in the internal organization. In reality, however, not all organizational members share the same way of thinking. It is not uncommon for developers, who promote innovation, and managers, who make decisions on resource mobilization, to differ in their thinking. Therefore, organizational uncertainty is a serious challenge that must be overcome for innovation proponents.

14.1.4 Virtuous and Vicious Cycles of Innovation

While it is difficult for a company to decide on a large-scale investment in the presence of various uncertainties, it is necessary to engage in new technology development to create new businesses. Therefore, companies will usually make small-scale investments in a number of exploratory developments in a diversified manner.

If such initial development investment decisions are successful and the technological development efforts of the developers bring about the expected improvement of technological level and, as a result, customers show a willingness to purchase the product and their expectations of profit are raised, the company will be able to make further additional investments based on the experience. In other words, if the virtuous cycle of "resource input → technology level → created value → profit level" continues, the company's innovation activities will proceed successfully, new business development will succeed, and new businesses will grow. This is a case of smooth sailing.

Figure 14.4 shows such a virtuous cycle as a graph. The key is whether the profit, R_1, obtained as a result of the input of resources, S_1, in the first cycle in this figure is commensurate with the input of resources, S_1. The desirable experience in the first cycle provides the basis for the investment decision in the second cycle, and the desirable experience in the second cycle provides the basis for the investment decision in the third cycle, which in turn justifies progressively larger development investments. In other words, the experience gained from the previous cycle provides new hypotheses about the relationship between the vertical and horizontal axes of each quadrant, and the resource input for the next cycle is determined based on these

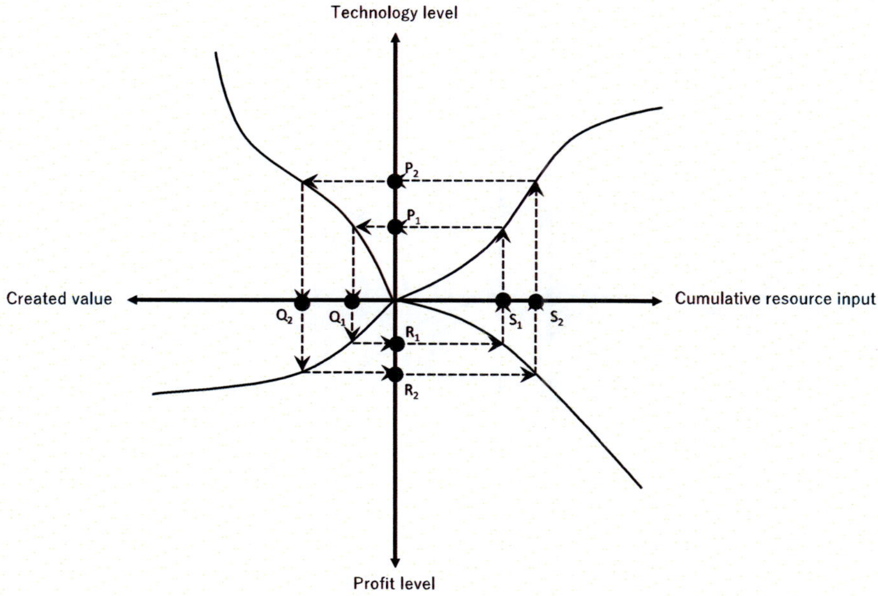

Fig. 14.4 Virtuous circle of innovation. *Source* Prepared by the authors

hypotheses. If the cycle proceeds smoothly, the development of new technologies and businesses will advance with a gradual increase in the input of resources, and the company will finally be able to establish its new target business.

However, it is rather rare that actual innovation activities proceed in such a successful manner. It frequently happens that the initial technological development does not proceed as planned and does not lead to the solution of technical problems. It is also often observed that the expected customers are not interested in the technology even when the technology level has been improved. In addition, it is not uncommon for a company to be exposed to unexpected competition even though the market has grown and to fall into a situation in which no profit remains for the company. In this way, when the investment of resources under uncertainty does not produce the expected results, the experience has a negative impact on the next investment decision, making it difficult to invest additional resources. This situation is illustrated in Fig. 14.5.

Figure 14.5 shows a situation in which the cumulative resource input of S_n improves the technological level to P_n, which creates value Q_n and leads to profit R_n. This is a case where the technological level did not improve as much as expected compared to the case in Fig. 14.4. Here, if the relationship between the profit level and the cumulative resource input is shown as a line in the graph in the fourth quadrant, then only the cumulative resource input of S_{n+1} can be accepted at the profit level of R_n. This means that additional resource input up to the level of S_n, which is necessary to advance technological development to the next stage, is not possible.

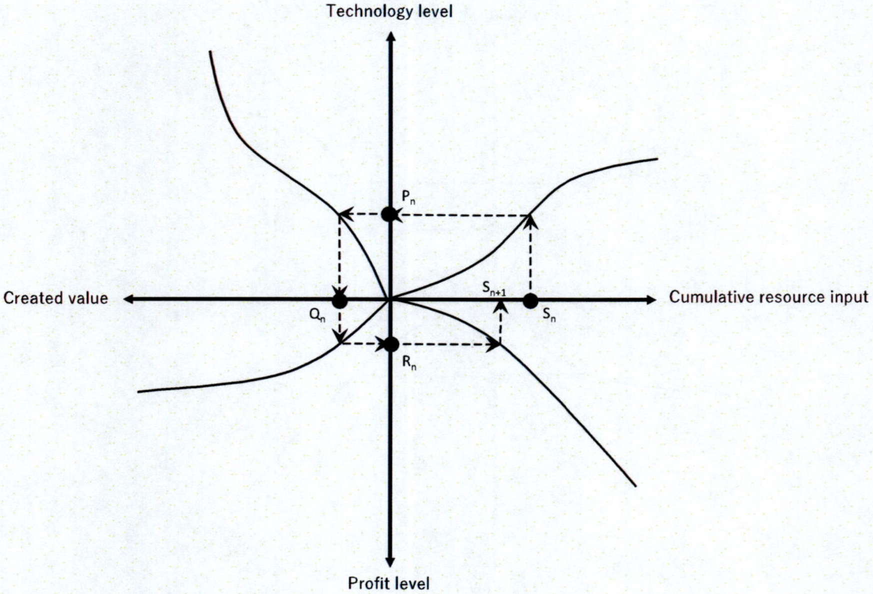

Fig. 14.5 Crisis faced by new technology and business development. *Source* Prepared by the authors

If this happens, firms may have to reduce the development of new technology or, in some cases, decide to withdraw from development.

New technology and business development often tend to fall into such a situation, and as a result, innovation activities can often face the danger of stopping. How can a company continue to develop despite such a situation?

The mechanism that makes this possible is the action of forces that change the situation of the above four uncertainties. The effect of these forces is to shift the expectations of firms toward a favorable direction in the four relationships: "resource input and technology level", "technology level and created value", "created value and profit level", and "profit level and resource input". The forces primarily acting on these relationships are "policy support" (Chap. 10), "initial markets" (Chap. 11), "emergence of dominant design" (Chap. 12), and "company-specific rationales" (Chap. 13), which we have discussed thus far.

As explained below, in the early RO industry, the government support in the presence of competitive and customer uncertainties raised the subjective profit expectations of companies and supported continued effort into new technology development. The new application markets that emerged under customer uncertainty encouraged continued development by raising expectations of the value created by the technology. In addition, the emergence of breakthrough technologies reflected in the '344 patent under technological uncertainty justified the continuation of further resource investment by raising the subjective expectations of companies regarding the possibility of solving difficult technological problems. Furthermore, the rationales specific to each firm discussed in Chap. 13 shifted the relationship between profit level and resource input in a direction favorable to continued development under social and organizational uncertainty. Firms had specific, historically cultivated philosophies that enabled them to mobilize additional resources even in the face of inadequate profits, as long as their development activities resonated with those philosophies.

14.1.5 Changing Expectations Under Uncertainty

As mentioned at the beginning of this chapter, uncertainty refers to a situation in which the probability of an event occurring as the result of a decision is unknown or in which the event itself is unknown. It is a situation in which "objective knowledge or empirical rules of causality" are lacking.

In such a situation, if companies invest their efforts in technological development and the causal relationships among technological elements and physical laws gradually become clear, technological uncertainty will be reduced. In addition, if the technology is introduced to the market as a product or service and interaction with customers and competing companies proceeds, the uncertainty in the market will be reduced. When uncertainty is reduced in this way, firms can make decisions about whether to invest resources.

However, the reduction of uncertainty does not always lead to the continuation of resource input. If the reduction of uncertainty reveals that additional development resources will not meet the company's profit targets, then innovation activities will be reduced or come to an end.

In contrast, even in a situation where uncertainty is high and the possible outcomes cannot be predicted accurately in terms of probability based on the accumulated knowledge and empirical rules, if the probability is "expressed in language as 'fairly high', 'fairly low', or 'fairly well'" (Takemura et al., 2004) and if an event occurs in which the expectation of the outcome is higher than before, such as, "it is probably going to happen this way," or, "it can be expected to go this way," additional resource input for development could be justified.

To clarify the role of expectation formation under uncertainty here, the position of each is conceptually depicted in Fig. 14.6(1)–(3). A state of high uncertainty is a situation where the relationship between variable x (e.g., cumulative resource input) and variable y (e.g., technology level) is not known and a graph cannot be drawn, as shown in (1). As the development activity progresses, the relationship between x and y gradually becomes clear as the values of variable y obtained in response to variable x are empirically accumulated, as shown in (2), and the uncertainty is reduced.

In contrast, the role played by expectations can be depicted as a change in the slope of the dashed graph shown in (3), i.e., a change in the expected value of the outcome. This is a case in which the relationship between x and y is not clear, but it is subjectively estimated that some event will cause a change in the relationship. If the slope of the graph is estimated to shift counterclockwise, then the resulting additional resource input would be justified.

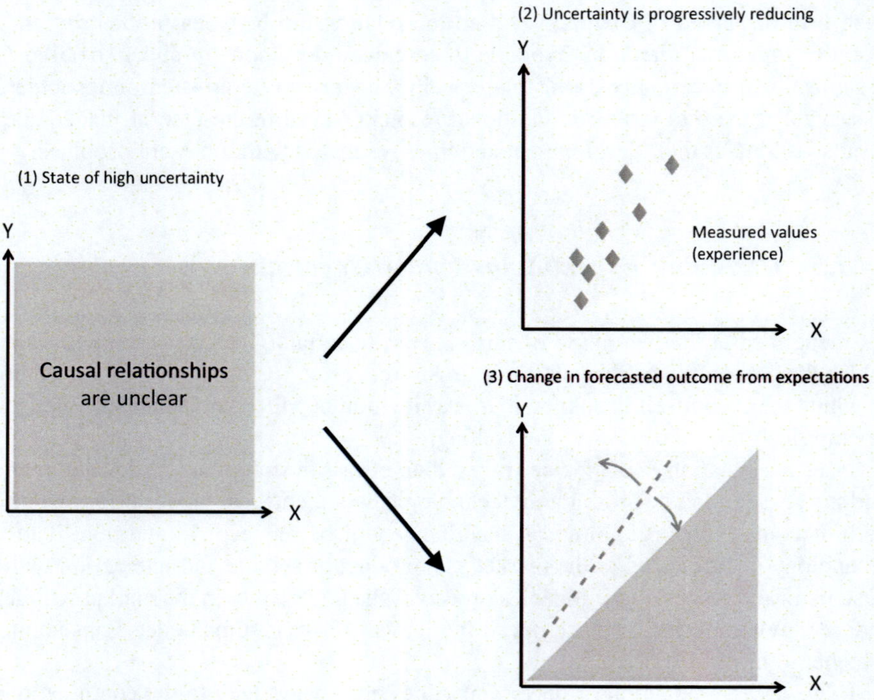

Fig. 14.6 The role of expectations under uncertainty. *Source* Prepared by the authors

14.1.6 Mechanisms to Continue Development Activities Under Uncertainty

When the initial investment in technological development does not produce sufficient results, market development is not achieved, and the profit level is not sufficient to justify further resource investment; then, innovation activities face the danger of discontinuation. For those who promote innovation in a company to overcome this situation, it is necessary for the expectations of the decision-makers in charge of resource mobilization to change in a favorable direction in relation to the vertical and horizontal axes of each quadrant in Fig. 14.5. In other words, a force that rotates the line of the graph expected in each quadrant counterclockwise is necessary.

The slope of the graph in the first quadrant, which represents the decision maker's expectation of the relationship between cumulative resource input and technology level, can change counterclockwise due to the emergence of dominant design and the convergence of technology approaches discussed in Chap. 12. As the emergence of dominant design means that promising technological approaches become clearer, developers will be able to narrow their development targets and use their development resources more efficiently than in the past. In addition, as technological approaches

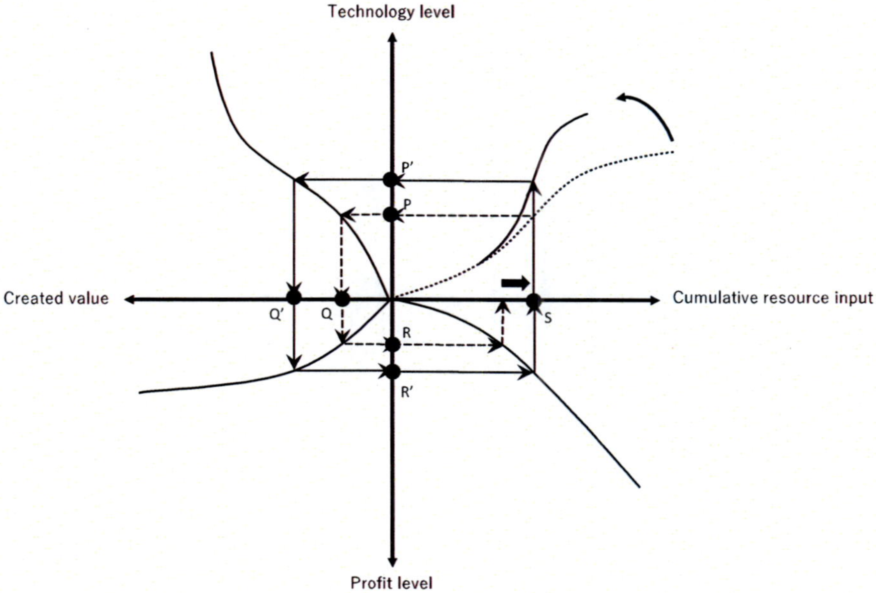

Fig. 14.7 Impact of the convergence of technological approaches. *Source* Prepared by the authors

are narrowed within the industry, developers will be able to refer to each other's development results, and mutual learning through technological spillover will increase. As a result of these external effects, decision-makers will believe that they can achieve development results more efficiently than before, even if the amount of resources invested remains unchanged. Such a change in expectations makes the slope of the graph rotate counterclockwise, as shown by the change from the dashed line to the straight line in the first quadrant of Fig. 14.7, and the expected technological level shifts from P to P′. As a result, the level of profit R′ commensurate with the amount of resource input S is expected, and the development can proceed to the next stage.

The slope of the graph in the second quadrant, which shows the expectation of the decision maker for the relationship between the level of technology and the value created, can change counterclockwise depending on the development of the initial market analyzed in Chap. 11. Even if the initial market to be developed is small in scale, the value expected to be created by the technology in the future is large because it suggests the future expansion of the market to which the technology will be applied. The discovery of a new range of applications raises the evaluation of the technology's potential and raises the estimate of its value, even if it is far from the initially targeted main market. This situation is depicted in Fig. 14.8 as a shift in the value created from Q to Q′, and as a result, the profit level R′ is expected to be commensurate with the resource input S.

The slope of the graph in the third quadrant, which shows the decision maker's subjective expectations of the relationship between the value created and the level of profits, could change counterclockwise with policy support. For example, if the

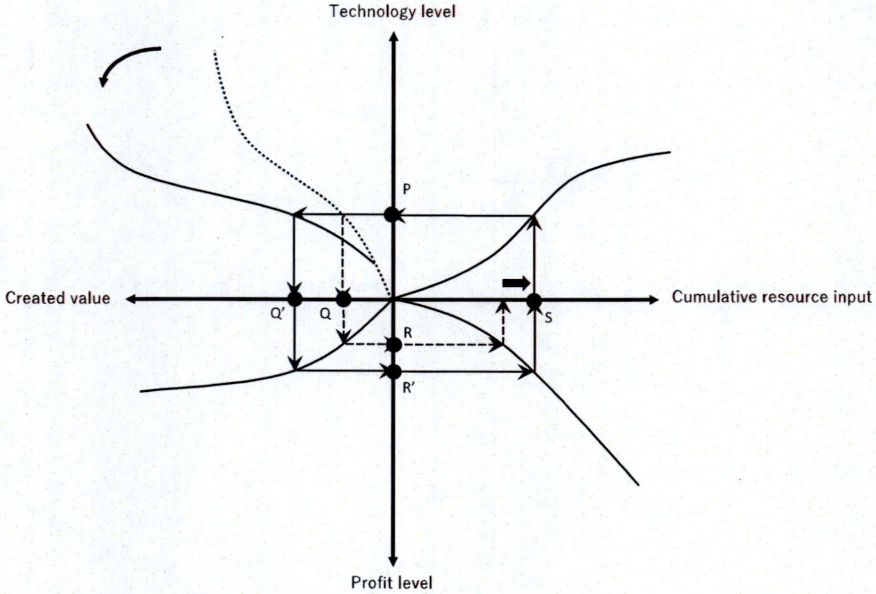

Fig. 14.8 Impact of initial market. *Source* Prepared by the authors

government takes policies that lead to entry restrictions, the profit expectations of existing firms will increase. In addition, subsidies for development activities may also increase the relative competitive advantage of subsidized firms, which may partially have a mitigating effect on competition and change the slope of the graph counterclockwise.

Furthermore, the injection of public funds reduces the investment burden of enterprises, lowers development costs, and thus increases their profits. In particular, since at least part of the R&D results of publicly subsidized projects are made public, companies can indirectly enjoy the technological knowledge of other companies. This is also expected to reduce the company's own development costs and increase the assumed profit level. Figure 14.9 shows how government support shifts the level of profit expected by the company to a profit level R′ that is commensurate with the amount of resource input S, and the development moves forward to the next stage.

Conversely, if the government's competition policy leads to excessive competition in the market, the graph will clockwise. Even if there is no change in the expected value of technology, the level of profit expected by companies will decline. We will discuss this point again in the next chapter.

Finally, the graph in the fourth quadrant shows the decision maker's expectation of profit level and cumulative resource input. Its slope can vary counterclockwise according to the company-specific rationales analyzed in Chap. 13. In the early stages of industrial development, the benefits derived from technology are often insufficient to justify further development investment. Even in such cases, however, company-specific rationales beyond mere economic rationality may drive development. If

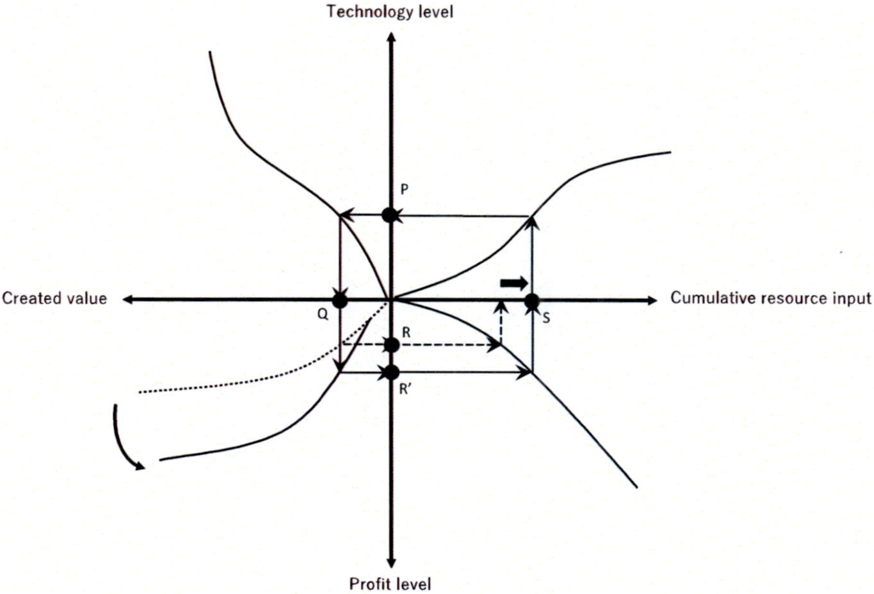

Fig. 14.9 Impact of policy support. *Source* Prepared by the authors

innovation activities are consistent with such company-specific rationales, they may induce further development investment even if they do not seem to generate high profits. In addition, socially responsible technologies such as RO membranes may be subject to more generous resource allocation even if the expected profit level is the same because companies and their employees find value in fulfilling their social responsibilities. Figure 14.10 illustrates how the slope of the graph changes counterclockwise due to conformity to the company's specific rationales and the social reasons, and how the amount of resource input S becomes commensurate with the amount of resource input and development can be continued.

Whether or not a company continuously invests resources in development activities depends on the multiplication of the slopes of the graphs in the four quadrants shown above. In reality, a single factor that changes the slope of the graphs in each quadrant is not sufficient to justify the resource input, and as shown in Fig. 14.11, the changes in the four slopes act in a complementary manner to enable the next additional resource input. Even if the change in the slope of each quadrant is small, this figure shows that the combination of these factors enables the input of resources necessary for continued development.

The meaning of the slopes of the graphs in the four quadrants may be expressed as "efficiency of technology development" (quadrant 1), "marginal market value of the technology" (quadrant 2), "ability of the firm to capture value" (quadrant 3), and "sensitivity to non-economic factors" (quadrant 4). As shown in the following equation, the amount of allowable additional resource input is defined as a result

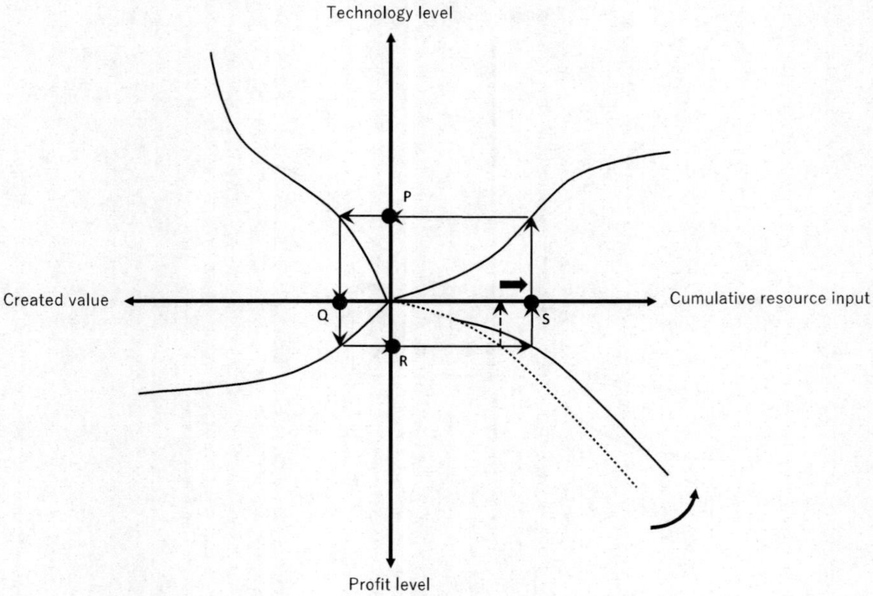

Fig. 14.10 Impact of society and company-specific rationales. *Source* Prepared by the authors

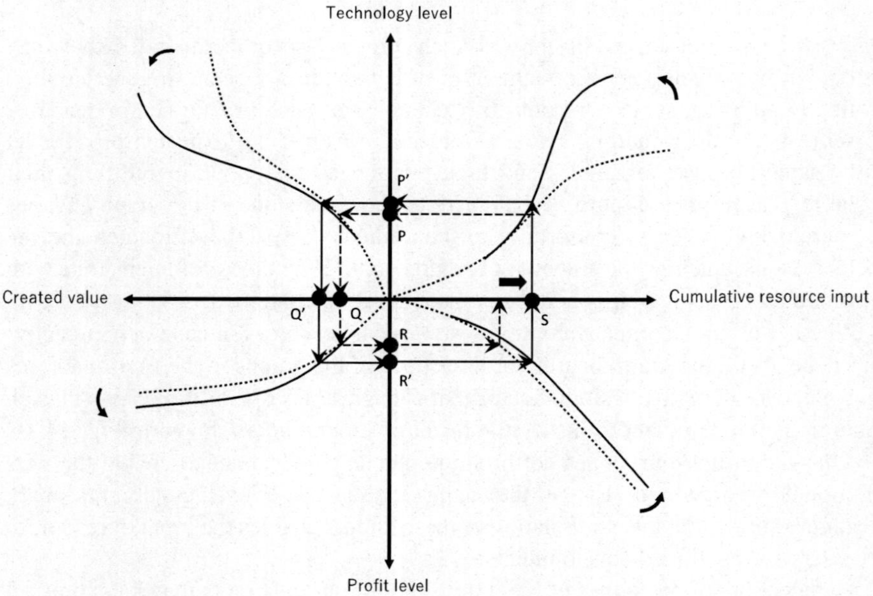

Fig. 14.11 Complementary effects of the four factors. *Source* Prepared by the authors

of the synthesis of these four factors. As long as the amount of the allowed additional resource input exceeds the amount of resources required by the developer, the continuation of development will be actualized.

Allowable additional resource input =
f(efficiency of technology development, marginal market value of the technology, ability of the firm to capture value, sensitivity to non-economic factors)

Thus far, we have explained the forces acting on the four factors corresponding to each quadrant of the graph, starting from the first quadrant, as "convergence of technological approaches", "search for initial markets", "policy support", and "society and company-specific rationales". In reality, however, these four forces operate across the other quadrants as well. For example, policy support not only has the effect of raising company profit expectations but also increases their expectations of future market expansion and may raise their expectations of efficient improvements in technology levels through spillover effects from other companies. The convergence of technological approaches not only increases the level of technology but also contributes to the increase in profit expectations by reducing the duplication of development activities. In addition, the emergence of breakthrough technologies may have the effect of raising expectations regarding future markets, just as the emergence of the '344 patent raised the prospect of practicality in desalination. Furthermore, apart from the expectation of improved economic efficiency, the fact that a dominant design has emerged may itself increase the legitimacy of resource mobilization.

Thus, the four forces described so far can have diverse effects. What is particularly important is that they play complementary roles in altering expectations about the slope of the graphs in the four quadrants and establishing the justification for additional resource input.

In the next section, in line with the model presented above, we interpret the mechanisms that enabled the continuous development of RO membrane companies.[3]

14.2 A Historical Interpretation of Continuous Development Under Uncertainty

14.2.1 The Impact of Early Policy Support in the United States

The development of RO membranes began at UCLA in response to the social problem of water shortages in Los Angeles, where the population was growing rapidly. Although the development of RO membranes was later pursued outside of California, no membrane material, which would demonstrate desalination performance

[3] We refer to actions across different quadrants in the next section.

for practical use, could be found until the invention of the L-S membrane at UCLA. Since the level of technology was not high enough to create a market, companies were reluctant to invest in development because they did not have enough profit expectations to justify the investment of resources. The active development support by the U.S. government was the key to breaking through this situation.

In the United States, the Saline Water Conversion Act was enacted in 1952, and governmental support by the OSW was promoted thereafter. In particular, as a result of a significant increase in public support under the Kennedy administration, many players started desalination research in the early 1960s, when the commercial potential of desalination technology was almost unknown. Dow, which was responsible for subsequent industrial development, also received government support for its development early on. DuPont, which had mainly developed internally, also began to receive support from the OSW in the 1970s. Furthermore, the Johnson administration, which succeeded the Kennedy administration, focused attention on desalination as a peaceful use of nuclear energy, which stimulated military companies such as Aerojet and GA to study desalination.

Such government support can be directly expressed as a change in the slope of the graph in the third quadrant, as shown in Fig. 14.9. When the cost of development decreases due to government support, the level of profit expected by companies increases. In addition, military companies, which are considered to have received preferential support from the government for the peaceful use of nuclear energy, may have thought that they could establish a market position as pioneers in desalination technology. These factors can be interpreted as a counterclockwise shift of the graph in the third quadrant, whereby the level of profit expected in relation to the value created increased so that the profit level became commensurate with the amount of resource input, and the power to proceed to the next stage of development was obtained.

The large-scale support by the U.S. government is thought to have had the effect of creating public interest in seawater desalination and raising the expectations of companies regarding future market expansion. This can be seen as a change in the slope of the graph in the second quadrant, as shown in Fig. 14.8. Furthermore, since a certain level of public disclosure is imposed on the results of government-supported research, spillover and mutual learning within the industry are expected to increase. This will lead to an increase in R&D efficiency for firms, which will be reflected in the change in the slope of the graph in the first quadrant shown in Fig. 14.7. Thus, the large-scale support by the U.S. government seen in the early stage of industrial development can be interpreted as supporting early technological development by enabling firms to invest resources under uncertainty through multiple reasons.

14.2.2 Early Development by Japanese Firms: U.S. Influence and Company-Specific Reasons

The huge amount of desalination support by the U.S. government not only had a direct impact on U.S. companies and research institutes but also indirectly affected the development behavior of Japanese companies. For Japanese companies that had followed American companies such as DuPont and Dow, the mere fact that those American companies were enthusiastically developing the technology may have been enough to justify their initial development investment, as it led to expectations of future market expansion.

Japanese public institutions also indirectly benefited from the U.S. government's support, which in turn influenced the development behavior of Japanese companies. Ishizaka of the GCIRIT, who supported Japan's desalination research, saw the U.S. desalination research on site and found a high affinity with abortive salt production research. As a successor to salt production research, Ishizaka became the leader of a large-scale project on seawater desalination. As a result of this series of events, the WRPC was established, and Japanese companies were able to conduct useful demonstration tests there. Through these tests, Toray learned the performance of PEC-1000, Nitto Denko learned the limitations of the tubular type, and Toyobo learned the practicality of the new kind of membrane. By utilizing the WRPC, Japanese companies were able to determine the possibility of solving their technical problems and set the direction of their subsequent development.

The initial development by Japanese companies triggered by the growing momentum of development in the U.S. can be understood as changes in the graphs in the second and third quadrants, as shown in Figs. 14.8 and 14.9. The increase in expectations for the future of the desalination market due to the information on the development trends in the United States can be understood as the change in the slope of the graph in the second quadrant. In addition, the demonstration tests using the WRPC dispelled some of the uncertainty about the relationship between the resource input and the level of technology and justified the resource input by raising expectations for the resolution of technical issues. At the same time, for the three companies, Toray, Nitto Denko, and Toyobo, participation in the national demonstration test directly reduced the development cost and suggested the establishment of a competitive advantage over other companies that were not selected for the demonstration test, which in turn raised the expectation of the level of profit to be obtained in the future. This is expressed as a change in the slope of the graph in the third quadrant.

Thus, although early-stage development in Japanese companies progressed under the influence of the U.S., promising markets still did not emerge, earnings were not expected to be sufficient to establish a business base, and the further investment of resources could be easily actualized. Even under these circumstances, the rationales unique to each company played an important role in continuing development without interruption. In the case of Toray, this was a technology-oriented philosophy of actualizing the world's best technology and doing it first. At Nitto Denko, it was realizing membranes business. At Toyobo, it was technology-related diversification. Each company's specific justification was critical to promote highly uncertain innovation activities (Takeishi et al., 2012).

By referring to such reasons specific to each company, it becomes possible to increase the amount of resources invested in development even if the profit level does not necessarily improve. Figure 14.10 depicts that the addition of company-specific reasons is expressed as a change in the slope of the graph in the fourth quadrant, and as a result, the additional resources required to continue development are justified.

Another factor that favorably affected the relationship between the level of profit and the amount of resource input is the social nature assigned to the RO membrane technology. In the case of a technology with a high public interest nature, the cause of solving social needs or social issues can be a justification for its development. The public interest nature of RO membranes is thought to have been one of the factors that encouraged the development of the technology in a situation where no immediate profit could be expected. This positive effect of the social nature on the continuation of development can be expressed as a change in the slope of the graph in the fourth quadrant, just as in company-specific reasoning.

14.2.3 Development of Application Markets

Despite the initial development supported by U.S. influence and the public support for demonstration testing, as well as company-specific rationales and social imperatives, the system fell far short of achieving the performance required by the desalination market, its original target, and took longer than expected to improve performance. Government support ceased, and the effect of indirect influence from the US disappeared. Although company-specific rationales might continue to operate, it is not acceptable for a commercial enterprise to invest resources endlessly in businesses with low profit expectations. In particular, as development progresses and the amount of development resources required increases, it becomes impossible to break through using only company-specific rationales.

In the midst of this period of uncertainty, developers in RO membrane companies were very interested in developing application markets that would allow them to apply their immature technology to maintain business viability for the time being. To sustain a highly public business that was not expected to be profitable immediately, it was important to search for an initial market where the existence and development potential of the market could be confirmed. In fact, although the technical hurdles for seawater desalination were high, it was possible to utilize the immature technology for other applications. For this reason, early RO membranes were developed for use in liquid waste and wastewater from factories, and in the 1970s, food concentration and medical applications were pioneered. In the 1980s, Japanese semiconductor companies, which were engaged in fierce competition in miniaturization, introduced RO membranes, even at high prices, to ensure high performance and the quality of their products. This gave the companies time to wait for the desalination market to blossom.

The effect of the development of such an application market is shown as the change in the slope of the graph in the second quadrant seen in Fig. 14.8. The emergence of a new applied market led to the expectation of the development of not only that market but also other various applied markets. As a result, even if the level of technology is the same, the expectation of the value created by the technology increases, and the level of profit considered by the company increases correspondingly, which have justified the introduction of additional resources for technology development. In particular, the expansion of demand for semiconductors made a significant contribution to the investment of additional resources in Toray and Nitto Denko.

14.2.4 The Convergence of Technological Approaches

The development of new application markets made an important contribution to the continuation of development in the late 1970s and throughout the 1980s. However, the size of those markets was too small to support the operations of large companies, and that alone did not guaranteed continued development. In the 1970s, various technologies, including asymmetric and composite membranes, were still being tested. There was no definitive technology that could convince the public of performance improvement for the full-scale expansion of desalination. Investing development resources for a long period of time without a clear direction for technology development is not an easy task, especially for a major company.

It was in this context that the industry saw the advent of Cadotte's '344 patent in 1981, which succeeded in achieving a variety of difficult-to-compare performances, all at a high level. It was a true breakthrough, a discontinuous jump in the course of the advancement of RO membrane technology.

With the appearance of the '344 patent, the superiority of polyamide-based flat composite membranes based on interfacial polymerization became apparent to the industry, and this gave a firm direction to the development of each company. Toray, which had been developing cellulose acetate-based membranes in parallel, shifted its focus to flat RO membranes using new polyamide materials in the 1980s. In 1985, Dow acquired FilmTec and took it under its umbrella and shifted to polyamide-based flat membranes. Nitto Denko, in addition to the results of its own development, concentrated on polyamide flat membranes by acquiring Hydranautics. This shift to polyamide-based materials was accompanied by a move toward the dominant flat membrane type as a membrane shape. After a patent dispute, it was decided that the '344 patent belonged to the U.S. government, and companies began to adopt Cadotte technology, leading to a rapid convergence in the technological approaches pursued by the industry.

The convergence of technological approaches triggered by the emergence of the '344 patent led to the reduction of uncertainty concerning the relationship between the input of development resources and the improvement of the technological level.

Companies were then able to concentrate their input resources on specific areas, implying an improvement in development efficiency that have led to the change in the slope of the graph in the first quadrant shown in Fig. 14.7. As a result, the expected profit with the same level of resource input increases, and the additional resources input can be justified.

Furthermore, once polyamide flat composite membranes are agreed upon by competing companies as the dominant design, technology spillover between companies is more likely to occur. Since learning in the industry is accelerated by companies referring to each other's development, companies can expect to reach a higher technological level with the same resource input. As a result, the graph in the first quadrant, which shows the expectation of the relationship between the amount of resource input and the level of technology, tilts further counterclockwise, which in turn justifies the additional resource input by firms.

However, the convergence of technological approaches due to the emergence of dominant design will result in the homogenization of technology among companies in the future. While this has the advantage of accelerating industrial development by stimulating performance and cost competition, the intensification of competition caused by homogenization also reduces the revenue that companies can earn from market expansion. This is indicated by the clockwise tilt of the graph in the third quadrant. As a result, the future profits that a firm can expect from the same customer base decrease, making it more difficult for the firm to invest additional resources. Existing studies on the relationship between competition and innovation point out that excessive competition inhibits product innovation (Beneito et al., 2015; Vives, 2008), and one rationale behind this is illustrated by this figure.

The above is an interpretation of how the development of RO membranes has continued and how the industry has been created and developed according to the framework of the basic model shown in Fig. 14.2.

14.3 Conclusion

Chapters 10–13 analyzed the mechanisms that enabled continuous new technology and business development activities in the RO membrane industry from four perspectives: government support, development of application markets, emergence of dominant designs, and company-specific reasons. In this chapter, we took these four perspectives and presented an integrated model to describe the mechanisms that enable innovation activities to continue under uncertainty. Uncertainties faced by innovation activities were classified into four categories: "technological uncertainty", "customer uncertainty", "competition uncertainty", and "social and organizational uncertainty", and it was argued that the continuation of innovation activities required dealing with each uncertainty. We argued that the factors influencing the subjective expectations plays an important role in coping with these uncertainties. These factors are the four factors analyzed in Chaps. 10–13.

The model presented in this chapter suggests that for innovation activities to continue, responses to different uncertainties are necessary, and that the action of complementary forces that affect them is key. This and other theoretical, practical, and policy implications of the analysis in this book are discussed in the next chapter.

References

Adner, R. (2002). When are technologies disruptive? A demand-based view of the emergence of competition. *Strategic Management Journal, 23*, 667–688.

Adner, R. (2011). *The wide lens: What successful innovators see that others miss*. Portfolio/Penguin.

Adner, R., & Levinthal, D. (2001). Demand heterogeneity and technology evolution: Implications for product and process innovation. *Management Science, 47*, 611–628.

Becchetti, L., Ciciretti, R., & Kobeissi, N. (2012). Corporate social responsibility and shareholder's value. *Journal of Business Research, 65*, 1628–1635.

Beneito, P., Coscollá-Girona, P., Rochina-Barrachina, M., & Sanchis, A. (2015). Competitive pressure and innovation at the firm level. *Journal of Industrial Economics, 63*, 422–457.

Christensen, C. M. (1997). *The innovator's Dilemma: When new technologies cause great firms to fail*. Harvard Business School Press.

Ellsberg, D. (1961). Risk, ambiguity, and the savage axioms. *Quarterly Journal of Economics, 75*, 643–669.

Foster, R. (1986). *The S-curve: A new forecasting tool*. Macmillan.

Jalonen, H. (2012). The uncertainty of innovation: A systematic review of the literature. *Journal of Management Research, 4*, 1–47.

Kellogg, R. (2014). The effect of uncertainty on investment: evidence from texas oil drilling. *American Economic Review, 104*, 1698–1734.

Knight, F. H. (1921). *Risk, uncertainty and profit*. Houghton Mifflin.

Leahy, J., & Whited, T. (1996). The effect of uncertainty on investment: some stylized facts. *Journal of Money, Credit and Banking, 28*, 64–83.

Malik, M. (2015). Value-enhancing capabilities of CSR: A brief review of contemporary literature. *Journal of Business Ethics, 127*, 419–438.

McGrath, R. G., & McMillian, I. C. (2000). *The Entrepreneurial mindset: Strategies for continuously creating opportunities in an age of uncertainty*. Harvard Business Press.

Meyer, R., & Johnson, E. (1995). Empirical generalization in the modeling of consumer choice. *Marketing Science, 14*, 180–189.

Peloza, J., & Shang, J. (2011). How can corporate social responsibility activities create value for stakeholders? A systematic review. *Journal of the Academy of Marketing Science, 39*, 117–135.

Pettigrew, A. (1973). *The politics of organizational decision-making*. Tavistock.

Takeishi, A., Aoshima, Y., & Karube, M. (2012). *Reasons for innovation: Creative legitimacy for resource mobilization*. Yuhikaku.

Takemura, K., Kikkawa, T., & Fujii, S. (2004). Taxonomy of uncertainties and risk evaluation: A proposal of the theoretical framework. *Sociotechnica, 2*, 12–20.

Thomas, R. (1994). *What machines can't do: Politics and technology in the industrial enterprise*. University of California Press.

Van Beurden, P., & Gössling, T. (2008). The worth of values: A literature review on the relation between corporate social and financial performance. *Journal of Business Ethics, 82*, 407–424.

Vives, X. (2008). Innovation and competitive pressure. *Journal of Industrial Economics, 56*, 419–469.

Chapter 15
Contributions and Future Research

Abstract In this chapter, we identify the contributions and implications of the book and then construct arguments for future research. One of the arguments concerns the contradiction between industry development and company profitability. Groundbreaking breakthroughs that led to the emergence of dominant designs, the convergence of technological approaches in the industry, and the creation of de facto standards for element shapes have certainly contributed to industrial development. However, they also led to fierce competition for homogeneity, which squeezed the profitability of companies. The question of how the balance between the convergence and standardization of technological approaches and the profitability of individual firms affects industrial development is beyond the scope of this book, but it is an important theme. In addition, the role of the commercial market as an experimentation ground and the possibility that the specific reasons justifying development within a company influence the direction of technological development are also presented as a possible future research theme.

15.1 Summary of this Book

15.1.1 Research Questions and Background

In this book, we have analyzed the development and commercialization process in the RO membrane industry based on the research question, "Why is it possible to continue new technology development and new business development for a long time under high uncertainty?".

It was only in the 2000s that RO membranes began to be used earnestly in the seawater desalination market, which had been the target from the beginning. Nearly 50 years had passed since development began in the United States, and more than 30 years had already passed since Japanese companies began developing. During that time, RO membrane companies faced a high degree of uncertainty in both technology and market development. For the purpose of supplying domestic water, the cost performance must be at least comparable to that of the evaporation method, if not as

M. Fujiwara and Y. Aoshima, *Mechanisms for Long-Term Innovation*,
Advances in Japanese Business and Economics 31,
https://doi.org/10.1007/978-981-19-4896-1_15

low as the filtration of river water or rainwater. The questions were: Is this technically feasible? Even if it is technically feasible, will there be a sufficient market for it? In wealthy Middle Eastern countries where the evaporation method had been the mainstream, a certain amount of alternative markets may be expected. However, many of the regions suffering from water shortages are in countries with a limited ability to pay. Therefore, the demand for RO plants, including in the Middle Eastern countries, will depend largely on the policies of each country, and therein lies much that cannot be predicted in advance.

Due to the nature of the industry, the history of the RO membrane industry has seen a variety of corporate consolidations, including mergers, acquisitions and withdrawals. On the other hand, however, the four companies that have led the way in developing the current market, Dow, Toray, Nitto Denko, and Toyobo, have consistently continued to develop technologies and cultivate business areas since the early stages of the industry. Why have they been able to continue to do so? This was the question pursued in this book.

The RO membrane is an innovation that was envisioned as a solution to the social problem of water shortage. Not only RO membranes but also other innovations with a social nature have been attracting particular attention in recent years due to environmental problems and population growth accompanying the development of the world economy. However, innovations that solve social problems are generally not viable as businesses and do not always generate high prices from customers due to their public mission.

On the other hand, the technological development of companies are becoming increasingly short-sighted. In an environment of high uncertainty, where it is difficult to foresee the future, it may be natural for companies to choose a sure-fire path with clear business potential. However, this makes it difficult for companies to promote innovations that will solve future social issues.

Thus, while the need for innovation that solves social issues is growing, it seems difficult for private companies, which are subject to profit pressure, to take on this task. So what should be done about it? This is the question that lies behind the analysis in this book.

The phenomenon of for-profit companies continuing to support technological and business development where high uncertainty exists in both technology and market, and where it takes decades for an industry to blossom is unlikely to be adequately explained by ordinary profit motives and market competition. This book, therefore, attempts to explain this phenomenon from a multifaceted perspective.

15.1.2 The Logic Beyond the Economic System

What has become clear from the analysis in this book is the importance of the complementary role of the operating rationales of various systems beyond the economic system. Innovation over time has been driven not only by the profit motive of firms but also by the existence of several other non-economic mechanisms.

According to the model presented in Chap. 14, the long-term commitment of resources to development under uncertainty was enabled by both the profit motive and the non-profit motives. The former corresponded to technological and market uncertainty while the latter corresponded to social and organizational uncertainty. Moreover, the factors that influenced the formation of expectations that made possible additional resource input based on the profit motive were not necessarily the result of rational or purposive behavior, as we will summarize again below. The logic of politics, the logic of engineers, the logic of compatibility between technology and market, and the logic of organization and society were at work.

15.1.2.1 The Logic of Politics: Policy Collaboration Effects

One of the mechanisms that drove innovation in RO membranes was the unintended collaboration of government policies and cross-border policy spillover. U.S. government's support was primarily aimed at developing technology and industry to solve water shortages in California. However, that reason alone would not have attracted the attention of the world, facilitated the activities of many companies and research institutes, and stimulated the development activities of Japanese companies and public institutions across the sea.

Because of President Eisenhower's political mission of "peaceful use of nuclear energy", desalination research has been justified as one of the peaceful uses of nuclear energy. In addition, President Kennedy's mission to "send man to the moon and make the desert bloom" led to a huge amount of policy support for desalination research, which was driven by the huge budget for space exploration.

Not all of those policy-supported companies achieved their intended results. Few of the military companies that received U.S. government support at the time have left their mark on history, except for GA's ROGA. GA also sold ROGA and could not continue the RO membrane business as a company. Therefore, whether the subsidies for individual projects were economically rational is highly questionable. However, DuPont and Dow, which had already started developing their own RO membranes, benefited from the large amount of money due to policy collaboration, and the results were transmitted across the sea to Japan, spreading to the development activities of Japanese companies. Furthermore, the choice of desalination research as a destination for salt production research, which had lost its place in Japanese public research institutes, can be seen as a kind of cross-border collaboration stimulated by the active development activities supported by the U.S. government.

This policy collaboration was the result of political circumstances separate from the rational behavior of firms within the economic system. However, because of its nature, it was a major force behind the early stages of the industry, when economic efficiency was not visible.

15.1.2.2 The Logic of Engineers

The second mechanism that supported the continuation of technological and business development was the emergence of dominant design in the evolutionary process of the technological system. Dominant design is the dominant product design or technology approach agreed upon within an industry, and its emergence has been cited as an opportunity to focus technology development that had been dispersed within an industry and to rapidly improve the price-performance ratio based on process innovation (Abernathy, 1978; Utterback & Abernathy, 1975). A similar phenomenon was identified in the RO membrane industry.

The dominant design in RO membranes was included in the ′344 patent by Cadotte. This led to a rapid convergence of development within the industry toward polyamide-based flat membranes by interfacial polymerization. By concentrating development activities on one technological approach, the performance of the membranes improved rapidly, and the market price fell substantially as competition increased. This led to the development of the seawater desalination market.

Such effects of dominant design on industries have already been robustly found in previous studies. However, the emergence process of dominant design has not always been sufficiently clarified. Unlike ordinary technology development activities by firms, the emergence of a dominant design cannot necessarily be explained by rational economic logic alone.

The ′344 patent became a dominant design because it was technologically superior and provided a path to simultaneously solve multiple technical issues that had been a concern. Since the term “technically superior” implies a high price-performance ratio, there is a certain rationality in the explanation that the emergence of the ′344 patent dispelled technological and market uncertainty and induced companies to continue investing in commercialization. This can be seen in the fact that Dow acquired FilmTec for the purpose of acquiring the ′344 patent.

However, ′344 patent attracted attentions of developers in other companies from a purely technological perspective, apart from such economic and business reasons. Although Cadotte’s achievements had been noticed by many parties before the ′344 patent, the ′344 patent was a distinct breakthrough. The challenge in the development of RO membranes was to achieve a high level of compatibility between basic performances, such as the desalination rate and water permeability, which are in a trade-off relationship. If one performance is outstanding but the others are inferior, the membrane cannot be used for seawater desalination. The key was to solve the trade-off problem across the entire technological system. In this situation, the ′344 patent provided a way to solve the trade-off problem that everyone had been struggling with. In this sense, the ′344 patent’s outstanding excellence or leap from the past surprised many engineers and made the shift to polyamide-based flat membranes decisive.

Although membrane performance actually improved and prices fell based on the ′344 patent, this was never promising ex ante. One can think of it as an ex post creation of economics, with a rapid convergence of technological approaches, as technological excellence attracted the attention of experts intent on pursuing the technology.

15.1.2.3 The Logic of Correspondence Between Technology and the Market

The third mechanism that enabled the continuation of development was the exploitation of strategic niche markets following the development of the technological system (Hoogma et al., 2002; Kemp et al., 1998; Schot & Geels, 2008). Continued development through the exploitation of niche markets was not necessarily a directly purposeful economic action, although it was the result of the rational activity of firms within the economic system. Rather, it is something that can be described as a means of creating a reason to mobilize the resources necessary to actualize the original objective.

As was the case with seawater desalination for RO membranes, many innovation activities are carried out with a final landing point in mind. However, there are many technical hurdles to overcome to reach the final goal, and it is not always possible to continue development until all of these hurdles are overcome. This is a barrier to innovation. Strategic niche markets are an important means of rescuing this situation. In the development of RO membranes, food applications, wastewater applications, and above all, ultrapure water applications for semiconductors served as such niche markets to justify the continuation of development.

The discovery of strategic niche markets is the result of the exploratory efforts of profit-seeking companies and developers, but the fruitfulness of these exploratory efforts is because the imperfections that inevitably arise during the development of technological systems have the potential to be linked to diverse applications. In seawater desalination, it is necessary to achieve a high level of compatibility between the desalination rate and permeability, and developers worked to achieve this, but in the process, imperfect technologies were created that could not achieve both performance levels. One example was Nitto Denko's NTR-7250. This technology was certainly incomplete for desalination, but it could be perfect for applications where desalination was not required to a great extent, such as in the back-end process of ultrapure water production.

These incomplete technologies that inevitably arise in the process of technological development suggest the possibility of being combined with various applications. If they are successfully explored, they justify the continuation of development for the time being, even if they differ from the original purpose. Technological development will go one step further and increase the feasibility of the original purpose by buying some time in this way.

Technology provides a bundle of functions, but the value of these functions is not uniquely determined. Technology evolves by being associated with various values with reference to concepts on the customer side (Clark, 1985). The value of a technology depends on its positioning in the customer's value network (Christensen & Rosenbloom, 1995). Hence, discovering the diverse positioning of a technology leads to the emergence of its potential value, which in turn enables the overcoming of high market uncertainty and the consequent continuation of development.

15.1.2.4 The Logic of Organization and Society

The fourth mechanism that enables companies to continue technological and business development is company-specific logic. Companies have unique philosophies, missions, and values that go beyond mere profit-making purposes. In the case of Toray, it was the value of the world's best technology. At Nitto Denko, it was the mission of making membranes a pillar of the business. At Toyobo, it was the mission of diversification. These values and missions provided them with the power to promote new schemes that could not be justified by economic rationality alone. Of course, it is not permissible to pursue these values and missions with complete disregard for profit. However, as long as they are compatible with inherent values and missions, they can avoid being judged hastily based on economics alone. That at least buys time for the proponents of innovation. While that time is being given, proponents can proceed with development and gradually eliminate technical uncertainty. In the meantime, they may be helped by the logic of politics, the logic of technology, the logic of strategic niche markets, or the logic of society, as described above and below.

As long as the company is a commercial enterprise, it must ultimately be subjected to the selection process via economic rationality. However, as was the case with the RO membrane industry, it often takes time for innovation activities to achieve their initial objectives and eventually form a large industry. Hence, the key to actualizing innovation is how to overcome the initial stage where uncertainty is high and sufficient economic rationality cannot be demonstrated. The company-specific logic plays a bridging role in this part.

Although not explicitly analyzed in this book, the development of RO membranes, which began as a solution to the social problem of water scarcity in the United States, has also been supported by social logic. As populations grow, water scarcity is still a problem in many parts of the world, with some predicting that by 2050, the population of river basins with severe water shortages will reach 40% of the world's population (OECD, 2012). Conflicts arising from water supply are also occurring in many parts of the world. Water is essential for industrial development, such as industry and agriculture, as well as for meeting fundamental social needs, such as sustaining people's lives and ensuring their safety. Hence, the development of RO membranes, which is expected to be a technology to solve water shortages, is socially driven for reasons beyond simple economics. Because there is no substitute for water, in situations where it is scarce, decisions that go beyond economics are required. Hence, at times, policy support has been instrumental in promoting development and stimulating demand.

However, it is difficult for profit-seeking companies to endlessly continue unprofitable business activities. Furthermore, since people suffering from serious water shortages often lack the ability to pay, it is not easy to draw up a reliable profit model.

Herein lies the difficulty of social problem-solving projects. On the other hand, for-profit companies, existing within a larger society, cannot continue to exist while ignoring the demands of society. Companies are allowed to pursue profit-making activities on the premise that they are in harmony with society. From this perspective, there is a reason for companies to engage in the development of RO membranes, even if they cannot directly achieve profit. Of course, they cannot continue to operate at a large loss. However, it can be said that at least the logic of society provided a certain amount of time for development activities under uncertainty before the business developed.

15.1.2.5 Mutual Complementation

No single one of the four logic rationales outlined above has been sufficient to ensure the long-term sustainability of RO membrane development. Looking back on history, we can see that long-term development under uncertainty was possible only when these rationales were combined and functioned in a complementary manner.

Government support through unintendedly collaborated policies certainly played an important role in driving early development. However, the abundant support ended after a certain period of time and was then left to private investment. Although the initial government support was a catalyst for private investment, it was not sufficient to develop technologies and markets that would allow for future industrial development. In fact, as it was more difficult to overcome the technological challenges than the U.S. government had thought, market formation did not proceed as expected. In the midst of lingering technological and market uncertainties, subsequent industrial development required a different mechanism to enable the private sector to continue development.

The continuation of development in Japanese companies was actualized by appealing to the inherent values and logic embodied in the company's mission and philosophy, while backed by the strong commitments of the developers. However, the company-specific logic alone does not fully justify the continuation of development without regard to profit. For decision-makers to make resource mobilization decisions, they needed some additional ingredient that would shift future profit expectations in a favorable direction. This was the emergence of a strategic niche market. At Nitto Denko, early development activities were supported by the fact that our engineers went directly to the market to develop various applications. The emergence of a market for the semiconductor industry, which arrived like a surprise boon, also supported the continuation of business from the 1980s onward. Similarly, Toray's success in establishing a small but constant revenue base in semiconductor applications with its general-purpose cellulose acetate RO membranes, which had been developed earlier, justified their continued development of composite membranes that had been unsuccessful in commercialization. In addition, Toray was able to focus on the development of next-generation composite membranes despite the two market failures of PEC-1000 and PEC-2000 because of the existence of the ′344 patent that was recognized as the dominant design. Although the engineers of each company had been developing technology independently, Cadotte's achievement dispelled some of the "fog" of technological uncertainty and gave the engineers direction and confidence in their development. As a result, the development of each company converged, performance improvement and cost reduction accelerated, and the path to industrial development became clear.

Although this technological convergence led to Dow's lawsuit against Nitto Denko, it was Nitto Denko's victory in the suit and the recognition of Cadotte's technology as U.S. state property that was decisive for the development of the industry. Without this institutional impetus, the industry might not have developed at the rate we have observed.

As we have discussed so far, the continuation of development with high uncertainty, which is common to the development of many social problem-solving technologies, cannot be explained by reducing the process only to the rational decision-making of profit-seeking companies. However, it also cannot be explained only by factors such as policy guidance, social pressure, or management vision. In the process of continuing innovation activities under uncertainty, the logic of various systems in society beyond the scope of the economic system overlaps and acts in a complementary manner. That is what this book has revealed through the history of RO membrane development.

15.2 Contributions and Implications

15.2.1 Understanding the Innovation Process Through an In-Depth Case Study

Innovation, unlike mere invention, is a long process from the development of innovative technology into a product or service to its widespread acceptance in the marketplace and the creation of significant economic value. The mainstream innovation research in recent years has been to measure the results and outcomes of innovation in terms of patent applications and productivity indices and to quantitatively clarify the various causal variables that bring about these results. Empirical research that clarifies the universal relationships among such variables is certainly important. However, the overall picture of innovation actualized over a long period of time cannot be fully grasped by reducing it to a simple relationship between variables. As this book has shown, the logic of various systems—economic, political, technological, organizational, and social—are at work in a single innovation story. Unfortunately, this book does not succeed in deriving a general statement about their interrelationships, but it does paint an overall picture of how they work in complementary ways to overcome the four uncertainties that act as barriers to mobilizing resources for innovation.

There have been few attempts to clarify in detail the history of industrial development from the early days to the present based on information obtained from interviews with people who have been directly involved in that history, as this book does. It is considered to be a contribution of this book to present various theoretical perspectives in understanding the innovation process based on the detailed historical analysis.

In addition, the findings from the analysis in this book have several important implications for intrapreneurs and corporate managers who seek to actualize innovation, as well as for policy makers. These are summarized below.

15.2.2 Implications for Innovators in a Company

What can innovators in a company do to move forward with innovation activities that face barriers due to high uncertainty? The analysis in this book provides some implications for the question. Innovators must not only immerse themselves in R&D activities and wait for external tailwinds to blow but must also proactively deal with the four uncertainties that stand in the way of innovation activities. The implications are as follows.

15.2.2.1 The Proactive Search for Initial Markets

The emergence of initial markets played an important role in terms of dealing with market uncertainty. This led to the demonstration of the existence of economic value created by the technology, even though it was different from the market initially envisioned by the developers.

Innovators often initiate and push forward innovation activities with unique ideas and goals. In the case of RO membranes, solving the social problem of water shortages was the ultimate goal for many developers. However, it is not clear whether the technology and the market will progress as expected to reach the final goal. Even if it does, it often takes many years to reach it. It is difficult for developers to continue to receive resource allocations within a commercial company without any visibility of market outcomes for a long period of time until the goal is reached. Therefore, initial niche markets that can be addressed by applying a technology that is still under development become an important means of persuading decision-makers within the company.

Initial niche markets may not generate enough revenue for the company to cover cumulative development costs. Nevertheless, it can be shown that such markets exist, and as long as there are customers, the immediate halting of development can be avoided.

If developers can gain time to proceed with development in this way, they can solve technical problems, open up further markets, and finally reach the goal that they initially set. Therefore, it is considered important to actively develop the initial market in the sense of building a "bridge" on the way to the goal, even if it deviates from the original objectives and thoughts of the innovators.

In the RO membrane industry, profits from ultrapure water production for semiconductors played an important role in justifying the continued development of technology by Toray and Nitto Denko. The market was far smaller than today's desalination market, but as Toray's Uemura mentioned, "Although we could not sell a lot of products because they were used for producing ultrapure water, we were able to feed dozens of researchers."[1] On the other hand, Toyobo's early development of membranes was underpinned by applications for small ships. Without such applications, the development of membranes could have ceased by the end of the 1990s.

Early niche market development also had the effect of providing a variety of learning opportunities through direct contact with customers. This effect has been pointed out as a recommended stepping-stone option in new business development with high uncertainty (McGrath & MacMillan, 2000).

When the required performance differs depending on the application, the expansion to various applications leads to technological improvement from multiple aspects. For example, the expansion of ultrapure water production applications led to the improvement of high water permeability under low pressure. Ohya and Niwa (1988) wrote, "Considering that more than half of the demand for RO membranes is for the semiconductor industry, it is natural that low-pressure RO membranes that can operate at an operating pressure of less than 1 MPa were developed."[2] Nakao (1990) also wrote, "In brine desalination, wastewater treatment, and ultrapure water production, the development of membranes that can be used at low pressure is an issue because the osmotic pressure of the original solution is low,"[3] indicating that low-pressure membranes had become popular for brine desalination, wastewater treatment, and ultrapure water production. On the other hand, when dirty water is used as raw water, such as in wastewater applications, the key will be to make the membranes resistant to contamination. In addition, for food enrichment applications, sanitation and chlorine resistance are also important in addition to contamination resistance. The technologies acquired through the development of these applications were later utilized in seawater desalination applications.

These initial markets in the RO industry did not necessarily emerge solely through the proactive search of the developers. However, the magnitude of the effect that the initial markets had suggests the high importance of the developers themselves proactively developing the application markets.

15.2.2.2 Strategic Use of Policy

Innovators can also strategically use government policy support in two ways to continue development in the face of uncertainty: acquiring complementary resources and utilizing tailwinds.

[1] See Chap. 7.

[2] Ohya and Niwa (1988), p.53.

[3] Nakao (1990) p. 235.

One of the reasons why it is difficult to continue innovation activities with high uncertainty is that profitability is not expected to match the development costs. Therefore, one way for innovators to persuade the continuation of development is to obtain supplementary resources from outside to reduce the burden of development costs within companies. In particular, if the company can obtain a government subsidy that will subsidize 100% of the development costs, including labor costs, it will be difficult for decision makers to reject the proposal to continue development for the time being. However, to obtain policy support, innovators may need to strategically position their development projects to align with policy intentions.

In the early stages of the RO industry, U.S. military companies applied for desalination research programs to meet the political goal of "peaceful uses of nuclear energy" and received public support from the OSW. When the support from the OSW ceased, some companies next continued to develop membranes while utilizing support from the Department of Energy, which was the predecessor of the U.S. Atomic Energy Commission. In this way, by positioning their development activities in a way that benefits the achievement of policy goals, they are able to successfully obtain external support and continue development, even if that positioning is only on the surface.

However, support from the government will eventually cease. At that time, proponents must be careful not to expose themselves to the danger of development discontinuation within their companies. In fact, many of the companies that received U.S. government support have since ceased development. Innovators need to work to obtain cooperation for commercialization within their companies while they are receiving government support. To obtain internal support for commercialization, they may want to avoid completely depending on government financial support for their R&D (Aoshima et al., 2013).[4]

Innovators in companies may also be able to use the policy stance of the government as a tailwind to continue development. For example, Ishizaka, who led desalination research at the AIST in Japan, won a large project from the MITI and promoted his own research activities by citing the views of the Ministry of Construction, which was concerned about future water shortages. Although the Ministry of Construction's view was probably to justify the construction of dams, Ishizaka justified his development activities by linking it to the necessity of desalination and obtained funds from MITI.

What Toyobo's developers used to survive within the company was their track record of technology application at the Seawater Desalination Center in Fukuoka, which was built under policy intervention. The company's technology development, which had been carried out in a low-key manner with a limited number of staff, began to attract attention thanks to this event. Although it was not a large-scale plant from a global perspective, the fact that their technology was selected for use in a new plant supported by government policy had a tremendous impact on the company. In this

[4] It should also be noted that resource acquisition strategies using surface level consent may become an encumbrance later on (Taniguchi, 2016; 2017; Tsuboyama, 2011).

way, by strategically using events with a large social impact, innovators can support their own development activities.

15.2.2.3 Reference to Other Companies and Inconspicuous Development Activities

The direct way to deal with technological uncertainty is to solve technological problems and improve the technological level. However, especially in the early stage of development, the solution to technological problems often does not proceed as expected. If the expectation that the technological level will not improve in proportion to the input resources is formed among the decision makers, the innovation activity is threatened with cancellation. Our analysis in this book suggested two ways to overcome these situations.

One is to refer to the development trend of other companies. When the company's own development is not progressing as planned, information on the successful development by other companies may lead to the formation of optimistic expectations by decision makers that it is possible to solve technological problems if resources are mobilized continuously. When fierce market competition is already underway, this is not desirable information because the success of other companies directly leads to competitive inferiority. However, the success of others in the early stages of an industry is more likely to serve as information about the potential of a technology than to determine its failure. It is considered that a series of successful developments by Cadotte raised expectations for polyamide composite membranes and provided a reason to support continued development at each company.

Here, the significance of referring to the technologies of other companies does not mean referring to or imitating their ideas. News of successful development by other companies is not good for developers. On the other hand, successes of other companies suggest the possibility that their company can also succeed and even surpass them. Unless the market is well developed where technological competition determines the market outcome, broadly referring to the technologies of other companies may actually form a justification for continuing development.

The case of Toyobo suggested another measure that could be taken by the innovators. If large-scale resource mobilization is required, decision-makers must be demanding the results. However, if the mobilization of resources is on a small scale, the continuation of development may be allowed even if there are no significant visible results. Developers at Toyobo obtained approval for business continuity in the difficult business environment of the 1990s by drastically reducing the size of the development team. Furthermore, by coming under the umbrella of the strong medical membrane business, they received informal flexibility in development personnel and were spared from being severely questioned about individual business results. In this way, innovators can avoid tough decisions on resource mobilization and overcome the challenges of uncertainty.

15.2.3 Implications for Corporate Strategy: The Dilemma of Industrial Development and Profit Acquisition

Since this book aims to elucidate the mechanisms by which the development of new technologies and businesses continues under uncertainty, there has been little analysis of corporate profits. However, firms pay attention to the creation of innovation because they expect it to bring them future revenues. Although they may not expect that innovation activities will immediately generate revenue, firms should expect to earn a return on their previous investment, at the very least, once the industry has developed sufficiently.

However, as explained below, our analysis suggests that the actualization of innovation and the acquisition of corporate profits may be incompatible. Corporate decision-makers will need to understand this and consider appropriate strategies.

To accelerate the technological progress for innovation, it is desirable for players in an industry to engage in competition in technological development on the same track (Numagami et al., 1992). Whether or not technology grows sturdily strongly depends on the concentration of development (Shimizu, 2016). The level of technology can be rapidly improved when developers share technological trajectories and engage in fierce technological race with each other.

In the history of the RO membrane industry, the convergence of membrane technology methods and membrane element shapes contributed to the development of the industry. In terms of membrane technology, the convergence of polyamide flat composite membranes became the dominant design in industry. This was largely due to the emergence of the Cadotte '344 patent and its attribution to the U.S. government. This event intensified the competition between companies with the same technology and membrane element shape, resulting in the expansion of the market and the development of the industry.

In terms of expanding the market by promoting competition, the standardization of element shapes also played an important role. As mentioned in Chap. 3, polyamide flat membrane RO elements have been standardized to a shape of 4 or 8 inches in diameter and 40 inches in length. The convergence of element shapes began in the 1970s, and by the 1980s, it had become almost a de facto standard (see the Column to Chap. 12). By the 2000s, when the desalination market began to expand in earnest, element shapes had already been standardized, and plant engineering companies were designing trains based on standard shapes.

RO membrane companies would like to lock in customers at the time of initial delivery and enjoy stable, long-term demand for renewals. However, because the polyamide group has standardized element shapes and the same membrane materials, competition has arisen with each renewal demand. This has been pushing down profitability. The convergence of technology approaches and the standardization of elements has led to a decline in membrane module prices, which has contributed to the widespread adoption of RO. This has, on the other hand, reduced the profits of membrane companies. Therein lies the dilemma between industrial development and corporate profits.

Toyobo, the only company that specializes in cellulose acetate hollow fiber membranes, has achieved high profitability despite its small sales scale. This is because Toyobo's element shapes are larger than those of polyamide flat membranes which has incurred a switching cost, allowing it to dominate the demand for renewal.

Normally, customers shun products that cannot be purchased from multiple companies because they lose the buying power on price. Despite this, Toyobo's membranes sell well in the Middle East because of their high chlorine resistance. The high chlorine resistance makes chlorine disinfection possible, which increases the plant's operating rate and lowers the cost of desalination. In the waters of the Middle East, which are prone to clogging due to the growth of microorganisms, these characteristics of Toyobo's membranes were exactly what was needed. Despite the high price of Toyobo membranes, the total cost of ownership (TCO), including running costs, was in some cases an advantage over the polyamide camp.[5]

Thus, there is a risk of discord between the expansion of an industry or market and a company's profits. This is where a strategy that considers the appropriate balance is necessary.

15.2.4 Implications for Policy Makers

15.2.4.1 A Grand Design for Long-Term Innovation

Not limited to the case of RO membranes, policy support from the government is often required for the development of budding and unknown new technologies. In recent years, as the importance of innovation has been emphasized, policy support is often provided even for development activities that are directly linked to commercialization. On the other hand, under severe financial conditions, there is an increasing demand for the efficient use of public funds. Therefore, policy makers are required to have the ability to envision how policy measures can lead to the creation of innovation. The comprehensive descriptive model presented in Chap. 14 may provide useful information in exercising such conceptual ability.

What the model in Chap. 14 showed was that sustaining innovation activity in a firm requires dealing with four types of uncertainty. The analysis of the RO membrane industry mainly positioned policy support as a force acting on competitive uncertainty. But in practice, policy instruments are not limited to that and can also work on a variety of other uncertainties.

There are various barriers that hinder innovation activities, such as a lack of technological prospects, a lack of market visibility, a lack of profit potential, and a lack of social conditions. Which barrier is more problematic may differ depending

---

[5] Toyobo's survival strategy of specializing in niche markets may be partly due to internal resource constraints. In the late 1990s, Toyobo managed to keep its RO membrane business profitable by drastically reducing the number of employees. In such a situation, it was impossible for the business to expand in a wide range of fields. they needed to focus on large-scale seawater desalination plants in the Middle East, where it could take advantage of the characteristics of its technology.

on the situation of the industry. Thus, policy makers are required to understand the specific situation and select appropriate policy measures to address the uncertainty.

For example, if the problem is the lack of technological prospects, a policy emphasis on basic R&D investment may be considered. In addition, if the proliferation of diverse technologies leads to a deterioration in the efficiency of R&D investment and reduces investment incentives, a policy response that standardizes technological approaches may be effective.

If the lack of market visibility is a barrier, then government procurement could be used to create an initial market. One could create a market by tightening regulations, as in the case of environmental regulations, or by relaxing regulations to expand the market. Furthermore, if fierce competition is hindering companies from making long-term investments, then policies to reduce competition may be desirable.

If innovation activities are stagnant due to organizational circumstances, it may be difficult to have a direct policy impact. However, if, for example, large firms subject to capital market pressures are hesitant to engage in highly uncertain development, policies targeting small and medium-sized firms that can make quick investment decisions, including technology transfer from large firms and human resource mobility, may be successful. It may also be possible to encourage the development of innovation that solves social problems, for example, by influencing the social climate. In this way, it may be possible to indirectly influence the decision-making of organizations.

It is important to understand that innovation processes over time are beset by different kinds of uncertainty and that the barriers to innovation vary across industries and time periods. It is also important to grasp the overall picture that there are both economically rational forces and other types of forces that act in complementary ways to tackle these uncertainties. Although it is difficult to predict the detailed causal effects of policies in advance, having a broad framework, such as one that this book has provided, is necessary.

15.2.4.2 The Significance and Risks of Government Intervention

As clarified in Chap. 10, the government has played a significant role in initiating the history of RO membrane development. Particularly in the case of a technology such as an RO membrane where the primary objective is to solve a social problem, it is difficult for society to invest sufficient development resources if it is left to private companies with a profit motive (Jaffe et al., 2005). Therefore, it was necessary for the government to trigger private investment by shaping favorable expectations regarding the state of technology, markets, and competition.

It seems that government support in the early stage was effective in developing human resources in the industry because the developers who were active in ROGA played an important role in the development of the industry by moving from one company to another. In addition, the fact that the ′344 patent of Cadotte was developed through government-supported activity shows that policy support has had an undeniable effect on the development of important technologies.

Looking back on history, however, it is true that many of the companies that relied heavily on U.S. government subsidies disappeared from the market and did not necessarily contribute to later industrial development. The companies that made significant contributions to the development of the industry at the commercialization stage were those, such as Dow and DuPont, that had already invested in their own development before the government support programs began. For these companies, government support was likely beneficial in accelerating demonstration testing and additional commercialization investments. On the other hand, the companies that intended to enter the market with the help of government support could not be in a position to lead the industry although they developed their products with government support. With the exception of ROGA, the military companies that received government support for the cause of the peaceful use of nuclear energy have not left their mark on history. Even ROGA could not grow as a business in its own right because of the outflow of human resources through repeated business sell-offs. In this light, while recognizing the role of large-scale government support in the early stages of industrial development, it is questionable whether it was sufficiently justified in terms of the cost-effectiveness.

As seen, the impact of policies is so diverse that it is not possible to make a general judgment on their merits. However, it should not be advised to abandon making anticipation in advance. Policymakers need to draw as far in advance as possible the direct or indirect effects of policy support.

In addition, policy support for the creation of new technologies and industries often combines a variety of policy objectives to ensure its legitimacy (Taniguchi, 2016).[6] This has been the case in the history of RO membranes. However, the more diverse the objectives are, the more ambiguous the relationship between the means and the ends becomes, and there is a risk that the originally intended policy objectives will not be achieved. Therefore, it is necessary for policy makers to be aware of the relationship between policy objectives and policy instruments. For example, if the objective is to accelerate commercialization, it is probably more effective to support companies such as Dow and DuPont that are committed to commercialization as part of their strategy (Matsushima, Aoshima, & Takada, 2016). On the other hand, if the objective is to develop general-purpose fundamental technologies that form the basis of the development of the industry as a whole, then support mainly from public institutions would be appropriate.

15.2.4.3 Cross-Border Spillover of Policy

In Chap. 10, we showed that U.S. government policy support in the early RO membrane industry period spilled over to Japanese firms across the ocean. This history provides an important perspective on policy effects.

[6] In the realm of social movement research, it has been noted that movements expand when multiple different frames phase in under an abstract master frame (Benford and Snow, 2000; Gerhards and Rucht, 1992).

Normally, the policy support of a country is intended to actualize the national interest of the country concerned. Sometimes, however, some of the policy effects spill over beyond national borders to players in other countries. As a result, the policy may produce different results from those originally intended. In particular, the greater the nature of the technology leans toward solving social problems, the less likely it is to find short-term profitability, the more important public support becomes, and the larger amount of support is provided. In such a case, the effects of the policy may spill over to companies and research institutions in other countries, even if not directly.

Former research on national innovation systems has tended to discuss the impact of policies with a focus on domestic innovation activities. Although the need to look at the relationship between domestic and international policies has been mentioned in the context of the globalization of corporate activities (Binz & Truffer, 2017; Corona-Treviño, 2016; Sharif, 2006), such an analysis has only just begun. In response to this situation, this book identifies one specific process through which the innovation policies of a country influence the innovation activities of other countries. In particular, it is an important finding that transnational policy spillover effects already existed in the 1960s, before the internationalization of economic activities.

Many such policy spillover effects have been observed in recent years. For example, Europe's environmental policies, particularly Germany's feed-in tariffs for renewable energy sources such as solar and wind power, have encouraged the rapid rise of Chinese companies but have also triggered the decline of their own companies. In Japan, the introduction of the eco-point system, which was intended to support electronics companies, led to a race to preempt demand and enriched overseas ODM companies, and caused the decline of Japanese companies (Aoshima & Shimizu, 2012). As business activities become more globalized, it will be easier for the policies of a particular country to have a spillover effect on other countries. At the same time, however, this will accelerate the development of the industry as a whole. The analysis in this chapter suggests that the early development of the RO membrane industry was the result of the domestic and international spillover of public support from the U.S. government.

It may also be possible to consider a strategy of free riding on the policies of other countries, assuming such cross-border policy spillovers. As technologies become more complex and investment in their development becomes large, it will be important to consider how much a country can utilize the policies of other countries rather than supporting them all on its own.

15.3 Further Research Topics

15.3.1 Application as a Testing Field

Finally, we would like to point out three research topics that should be promoted in the future that could not be sufficiently addressed in this book: (1) analysis of the application market as an experimentation field, (2) analysis of the impact of development reasons on development activities, and (3) clarification of micro-mechanisms caused by innovation policy.

The first topic is the analysis of the application market as a experimentation field. The search and creation of an initial market is important to overcome market uncertainty and to enable the continuation of development. It was also pointed out that the search for initial markets may lead to diverse learning in application markets with different performance requirements. What effect, then, will such diverse learning have on the direction of technological progress in the industry as a whole?

If the application markets that emerge show different technological challenges to developers, then technological progress may be influenced by the nature of application markets, whether by deliberate search or accidental discovery. Such an argument may add the new perspective to the existing model of technological evolution.

If there are unique performance requirements that are outstandingly required by each application, the emergence of an application-specific market will focus the development on the performance requirements of that application. In this case, it is not necessary to satisfy all the performance requirements at the same time at a high level, which tends to cause a trade-off problems, but rather to overcome technical issues focusing on specific performances. Then, as new applications emerge again, challenges will be overcome by focusing on other distinctive performance factors.

This process of problem solving that occurs sequentially in the flow of industrial development could possibly be compared to the experimental process of science. Just as scientific research accumulates research results one by one by focusing on specific factors while controlling various experimental conditions, it may be said that responding to the various applications that arise in the process of industrial development has an experimental aspect in which specific performance conditions are relaxed and the focus is on unique performance. When a specific application emerges, development is focused on the performance requirements required by that application, and the knowledge gained from that development is utilized in the next development. As a result of the accumulation of responses to various applications in this way, it is likely that a technology will be found that will eventually achieve a higher level of compatibility with all performance requirements.

It has been observed in other cases that gradual technological progress occurs when the focus of development differs depending on the application. For example, Fujiwara (2004) analyzed the process of converting the precision press processing technology accumulated in Seiko Epson's watch business to the inkjet printer business and pointed out the importance of learning that occurs in relay applications. Since the nozzle hole that discharges ink required extremely high punching accuracy and hardness, it was thought to be difficult to transfer the pressing technology of the watch business as it was. Nevertheless, the watch engineers were able to successfully apply the technology because they had previously applied the technology to a different type of printer (the dot matrix printer). The engineers had learned about some requirements for producing inkjet printer nozzles through their previous experience with dot matrix printers, which led to a smooth transfer of technology.

In this book, we have introduced some circumstances that suggest that the requirements of different applications have affected the direction of technological progress, such as the development of low-pressure technology due to the emergence of applications for semiconductors, but we have not been able to conduct an in-depth analysis. In addition, it is hypothesized that different companies with different backgrounds may focus on different applications, and that the direction of technological development and the resulting technological results will be affected. Research on such a hypothesis is recognized as an issue to be pursued in the future.[7]

15.3.2 The Influence of Development Reasons on Development Activities

As discussed in Chap. 13, one effective way to justify the continuous mobilization of resources for innovation activities is to create reasons that match the unique values and orientations of the organization. The reasons created in this way may be different from those originally intended by the proponents of the innovation. However, in some cases, the reasons created as a pretext for acquiring resources may eventually take on substance and influence the direction of subsequent development activities.[8] In other words, the values and orientation of an organization may induce a particular reason for development, and they may even define the content of development activities.

In relation to this question, in recent years, there has been an accumulation of research that clarifies the impact of organizational orientation (Kohli & Jaworski, 1990; Narver & Slater, 1990) on innovation activities (Lukas & Ferrell, 2000; Narver, Sater, and MacLachlan, 2004: De Luca et al., 2010).[9] In light of this previous research, clarifying the causal pathways by which differences in organizational values and

[7] The importance of innovation activities through the interaction of demand pull and technology push has also been pointed out by Di Stefano, Gambardella, and Verona (2012).

[8] If the development is expected to be completed within a short period of time, the inconsistency between reasons and actions may not be so much of a problem and subterfuge may be used because it is only necessary to mobilize resources for the time being. However, if the development is prolonged, the duplicitous strategy will damage the credibility of the individual developer or the project, and it will be difficult to mobilize resources continuously for a long time. Therefore, when development is expected to be prolonged, a high degree of consistency between reasons and actions is even more needed.

[9] As presented by Kato (2013), strategic orientation is a concept that was originally discussed in relation to management outcomes. In addition, Kohli and Jaworski (1990) developed the concept of "strategic orientation" in a different manner from Venkatraman (1989) by adding technology orientation instead of interfunctional coordination among the three original dimensions of market

orientations produce differences in development behavior, which in turn affect the direction of technological progress and technological outcomes, will be one of the important future research themes in innovation research.

With regard to this theme, in the Appendix of Chap. 13, we have taken up Toray and Nitto Denko, which were in the same polyamide group but whose continued development was justified for different reasons. We confirmed through their applied patents that the technological development areas of the two companies differed. Although the analysis is only an initial attempt and no conclusion can be drawn with sufficient rigor, it does point the way toward future research.

15.3.3 The Dynamism of Industrial Development and Profit Acquisition

Identifying the mechanisms that work behind the innovation activities of firms and the development processes of new industries is also important for the economic development of a country. This is because, in many cases, the creation of new industries is necessary for a country to achieve sustained economic growth. In fact, looking at the economic growth of postwar Japan, for example, the driving industries have a history of being replaced in various ways, from shipbuilding and steel to electronics and automobiles (Itami, 1998).

In today's Japan, however, it is difficult to create a new leading industry. One reason for this may be the low profitability of R&D activities (Sakakibara, 2005; Yonekura et al., 2010). Further investment in innovation can be made only when companies are able to earn profits, and the subsequent industry can be newly created. If a company does not generate profits, it will not be able to invest in the next innovation. Therefore, it is particularly important to clarify the profit-earning mechanism of firms as well as the development of new industries through innovation, if the government aims to achieve economic growth based on innovation as a policy objective.

RO membranes have attracted attention as a leading material business because of the high market share of Japanese companies and have often been introduced as an exemplary case for Japanese industry to aim for in the future. Unfortunately, however, at least at this point in time, many RO membrane companies do not appear to be securing ample and stable profits. One of the reasons for this is the dilemma between industry development and profit acquisition, as discussed in the previous section on strategic implications. If industries do not develop, companies will not be able to generate profit opportunities. However, if the commonization, standardization, and modularization of technologies are expanded to speed up the development of industries, it becomes difficult for companies to obtain profits. In this dilemma, how

orientation (customer orientation, competition orientation, and interfunctional coordination) and discussed the relationship between their three orientation dimensions and innovation (Gatignon and Xuereb, 1997; Spanjol et al., 2012; Spanjol et al., 2011).

can companies obtain profits? This is a theme that has been studied in fields of strategy theory, such as platform theory, in recent years (e.g., Tatsumoto, 2017). However, there still remains much that needs to be clarified. Industrial development and profit-earning mechanisms through innovation will continue to be important research themes in the future.

15.4 Conclusion

Innovation, defined as "inventive creation that brings economic value", is a phenomenon fraught with contradictions (Takeishi et al., 2012). Since it is an inventive creation that is a break with the past, there is a great deal of uncertainty attached to it. Moreover, to create economic value, resources such as people, goods, money, and information that exist in society must be poured into it continuously. This is like patiently applying water and fertilizer to a new species of plant that does not sprout easily.

Because innovations are expected to generate economic value, activities to actualize them are often driven by commercial firms guided by the profit motive. However, the analysis in this book suggested that innovation is not necessarily actualized only by endogenous corporate activities in the economic system. For innovation to emerge, excess resources in society need to be combined with innovative ideas. However, uncertainty, which is an inevitable part of the process of innovation, often prevents that combination. This is because the criterion of economic rationality, which directs the development of the economic system, cannot justify the combination.

To overcome this problem and move forward with the actualization of innovation, forces and rationales outside the economic system must be used. In the case of this book, these were politics, technology, organizations, and society. Innovation activities move forward step by step and approach the final goal through the complementary action of the various justifying rationales brought about by these different systems, including the economic system, across time and place. This is the picture of the innovation process derived from this book.

We do not believe that such conclusions can be generalized from the analysis of just one case study. The innovation of RO membranes has been pursued mainly as a part of new business development within large companies. This is not the representative case of innovation in general. The question raised in this book is why the excess resources accumulated within companies are continuously poured into innovative activities. This is different from the innovation realized by combining ideas and resources inside and outside the organization, as seen in open innovation, or innovation activities promoted through angel funds and venture capital as the main source of resources.

However, if high uncertainty is inevitable in the process of innovation, any case of combining innovative ideas and excess resources will require the help of forces beyond the economic system. Further research on this issue is awaited in the future.

References

Abernathy, W. J. (1978). *The productivity dilemma: Roadblock to innovation in the automobile industry*. Johns Hopkins University Press.

Aoshima, Y., & Shimizu, H. (2012). A pitfall of environmental policy: An analysis of 'eco-point program' in Japan and its application to the renewable energy policy. *International Journal of Global Business and Competitiveness, 7*, 1–13.

Aoshima, Y., Matsushima, K., & Eto, M. (2013). Effects of government funding on R&D performance leading to commercialisation. *International Journal of Environment and Sustainable Development, 12*, 22–43.

Benford, R., & Snow, D. (2000). Framing processes and social movements: An overview and assessment. *Annual Review of Sociology, 26*, 611–639.

Binz, C., & Truffer, B. (2017). Global innovation systems: A conceptual framework for innovation dynamics in transnational contexts. *Research Policy, 46*, 1284–1298.

Christensen, C., & Rosenbloom, R. (1995). Explaining the attacker's advantage: Technological paradigms, organizational dynamics, and the value network. *Research Policy, 24*, 233–257.

Clark, K. (1985). The interaction of design hierarchies and market concepts in technological evolution. *Research Policy, 14*, 235–251.

Corona-Treviño, L. (2016). Entrepreneurship in an open national innovation system (ONIS): A proposal for Mexico. *Journal of Innovation and Entrepreneurship, 5*, 1–13.

De Luca, L. M., Verona, G., & Vicari, S. (2010). Market orientation and R&D effectiveness in high-technology firms: An empirical investigation in the biotechnology industry. *Journal of Product Innovation Management, 27*, 299–320.

Di Stefano, G., Gambardella, A., & Verona, G. (2012). Technology push and demand pull perspectives in innovation studies: Current findings and future research directions. *Research Policy, 41*, 1283–1295.

Fujiwara, M. (2004). Technological innovation triggered by manufacturing technology transfer between divisions: The process of technological change in the printer business of Seiko-Epson. *Journal of Business Management, 12*, 32–44.

Gatignon, H., & Xuereb, J. (1997). Strategic orientation of the firm and new product performance. *Journal of Marketing Research, 34*, 77–90.

Gerhards, J., & Rucht, D. (1992). Mesomobilization: Organizing and framing in two protest campaigns in West Germany. *American Journal of Sociology, 98*, 555–596.

Hoogma, R., Kemp, R., Schot, J., & Truffer, B. (2002). *Experimenting for sustainable transport: The approach of strategic niche management*. Routledge.

Itami, H. (1998). *Nihon sangyō 3-tsu no nami (The three waves of Japanese industry)*. NTT Publishing.

Jaffe, A. B., Newell, R. G., & Stavins, R. N. (2005). A tale of two market failures: Technology and environmental policy. *Ecological Economics, 54*, 164–174.

Kato, T. (2013). Strategic orientation in Japanese firms: Examining measurement of the construct. *Hitotsubashi Shōgaku Ronsō (hitotsubashi Review of Commerce and Management), 8*, 2–15.

Kemp, R., Schot, J., & Hoogma, R. (1998). Regime shifts to sustainability through processes of niche formation: The approach of strategic niche management. *Technology Analysis and Strategic Management, 10*, 175–196.

Kohli, A. K., & Jaworski, B. J. (1990). Market orientation: The construct, research propositions, and managerial implications. *The Journal of Marketing, 54*, 1–18.

Lukas, B. A., & Ferrell, O. C. (2000). The effect of market orientation on product innovation. *Journal of the Academy of Marketing Science, 28*, 239–247.

McGrath, R. G., & McMillian, I. C. (2000). *The entrepreneurial mindset: Strategies for continuously creating opportunities in an age of uncertainty*. Harvard Business Press.

Nakao, S. (1990). Reverse osmosis membrane. *Bulletin of the Society of Sea Water Science Japan, 44*, 235–247.

Narver, J. C., & Slater, S. F. (1990). The effect of a market orientation on business profitability. *Journal of Marketing, 54*, 20–34.

Narver, J. C., Slater, S. F., & MacLachlan, D. L. (2004). Responsive and proactive market orientation and new-product success. *Journal of Product Innovation Management, 21*, 334–347.

Numagami, T., Asaba, S., Shintaku, J., & Amikura, H. (1992). Competition as an interaction: Reinterpretation of competitive behavior in electronic calculator industry. *Soshiki Kagaku (organizational Science), 26*, 64–79.

OECD. (2012). *Environmental Outlook to 2050: The Consequences of Inaction*. https://www.oecd.org/env/indicators-modelling-outlooks/49884270.pdf. Accessed 17 Jan 2019.

Ohya, H., & Niwa, M. (1988). *New Polymeric Materials ONE POINT: High-Performance Separation Membrane*. Tokyo: Kyoritsu Shuppan.

Sakakibara, K. (2005). *Innovation no shūekika: Gijutsu keiei no kadai to bunseki (Monetization of innovation: Issues and analysis of technology management)*. Tokyo: Yuhikaku.

Schot, J., & Geels, F. (2008). Strategic niche management and sustainable innovation journeys: Theory, findings, research agenda, and policy. *Technology Analysis & Strategic Management, 20*, 537–554.

Sharif, N. (2006). Emergence and development of the national innovation systems concept. *Research Policy, 35*, 745–766.

Shimizu, H. (2016). *Innovation in General Purpose Technology: Technological Development of Semiconductor Lasers in the U.S. and Japan*. Tokyo: Yuhikaku.

Spanjol, J., Qualls, W. J., & Rosa, J. A. (2011). How many and what kind? The role of strategic orientation in new product ideation. *Journal of Product Innovation Management, 28*, 236–250.

Spanjol, J., Muhlmeier, S., & Tomczak, T. (2012). Strategic orientation and product innovation: Exploring a decompositional approach. *Journal of Product Innovation Management, 29*, 967–985.

Takeishi, A., Aoshima, Y., & Karube, M. (2012). *Reasons for innovation: Creative Legitimacy for resource mobilization*. Tokyo: Yuhikaku.

Taniguchi, R. (2016). Successful resource acquisition-induced-resource allocation failures: Case study of symbolic management on "biomass nippon strategy." *Soshiki Kagaku (organizational Science), 50*, 66–81.

Taniguchi, R. (2017). *Formation and dispersion of chaotic visioned projects: A case study of Biomass-Nippon Strategy*. Ph.D. dissertation.

Tatsumoto, H. (2017). *Platform Strategy for Global Markets: The Strategic Use of Open Standards and the Management of Business Ecosystems*. Tokyo: Yuhikaku.

Tsuboyama, Y. (2011). Intra-organizational processes of making organizational facade: A case study of the first restructuring plan of Japanese National Railways. *Soshiki Kagaku (organizational Science), 44*, 87–106.

Utterback, J. M., & Abernathy, W. J. (1975). A dynamic model of process and product innovation. *Omega, 3*, 639–656.

Venkatraman, N. (1989). Strategic orientation of business enterprises: The construct, dimensionality, and measurement. *Management Science, 35*, 942–962.

Yonekura, S., Nobeoka, K., & Aoshima, Y. (2010). No more lost decade for Japan. *Hitotsubashi Business Review, 58*, 12–31.

Appendix
Procedure of the Patent Analysis

This appendix explains the procedure of the constructing dataset for the patent analysis in this book and presents supplementary explanations for each chapter.

A.1 The Procedure of Constructing a Patent Dataset

A.1.1 F-Terms

In the analysis of Chaps. 11–13 of this book, we conducted a patent analysis on three companies, Toray, Toyobo, and Nitto Denko, based on the patent database UltraPatent. In the analysis, we mainly searched for and extracted patents based on their F-terms because the F-term includes a classification along the expected use of each patent. Since International Patent Classification (IPC) code only classifies the technical field, it is difficult to determine the expected use of the patent. Since this book aims to trace the history of RO membranes, which have been developed through the growth of various different applications, we judged that the F-term, which includes the application classification, would be more suitable.

However, the reliability of the F-term is questioned in some cases due to the roughness of the process of assigning classification codes. Therefore, we checked the consistency between the application trends of patents based on the F-term classification and the case descriptions based on our own interviews. As a result, we concluded that the two are consistent and that there is no problem in employing the F-term classification.

The F-term has a kind of tree structure under a large technological classification called a theme code. For example, the theme code used in the analysis of this book is 4D006 "Separation using semipermeable membranes", which is divided into several "viewpoints", such as MA, MB, MC, and NA. For each of these viewpoints, a major F-term is assigned, such as MA00 (shape and structure of the membrane),

M. Fujiwara and Y. Aoshima, *Mechanisms for Long-Term Innovation*,
Advances in Japanese Business and Economics 31,
https://doi.org/10.1007/978-981-19-4896-1

MB00 (properties of the membrane), MC00 (material of the membrane), and NA00 (manufacturing method of the membrane). The number of subordinate terms held by a particular F-term of a major category varies from one to another. For example, under MA00, there are a total of 31 subordinate F-terms across the three levels, while under MB00, there are a total of 19 subordinate F-terms across the three levels. As an example, the structure of the F-term for viewpoint MA is shown in Table A.1 .

We constructed two kinds of patent datasets based on F-terms. The difference between the two datasets is that Dataset A extracts patents from only the three Japanese companies, while Dataset B extracts patents from the whole industry.

A.1.2 Dataset A: The Procedure of Dataset Construction for Three Japanese Companies

The extraction of the patents of the three Japanese companies (Toray, Toyobo, and Nitto Denko) was carried out in two stages.

First, we broadly extracted the patents of each company related to separation membranes by using only the F-term theme code 4D006 (FTERMH, including subordinate classes) that covers separation membranes as the search term. This is because we were concerned that there might be omissions if we initially narrowed the search to only F-terms related to RO membranes.[1] The search date was June 24, 2014. To trace the historical path of development, the search period covered the entire period of 43 years from 1971, the first UltraPatent date searchable for our research, to June 19, 2014 (the date of publication). The year of the start of development for each company is 1968 for Toray, 1973 for Nitto Denko, and 1971 for Toyobo. This means that the development for the three companies can be almost completely covered from the beginning. As a result of the search, the extracted patents were 1,768 for Toray, 1,508 for Nitto Denko, and 553 for Toyobo.

Here, the term "company" refers to each corporate group. For example, "Company A" includes its group companies that the company put on its website as of the date

[1] We confirmed that there was no need to be concerned about omissions by this method. Thus, when constructing a new Dataset, which is explained in the next section, the patents were extracted using 4D006GA03 (FTERMH, including subordinate classes) as a search target from the initial search stage.

Table A.1 Examples of viewpoints and F-terms

Viewpoint	F-terms			
MA	MA00 shape and structure of the membrane	MA01—hollow fiber membrane		
		MA02—tubular membrane		
		MA03—flat sheet membrane		
		MA04—other shapes		
		MA06—multiple layer structure (composite membrane)	MA07—ultra-thin separation layer	
			MA08—multiple separation layers	
			MA09—support layer is distinctive	
			MA10—has covering layer	
		MA11—charged membrane	MA12—Ion exchange membrane	MA13—cation exchange membrane
				MA14—anion exchange membrane
				MA15—amphoteric ion exchange membrane
		MA16—dynamic membrane		
		MA17—molecular membrane		
		MA18—liquid membrane	MA19—emulsion liquid membrane	
		MA21—pores are distinctive	MA22—pore diameter (limited numerical value)	
			MA23—surface aperture ratio	

(continued)

Table A.1 (continued)

Viewpoint	F-terms			
			MA24—void ratio (limited numerical value)	
			MA25—anisotropic, asymmetric (single layer)	
			MA26—pore distribution (MA25 is prioritized)	
			MA27—pore shape (MA25 is prioritized)	MA28—mesh
			MA30—nonporous	
		MA31—membrane thickness is distinctive		
		MA33—hollow fiber, tubular membrane with distinctive diameter		
		MA34—hollow fiber, tubular membrane with distinctive cross-section		
		MA40—other characteristics are distinctive (*)		

Source Prepared by the authors

of the search.[2] We included group companies because inventors in a corporate group could be dispatched to its group companies and file patent applications there.

There is a concern that the number of patent applications will be counted unfairly high by including patents filed independently by group companies as well. However, the number of such patent applications was only three at Toray (all Suido Kiko), zero at Nitto Denko, and nine at Toyobo (including one at Toyo Cloth and eight at Japan Exlan), confirming that there was no problem.

Based on the data extracted in this way, we next narrowed down the search to patents with any one of the F-terms AA01 (RO membrane), GA03 (RO membrane), GA04 (loose RO), or GA05 (low pressure RO) in the theme code of 4D006. Then, we checked whether there were any extraction omissions in the patent data. As a result, the number of patents that were extracted through this narrowing down process was 786 for Toray, 808 for Nitto Denko, and 225 for Toyobo. Figure A.1 shows the number of patent applications filed by each company. After that, the names of applicants and inventors were identified, and the dataset concerning the three Japanese companies was constructed.

The reason why AA01 and GA03 have the same classification for "RO membrane" is that the F-term of AA-FA is an old term assigned to patents filed up to 1988, while the F-term of GA-PC is a new term that was established after 1989. This new F-term is also retroactively assigned to patents filed up to 1988. Therefore, patents filed after 1989 are assigned only GA03, while patents filed before 1988 are assigned GA03 in addition to AA01 in principle.[3] However, there were a small number of patents to which AA01 was assigned but GA03 was not assigned and patents to

---

[2] The following is a list of specific group companies that were subject to the search.

The Nitto Denko Group: Nitto Shinko Corporation, NITOMS Corporation, Nissho Corporation, Nitto Lifetech Corporation, El Nissho Corporation, Matex Kakoh Corporation, Nitto Medical Corporation, Nitto Seiki Corporation, Fukushima Nitto Shinko Corporation, Nitto Business Expert Corporation, Nitto Logi-Com Co. Ltd., Nitto Analysis Center Corporation, and Optmate Corporation (Note: Hitachi is excluded.).

The Toray Group: Ichimura Sangyo, Co., Ltd., Ogaki Fuso Boseki Co., Ltd., Sowa Textile Co. Ltd., Maruichi Fiber Co. Ltd., Marusa Co. Ltd., Soda Aromatic Co. Ltd., Suido Kiko Kaisha Ltd., Panasonic Plasma Display Co., Ltd., Sunrich Mode Inc., Japan Apparel System Science Co., Ltd., Eitopia Co., Ltd., Kamakura Techno Science Corporation, Toyo Jitsugyo Corporation, Toyo Business Support Corporation, Chori Corporation, Ishikawa Shokusan Corporation, Okazaki Shokusan Corporation, Gifu Shokusan Corporation, Shiga Shokusan Corporation, Chiba Shokusan Corporation, Tsuchiura Shokusan Corporation, Toyo Service Co., Mishima Shokusan Corporation, and Meinan Service Co.

The Toyobo Group: Kingo Shoji Co., Ltd., Kizu Kasei Kogyo Co., Ltd., Kureha Elastomer Co. Ltd., Kureha Ltd., Goudou Shouji, COSMO Electronics Co.,Ltd., Tsuruga Film, TNC Corporation, Toyo Cloth Co., Ltd., Toyo Knit Co., Ltd., Toyoshina Films Co.,Ltd., Japan Exlan Co., Ltd., Nippon Dyneema Co. Ltd., Japan Unipet Co. Ltd., Japan U-PICA Company. Ltd., Nippon Ryutan Kogyo Co., Ltd., Mizushima Aroma Co. Ltd., Mitsumoto Kasei Corporation, Miyuki Keori Co.,Ltd., and Yuho Co.

[3] The number of patents for each company in the former F-term only is 351 for Toray, 576 for Nitto Denko, and 186 for Toyobo.

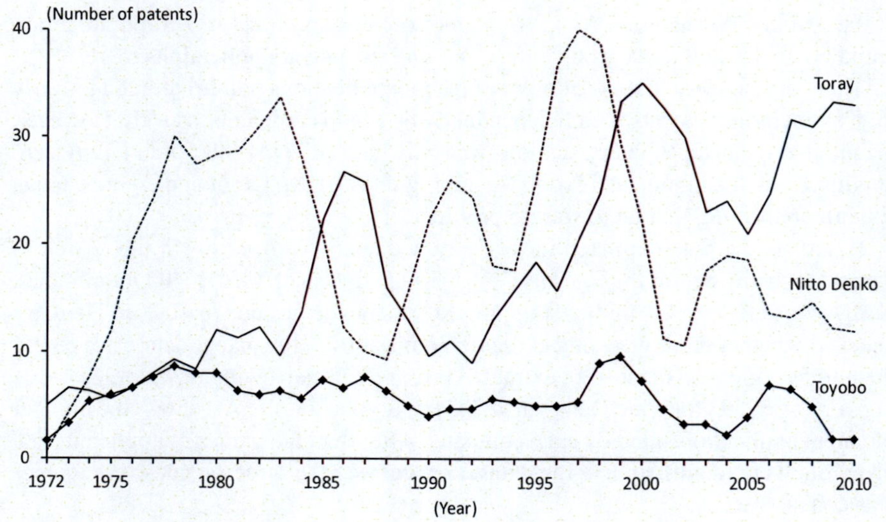

Fig. A.1 Filing trend of the three Japanese companies. *Source* Prepared by the authors based on data from UltraPatent. *Note* Three-year moving average deviation

which AA01 was not assigned but GA03 was assigned. For these patents, the above-mentioned extraction rule of "patents with one of the F-terms" is applied, and they are incorporated into the dataset without omission.

A.1.3 Dataset B: Procedure for Constructing an Industry-Wide Dataset

After developing the patent dataset for the three Japanese companies, we then extracted the patent data for the entire industry again on August 24, 2017. This was done to conduct the analysis in Chaps. 11 and 12. In this case, we selected GA03 of 4D006 initially to extract the applied patents. Since the patents were extracted in a way that included the subclassifications, the patents with GA04 (loose RO) and GA05 (low pressure RO), which are the subclassifications of GA03, were also captured. The number of patents obtained as a result is 9,296.

There are three differences from the dataset in the previous section: (1) the search target is not limited to the three Japanese companies, (2) the old term AA01 is not included in the search target, and (3) more recent data are added to the dataset. Although we first thought that the patent information of the three Japanese companies could be updated using these 9,296 new patents, we later judged that there was no necessity to conduct a huge amount of name identification work again for the purpose of analysis. Therefore, we used Dataset A for the analysis of the three Japanese companies and Dataset B for the analysis at the industry level. To be sure,

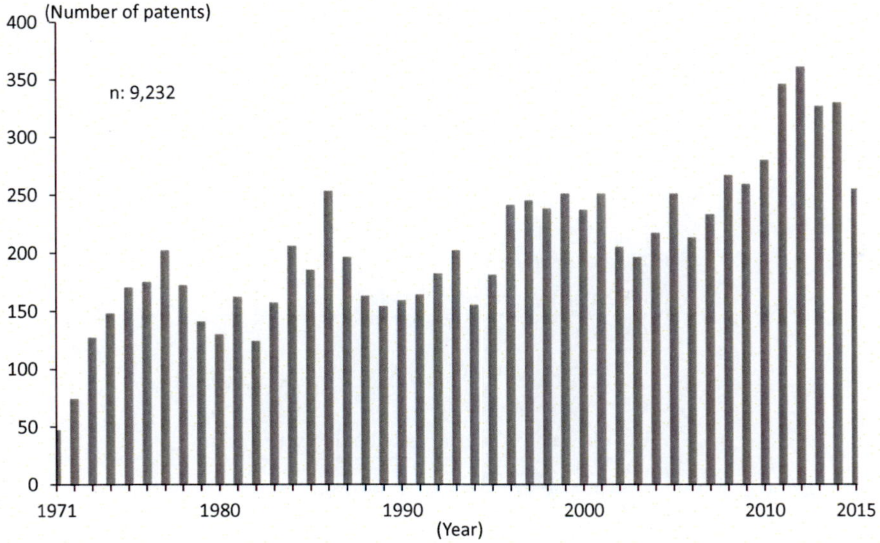

Fig. A.2 A trend of patent application by year. *Source* Prepared by the authors based on data from UltraPatent.

we compared the patent information of the period in common with dataset A and confirmed that there was no difference whatsoever.

Figure A.2 shows the trend of 9,232 RO membrane patents filed by 2015 out of 9,296 RO membrane patents granted in Japan.[4] The long-term trend in the development of RO membranes at the industrial level is an increasing one with several peaks in the late 70s and the mid 80s, around 2000, and after 2010.

A.2 Supplement for Chap. 11

A.2.1 F-Terms for Raw and Produced Water

This section is provided as supplementary material with respect to the application development activities analyzed in Chap. 11.

There are two major applications of RO membranes: the development of applications on the raw water side, which corresponds to the input, and the development of applications on the produced water side, which corresponds to the output. Application development pertaining to the raw water side refers to what kind of water is treated, while application development pertaining to the produced water side refers to where the water is to be used. As described in Chap. 11, while desalination pertains

[4] There are 64 published patents filed in 2016 and 2017, but it is thought that there are many other under examination and have not been published.

Table A.2 F-Terms by application for raw and produced water

List of F-terms for raw water		
PB01 treated liquids	PB02 usable water	PB03 seawater
		PB04 surface stream water river water, lake and marsh water
		PB05 subterranean water, well water
		PB06 tap water
	PB07 circulating water	
	PB08 liquid waste and wastewater	
	PB09 blood	
	PB12 aqueous solutions	
	PB13 non-aqueous liquid	
	PB 14 oil-water system, organic liquid-water system	
	PB15 slurry, emulsion	
	PB17 air	
	PB18 fuel gas, natural gas	
	PB19 emission gas, waste gas	
	PB20 other treated liquids	

to raw water ultrapure water production for semiconductors pertains to where the produced water is to be used.

Both raw water and produced water have a wide variety of applications. In addition to the desalination of seawater, raw water can be used to treat factory wastewater and blood. Produced water can be used in a variety of ways, such as supplying ultrapure water for semiconductor manufacturing, developing pure water for hospitals, and supplying drinking water to society at large.

Table A.2 lists the F-Terms assigned to each of these raw and produced water uses within Theme Code 4D006. Unfortunately, there is no F-term corresponding to drinking water use.

Using these F-terms as a clue, Chap. 11 analyzes how Toray, Nitto Denko, and Toyobo searched for raw water and produced water until desalination, mainly for drinking purposes, reached a practical level. Since the purpose of the analysis was to compare the three companies, we used Dataset A.

List of F-terms for produced water			
PC00 fields of use and applications	PC01 electronics and semiconductor industry	PC02 ultrapure water production	PC03 primary pure water production
			PC04 secondary pure water production
		PC05 clean rooms	
	PC11 food industry	PC12 fermentation industry	
		PC13 soybean processing	
		PC14 dairy industry	
		PC15 egg whites	
		PC16 seafood processing, edible meat	
		PC17 fruit and vegetable juice	
		PC18 sugar production, starch	
	PC21 electrodeposition coating		
	PC22 plating industry		
	PC23 oil-containing wastewater		
	PC24 degreasing wastewater		
	PC25 paper pulp		
	PC31 water for boilers (thermal power plants)		
	PC32 nuclear power plant condensate treatment		
	PC33 nuclear power plant wastewater treatment		
	PC36 isotope separation, recovery and purification of uranium		
	PC38 analysis		
	PC41 medical treatment	PC42 pharmaceutic production water	
		PC43 hospital water	
		PC44 dialysis water, dialysis fluid	
		PC45 transfusion fluid, injection solutions, and treatment filters	
		PC46 dialysis fluid for peritoneal dialysis	
		PC47 artificial kidneys	
		PC48 artificial lungs	
		PC49 contact lens cleaning fluid	
	PC51 water purification	PC52 household water purifiers	

(continued)

(continued)

List of F-terms for produced water			
		PC53 emergencies, marine emergencies	
	PC54 greywater, miscellaneous water		
	PC55 pools		
	PC56 baths		
	PC61 human waste, stored waste treatment		
	PC62 activated sludge treatment	PC63 supernatant water separation (PC65 is prioritized)	
		PC64 activated sludge condensation	
		PC65 household septic tanks	
	PC67 bioreactors		
	PC69 membrane reactors		
	PC71 oxygen enrichment		
	PC72 dehumidification		
	PC73 air-purifiers		
	PC77 gasoline vapor		
	PC80 other fields of use and applications		

Source Prepared by the authors based on the JPO website
Note F-term classification within theme code 4D006

A.2.2 Focus Areas of the Three Japanese Companies (Raw Water)

In analyzing the focus areas of the three Japanese companies pertaining to raw water, we focused on PB01 "Treated liquid" and its sub-terms, as shown in Table A.2, to confirm the application search activities of the companies. For the sake of convenience, we refer to these F-terms collectively as PB01 group F-terms. In Chap. 11, we extracted only the areas in which the companies were particularly focusing and did not present the whole picture. Therefore, in this section, the patents obtained by Toray, Nitto Denko, and Toyobo are organized according to the classification of the F-term group, and an overview of the application areas assumed by each company for raw water is presented.

The number of patents with the PB01 group F-term was 239 for Toray, 493 for Nitto Denko, and 177 for Toyobo. When conducting F-term analysis of these patents, it is important to note that some patents are assigned only one PB01 group F-term, while others are assigned multiple PB01 group F-terms. Since the F-term is assigned by a third party outside the company, a patent with only one specific F-term is

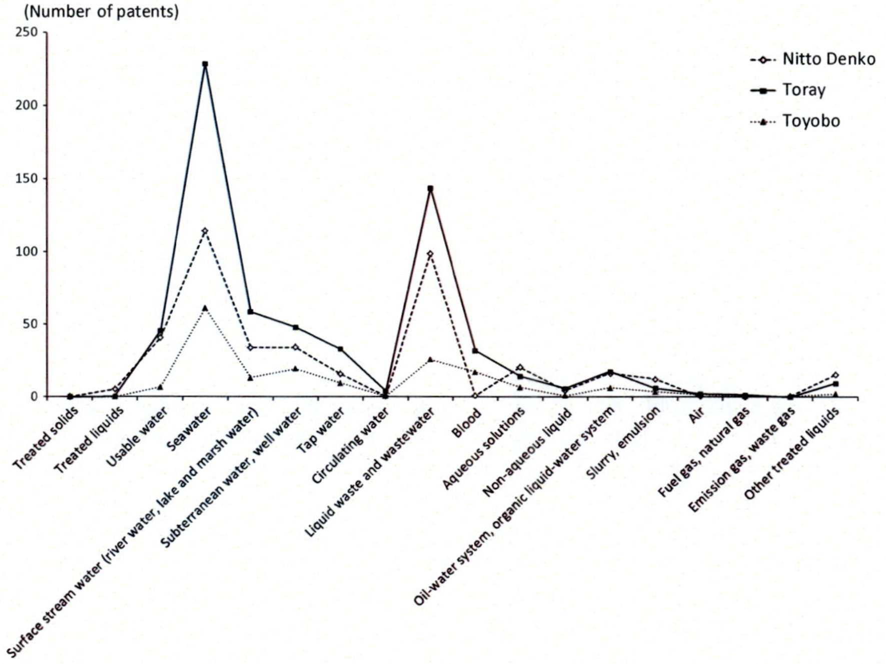

Fig. A.3 Comparison of three companies on expected use (raw water). *Source* Prepared by the authors based on data from UltraPatent

considered to be a technology developed for a more narrowly focused application in the eyes of a third party.

Therefore, we quantified the application-relatedness of each patent by weighting it with the reciprocal of the number of PB01 group F-terms assigned to it. Specifically, if the number of PB01 group F-terms assigned to a patent is n, 1/n is assigned to each PB01 group F-term. For example, in the case of a patent that has only one PB01 group F-term, we assigned 1 (=1/1) to the F-term. On the other hand, in the case of a patent that has three PB01 group F-terms, each of the three F-terms is assigned 1/3.

Based on these methods, we calculated which PB01 group F-terms are assigned to each patent filed by each company (Toray: 239 patents, Nitto Denko: 493 patents, and Toyobo: 177 patents). The results are shown in Fig. A.3. From the figure, it can be seen that all three companies put the most effort into the development of "seawater" as raw water, followed by the treatment of "liquid waste and wastewater". From this, we can confirm that all three companies have developed membranes with a strong focus on "seawater" and "liquid waste and wastewater" as raw water.[5]

[5] In the Figure, some readers may wonder why the value given to water (PB08), which is the upper level of the F-term, is low even though many values are given to seawater and surface water. This is because, based on the original F-term assignment rules, this higher level F-term is assigned to patents of other water that does not apply to seawater, surface water, groundwater, and tap water.

Looking at the differences among the companies, Toray's specialization in "seawater" is clearly evident. For Nitto Denko, "seawater" and "liquid waste and wastewater" are in close competition. This partially confirms that Nitto Denko, which claims to be a department store for membranes, has been developing its products for water treatment other than seawater. However, Nitto Denko's value for "blood" was low, perhaps suggesting that the company was not focusing on development for hospitals.

A.2.3 Focus Areas of the Three Japanese Companies (Production Water)

Next, we will focus on the PC00 group F-term for produced water and show the overall picture of the search activities for produced water use by the three Japanese companies in the same way as in the case of raw water. However, unlike the PB01 group for raw water, the PC00 group for produced water contains as many as 51 F-terms, including the PC00 term itself. Since it is difficult to see all of them in a graph, we grouped them into the major categories.

The major categories of the PC00 group are as follows: electronics and semiconductor industry (PC01), food industry (PC11), electrodeposition coating (PC21), plating industry (PC22), oil-containing wastewater (PC23), degreasing wastewater (PC24), paper pulp (PC25), water for boilers (thermal power plants) (PC31), nuclear power plant condensate treatment (PC32), nuclear power plant wastewater treatment (PC33), isotope separation, recovery and purification of uranium (PC36), analysis (PC38), medical use (PC41), water purification (PC51), greywater, miscellaneous water (PC54), pools (PC55), baths (PC56), human waste, stored waste treatment (PC61), activated sludge treatment (PC62), bioreactors (PC67), membrane reactors (PC69), oxygen enrichment (PC71), dehumidification (PC72), air purifiers (PC73), gasoline vapor (PC77) and other fields and applications (PC80). As mentioned earlier, potable water applications are not included in this category and are likely to be included in other fields of use and applications.

Some of these terms have more detailed sub-terms. For example, ultrapure water production (PC02) and clean rooms (PC05) branch out from the electronics and semiconductor industry (PC01). Primary pure water production (PC03) and secondary pure water production (PC04) branch out from ultrapure water production (PC02). Thus, when a term has subordinate terms, no matter which hierarchy of subordinate terms is given, it is assigned an F-term in the large category, so we decided to assign a value to the F-term in the large category. For example, if a patent is assigned to secondary pure water production (PC04), we assigned a value of 1/n to PC01, considering that electronic industry and semiconductor industry (PC01) were assigned to the patent. Here, n is the number of the large classification PC group F-term assigned to the patent.

Figure A.4 shows the results of the three-company comparison on the uses of produced water based on the above procedures. The top five applications for

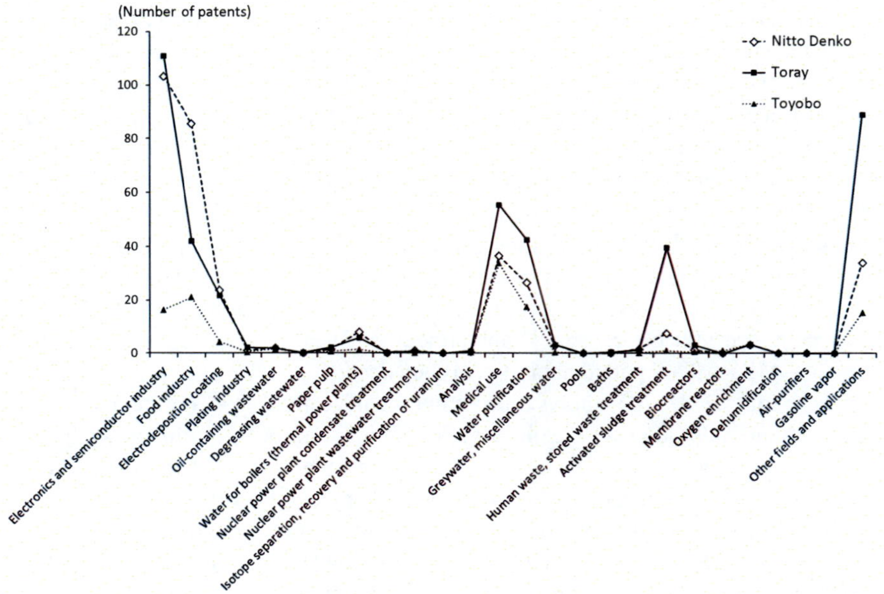

Fig. A.4 Comparison of three companies regarding expected applications Source: Prepared by the authors based on UltraPatent.

all three companies are the same: electronics and semiconductor industry, food industry, medical use, water purification, and other fields and applications. When narrowing the list to the top three applications, Toray's top three applications were electronics and semiconductor industry, other fields and applications, and medical use. Nitto Denko's top three applications were electronics and semiconductor industry, food industry, and medical use. Toyobo's top three applications were medical use, food industry, and electronics and semiconductor industry. Given the large number of "other fields and applications" for Toray, potable water applications assumed to account for the majority of this category.

A.3 Supplement for Chap. 12

A.3.1 Distribution of Developed Materials

As clarified in Chap. 12, the major materials used in RO membranes were converted from cellulose acetate to polyamide, and the membrane shapes converged to the flat membrane type. The procedure taken to depict these conversion processes shown in Fig. 12.1 is as follows. Since this is an analysis of the entire industry, Dataset B was used.

First, out of 9,296 total patents, we extracted 3,527 patents that are assigned to the MB group, which is the F-term related to "properties of membranes" in the theme code 4D006 "separation using semipermeable membranes". Next, we extracted patents with the F-term "polymer organic material" in group MC09 with its subclassifications, which is included in group MC00 that indicates "material of the membrane".[6] Then, since the specific names of membrane materials such as cellulose and polyamide appear in the two lower levels of the MC09 group, we extracted 2,797 patents that can be organized by the F-term of that level and classified them by the materials used.[7] The patent data organized in this way are called the MB × MC09 patent data for convenience. Figure 12.1 shows the ratio of the number of patents assigned to the MC11 group "cellulose", the MC54 group "polyamide", and the MC62 group "polysulfone" to the total MB × MC09 patent data filed in each decade.

In addition, Table A.3 shows the comprehensive transition of the materials, including materials other than cellulose, polyamide, and polysulfone as well. The reason why the total ratio of each F-term group is much higher than 100% is that more than one F-term group is assigned to a single patent in many cases.

This table shows that the largest number of patents still concern polyamide (1,621), polysulfone (1,515), and cellulose (1,291). The reason for the large number of polysulfone patents is that, as noted in Part 4, the development of the support layer progressed as the composite structure became popular.

Table A.3 shows that 62.8% of the MB × MC09 patent data filed in the 1970s had the cellulose F-term. However, this percentage dropped to 36.51% in the 1990s. Conversely, polyamide has been increasing. In the 1970s, only 38.82% of MB × MC09 patent applications were assigned the F-term for polyamide, but this percentage has since increased to 72.94% in the 2010s, meaning that more than 70% of MB × MC09 patent applications filed in the 2010s were assigned the F-term for polyamide. Polysulfone, which is responsible for the support layer, also increased from only 27.67% in the 1970s to 66.67% in the 2010s

A.3.2 *From Cellulose Acetate to Polyamide Flat Sheet Membranes*

As the next step, we add information on the membrane shape to that on the membrane material to indicate how the technological domains converged.

Figure A.5a shows decade trends categorized by technological domain for 805 patents for which the material and shape of membrane were identified out of 9,296

[6] In other words, the target ranges up to MC69.

[7] The reason why we did not limit the analysis to the patents belonging to each group only (696 patents) is that, as a result of the progress in the composite of membranes, there are cases where the F-term of the membrane for the support layer is given at the same time even when the membrane only for the separation layer is developed. If the patents belonging exclusively to each group were to be analyzed, the number of patents related to composite membranes would be unreasonably small.

Table A.3 Number of patents by F-term group for each decade

4th layer	F-term		Total	Breakdown by decade				
				1970s	1980s	1990s	2000s	2010s
MC11 group	Cellulose	Patent count	1291	309	294	207	227	254
		%	46.16	62.80	42.12	36.51	42.83	49.80
MC22 group	Polyolefin	Patent count	845	94	203	178	151	219
		%	30.21	19.11	29.08	31.39	28.49	42.94
MC24 group	Polyvinyl aromatics (polystyrene, etc.)	Patent count	331	94	109	60	27	41
		%	11.83	19.11	15.62	10.58	5.09	8.04
MC25 group	Polyalkenyl halogen	Patent count	1029	125	240	166	204	294
		%	36.79	25.41	34.38	29.28	38.49	57.65
MC32 group	Polyalkenyl alcohol and derivatives	Patent count	655	112	191	131	102	119
		%	23.42	22.76	27.36	23.10	19.25	23.33
MC35 group	Unsaturated acid polymer and derivatives	Patent count	959	182	259	176	147	195
		%	34.29	36.99	37.11	31.04	27.74	38.24
MC40 group	Nitrogen heterocycles in side chain	Patent count	245	62	66	35	35	47
		%	8.76	12.60	9.46	6.17	6.60	9.22
MC42 group	Polyacetylenes	Patent count	22	1	13	5	1	2
		%	0.79	0.20	1.86	0.88	0.19	0.39
MC45 group	Polyether	Patent count	546	95	136	99	88	128
		%	19.52	19.31	19.48	17.46	16.60	25.10
MC48 group	Polyester	Patent count	636	107	159	90	103	177
		%	22.74	21.75	22.78	15.87	19.43	34.71
MC49 group	Polycarbonate	Patent count	252	39	87	27	31	68
		%	9.01	7.93	12.46	4.76	5.85	13.33
MC50 group	Epoxy resin	Patent count	73	9	16	6	6	36
		%	2.61	1.83	2.29	1.06	1.13	7.06
MC52 group	Polyurea	Patent count	106	19	36	26	13	12
		%	3.79	3.86	5.16	4.59	2.45	2.35

(continued)

Table A.3 (continued)

4th layer	F-term		Total	Breakdown by decade				
				1970s	1980s	1990s	2000s	2010s
MC53 group	Polyurethane	Patent count	120	20	26	21	18	35
		%	4.29	4.07	3.72	3.70	3.40	6.86
MC54 group	Polyamide	Patent count	1621	191	333	359	366	372
		%	57.95	38.82	47.71	63.32	69.06	72.94
MC57 group	Has nitrogen heterocycles	Patent count	774	98	219	159	118	180
		%	27.67	19.92	31.38	28.04	22.26	35.29
MC60 group	Polyamine	Patent count	184	49	67	37	12	19
		%	6.58	9.96	9.60	6.53	2.26	3.73
MC62 group	Polysulfone	Patent count	1515	137	425	338	275	340
		%	54.17	27.85	60.89	59.61	51.89	66.67
MC66 group	Also has nitrogen/sulfur in main chain	Patent count	18	2	7	3	1	5
		%	0.64	0.41	1.00	0.53	0.19	0.98
Total number of patents					2797	492	698	567

Source Prepared by the authors based on data from UltraPatent.
Note Because some patents overlap in multiple F-term assignations, the sum of the patent count does not match the total number of patents.

patents filed in Japan.[8] Here, the technological domain is shown by combining the material and shape of the membrane. Figure Appendix 5b shows the trends categorized by technological domain for 351 patents (patents considered to be related to composite membranes) with F-terms related to composite structures out of 805 patents.

Figure A.5a shows that in the 1970s, tubular, hollow fiber, and flat sheet membranes were developed relatively dispersedly using cellulosic materials. The development of tubular membranes, which was the mainstream at first, was almost over in the 1970s, and hollow fiber membranes became the mainstream in the 1980s. On the other hand, development using polyamide materials rapidly increased after the 1980s, for which development of flat sheet membrane is the mainstream.

[8] Of these 9296 patents, we excluded the patents with F-terms across the technological domains from the dataset and extracted 835 patents which F-terms indicate only a certain technological area. Of these 835 patents, 805 patents had F-terms related to either tubular, hollow fiber, or flat membranes, rather than "other" membrane shapes. We classified these 805 patents into six combinations of materials and shapes (called technological domains).

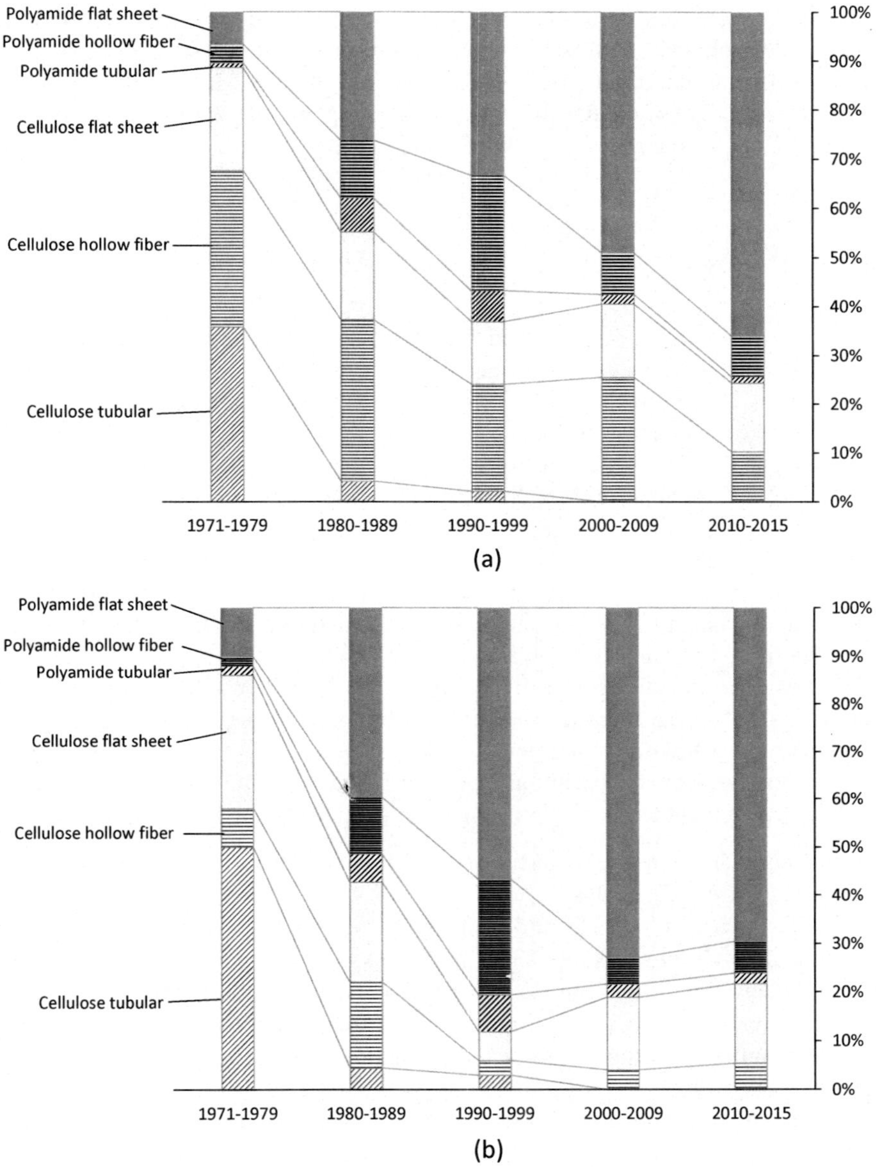

Fig. A.5 Share of patent applications by technological domain. **a** Share of patents by technological domain (total: n = 805). **b** Share of patents by technological domain (limited to the patents in the composite membrane F-term MA06 group: n = 351). *Source* Prepared by the authors based on UltraPatent

Figure A.5b, which shows the patents related to composite membranes, shows that the development of tubular membranes was significant in the 1970s but quickly shifted to the development of flat sheet membranes. Many cellulose-based membranes are also of the flat sheet type, suggesting that flat sheet membranes are suitable for composite sheets regardless of the material.

A.3.3 *Procedure for Each Area in the Value Chain*

In Figs. 12.8 and 12.9, we classified the patents into three areas along the value chain: the membrane system, membrane module system, and operation system. The same categorization was also performed in Fig. 13-Appendix. 1a&b. The procedure adopted in these cases is as follows.

Looking at the viewpoints starting from GA in the theme code 4D006, they are in the order of membrane area (MA, MB, MC, NA), membrane module area (HA, JA, JB), and operation area (KA, KB, KC, KD, KE, LA). Using the F-terms belonging to these viewpoints, we classified which area of the value chain each patent is related to. However, as we have noted in the past, there are many cases in which multiple F-terms are assigned to a single patent, so we decided to adopt the following rules when classifying patents into each area.

The basic classification rule is to identify which of the three areas is most heavily assigned the F-Term and then to designate that area as the subject area of the patent. As mentioned above, there are four F-terms in the membrane area, three in the membrane module area, and six in the operation area as the first level, as the major classification. We call them "parent terms" for convenience and then take the following five steps.

(1) Patents that are not assigned a single F-term for any of the three areas are excluded from the analysis.
    In the case of Nitto Denko and Toray in Dataset A, two patents of Nitto Denko (Patent Application 2008-245372 and Patent Application Hei08-113407) were excluded from the analysis. In Dataset B, 41 patents fit this case.
(2) If a patent is assigned one or more F-Terms that are in the lower layer of the parent term of a certain area, the corresponding parent term is assumed to have been assigned.
    Example: If MA06 from the second level is assigned to a certain patent, we consider that MA has been assigned to the patent and recognize it as an MA-related patent. Also if a patent is assigned MA08 from the third level, it is recognized as an MA-related patent in the same way. At this time, we do not distinguish how many F-terms belonging to MA are assigned to the patent in question.
(3) We calculate the percentage of parent terms assigned to each of the membranes area, the membrane modules area, and the operation area and refer to this as the "patent area ratio" in each area for convenience.

For example, there are four parent F-terms, MA, MB, MC and NA, in the membrane area. If the F-term for MA and MC is assigned to a patent, the ratio for the membrane area is 50% (2/4). Similarly, if only HA is granted in the patent among the three terms HA, JA, and JB, which belong to the area of the membrane module, the patent area ratio for the membrane module is 1/3, or 33.3%.

(4) We compare the area ratios calculated for the membrane area, the membrane module area, and the operation area and identify the area with the highest ratio as the subject area of the patent.
Example: If a patent has two parent terms for the membrane area, one parent term for the membrane module area, and two parent terms for the operation area, the ratio for each area is 50%, 33%, and 33%, respectively, and the patent is judged to be in the membrane area.

(5) If the ratio is the same in several areas and is the highest value, the patent is identified as a multi-area one, and a value of 0.5 is given to each area.
For example, if the ratio is 50% in the membrane area, 33.3% in the membrane module area, and 50% in the operation area, the patent is considered to be a composite area patent of the membrane and operation areas, and 0.5 is given to both areas.

Since the number of patent applications fluctuates greatly from one year to the next, the patents analyzed in the Appendix of Chap. 13 are grouped by decade, and the ratios of membrane-related patents, membrane-module-related patents, and operation-related patents to the total number of patent applications in each decade are calculated to mitigate the effects of short-term fluctuations.

Interview and Visit List

The following industry professionals have kindly agreed to be interviewed. We would like to express our sincere appreciation for their cooperation in spite of their busy schedules. The affiliations are current at the time of the interviews.

U.S. Companies

Name	Company name/institution name (as of interview date)	Date	Place
Baker, Richard	Membrane Technology and Research	March 3, 2015; June 21–24, 2018	California; e-mail
Riley, Robert (Bob)	Separation Systems Technology, Inc	March 5, 2015; June 21–25, 2018	California; e-mail
Truby, Randy	RL TRUBY & Associates	March 4, 2015	California
Eriksson, Peter	GE Power & Water, GE Water & Process Technologies	March 6, 2015	California
Maeda, Yasushi	Dow Chemical Japan Limited	November 13, 2014	Tokyo
Peery, Martin	Dow Water & Process Solutions, The Dow Chemical Company	March 2, 2015	Minnesota
Rosenberg, Steve	Dow Water & Process Solutions, The Dow Chemical Company	March 2, 2015	Minnesota

M. Fujiwara and Y. Aoshima, *Mechanisms for Long-Term Innovation*, Advances in Japanese Business and Economics 31,
https://doi.org/10.1007/978-981-19-4896-1

Institutions in Japan

Name	Company name/institution name (as of interview date)	Date	Place
Ohya, Haruhiko	Formerly of Yokohama National University	October 15, 2014; May 19–21, 2018	Tokyo; Letter
Hirai, Mitsuyoshi	Water Reuse Promotion Center	March 16, 2015	Tokyo
Taniguchi, Yoshio	Water Reuse Promotion Center	March 16, 2015	Tokyo

Desalination Plants

Name	Company name/institution name (as of interview date)	Date	Place
Higa, Yoshio	Okinawa Prefectural Enterprise Bureau	June 22, 2012	Okinawa
Nakasone, Tomonori	Okinawa Prefectural Enterprise Bureau	June 22, 2012	Okinawa
Isa, Tomoaki	Okinawa Prefectural Enterprise Bureau	June 22, 2012	Okinawa
Matayoshi, Takayuki	Okinawa Waterworks Control Center Co., Ltd	June 22, 2012	Okinawa
Morita, Yukio	Fukuoka District Waterworks Agency	August 28, 2013	Fukuoka
Abdulhadi Al Sheikh	Saline Water Conversion Corporation	February 2, 2016	Jeddah
Mohammed Al Thubaiti	Saline Water Conversion Corporation	February 2, 2016	Jeddah
Yaser Zaki Al-Jehani	Saline Water Conversion Corporation	February 3, 2016	Jeddah

Toyobo Co., Ltd.

Name	Company name/institution name (as of interview date)	Date	Place
Ariji, Akihiro	Toyobo Co., Ltd	February 1, 2016; February 2, 2016	Jedda; Rabigh
Ikeda, Kazuhito	Toyobo Co., Ltd	June 16, 2015	Yamaguchi
Ukai, Tetsuo	Formerly of Toyobo Co., Ltd	December 13, 2013; January 24, 2014	Osaka; Shiga
Uchimura, Tsuyoshi	Arabian Japanese Membrane Company	February 1, 2016; February 2, 2016	Jedda; Rabigh

(continued)

(continued)

Name	Company name/institution name (as of interview date)	Date	Place
Oita, Yuji	Toyobo Co., Ltd	August 4, 2016	Osaka
Konagaya, Shigetsugu	Nagoya University (Formerly of Toyobo)	January 15, 2015	Aichi
Katsube, Mikio	Toyobo Co., Ltd	June 16, 2015	Yamaguchi
Kumano, Atsuo	Toyobo Co., Ltd	October 11, 2013; November 16, 2013; December 13, 2013; January 24, 2014; January 24, 2014; March 26, 2014; March 27, 2014; December 17, 2014; June 16, 2015; April 8, 2016; April 3–4, 2018; May 3–4, 2018; June 23–25, 2018	Osaka, Shiga, Tokyo, Yamaguchi, and e-mail
Sakamoto, Ryuzo	Toyobo Co., Ltd	April 8, 2016; June 16, 2016; August 4, 2016; October 31, 2016	Osaka
Sato, Hiroyuki	Toyobo Co., Ltd	April 8, 2016	Osaka
Shigekiyo, Masahiko	Toyobo Co., Ltd	June 16, 2015	Yamaguchi
Tanaka, Satoshi	Arabian Japanese Membrane Company	February 1, 2016; February 2, 2016	Jedda; Rabigh
Takeuchi, Ikuo	Toyobo Co., Ltd	January 27, 2017	Tokyo
Tsumura, Junji	Toyobo Co., Ltd	October 31, 2016	Osaka
Narahara, Seiji	Toyobo Co., Ltd	April 8, 2016; August 4, 2016	Osaka
Fujiwara, Nobuya	Toyobo Co., Ltd	October 11, 2013; November 16, 2013; December 17, 2014; June 16, 2015; February 2, 2016; April 8, 2016; June 16, 2016	Osaka, Tokyo, Yamaguchi, and Rabigh,
Fuse, Tomonori	Toyobo Co., Ltd	June 16, 2015	Yamaguchi
Saeed Saad AlHarthi	Arabian Japanese Membrane Company	February 2, 2016	Jeddah
Suzuki, Toshitake	Toyobo Co., Ltd	September 12, 2016	Fukui

Toray Industries, Inc.

Name	Company name/institution name (as of interview date)	Date	Place
Inoue, Takeharu	Toray Industries, Inc	August 26, 2010	Ehime
Ueda, Fujio	Toray Industries, Inc	August 26, 2010	Ehime
Ueno, Kenji	Toray Industries, Inc	July 15, 2010; August 26, 2010	Tokyo; Ehime
Uemura, Tadahiro	Toray Industries, Inc	July 15, 2010; August 30, 2010; July 14, 2015; November 18, 2015; April 6, 2018; May 3, 2018	Tokyo, Shiga, and e-mail
Kawabata, Tatsuo	The Democratic Party (Formerly of Toray)	April 25, 2014; March 15, 2016	Shiga
Kurihara, Masaru	Toray Industries, Inc	August 30, 2010; February 6, 2018	Shiga
Toyama, Motoyuki	Toray Industries, Inc	August 26, 2010	Ehime
Nishioka, Eiji	Toray Industries, Inc	August 26, 2010	Ehime
Himeshima, Yoshio	Toray Industries, Inc	August 19, 2010	Tokyo
Fukui, Fumiaki	Toray Industries, Inc	August 26, 2010	Ehime
Fusaoka, Yoshinari	Toray Industries, Inc	July 15, 2010; August 30, 2010	Tokyo; Shiga
Mitsuiri, Seiji	Toray Industries, Inc	August 26, 2010	Ehime
Miyata, Kazuhiro	Toray Industries, Inc	November 18, 2015	Tokyo
Yamamoto, Masanori	Toray Industries, Inc	August 26, 2010	Ehime

Nitto Denko Corporation

Name	Company name/institution name (as of interview date)	Date	Place
Ikeda, Kenichi	Research Institute of Innovative Technology for the Earth (Formerly of Nitto Denko Corporation)	November 11, 2013; December 27, 2013; January 24, 2014; February 7, 2014; February 8, 2014; May 3–10, 2018	Kyoto, Tokyo, and e-mail
Kamiyama, Yoshiyasu	Formerly of Nitto Denko Corporation	August 25, 2014	Okayama
Shintani, Takuji	Nitto Denko Corporation	December 12, 2014	Shiga

(continued)

(continued)

Name	Company name/institution name (as of interview date)	Date	Place
Nakagome, Keisuke	Formerly of Nitto Denko Corporation	April 11, 2014; April 13, 2014; July 23, 2014; September 23, 2014	Tokyo and e-mail
Hachisuka, Hisao	Hydranautics	March 4, 2015	California
Yamamoto, Hideki	Nitto Denko Corporation	August 29, 2014; December 12, 2014	Osaka
Yoshioka, Noriaki	Nitto Denko Corporation	August 29, 2014; December 12, 2014	Osaka

Visit List

Name	Date	Place
Okinawa Seawater Desalination Center	June 22, 2012	Okinawa
Fukuoka Seawater Desalination Center	August 28, 2013	Fukuoka
45th Autumn Meeting, Society for Chemical Engineers, Japan	September 17, 2013	Okayama
FY2014 activity report meeting, Creating green innovation in the water business with innovative membrane engineering at its core	August 22, 2014	Hyogo
Saline Water Conversion Corporation	February 3, 2016	Jeddah
NEWater Visiting Centre	May 4, 2018	Singapore

A Chronological Table of RO Membrane Development (1949–2014)

M. Fujiwara and Y. Aoshima, *Mechanisms for Long-Term Innovation*, Advances in Japanese Business and Economics 31,
https://doi.org/10.1007/978-981-19-4896-1

The United States				Japan					
Public sector		Private company		Toray		Toyobo		Nitto Denko	
1949	G. Hustler (UCLA) "The sea as a source of fresh water"								
1952	The Saline Water Act								
1955	Office of Saline Water								
1959	C. E. Reid (Univ. of Florida) Desalination demonstration								
1960	Loeb and Sourirajan (UCLA) Cellulose acetate-based asymmetric membrane								
		1961	GA began development						
		1962	DuPont began development						

(continued)

(continued)

The United States				Japan					
Public sector		Private company		Toray		Toyobo		Nitto Denko	
1964	North Star Cellulose acetate-based composite membranes								
1967	North Star Non-cellulose support layer	1967	DuPont B-5	1967	Started survey				
				1968	Officially started R&D (CAT project)				
		1969	DuPont B-9						
1970	North Star NS-100								
				1971	Integrated three research groups	1971	Started survey		
1972	North Star NS-200	1972	DuPont B-10	1972	Formed PEC Group	1972	Moved from survey to exploration		
								1973	Officially started R&D
		1974	UOP acquired GA and formed Fluid Systems					1974	S. Hijikata appointed president

(continued)

(continued)

The United States				Japan					
Public sector		Private company		Toray		Toyobo		Nitto Denko	
		1975	Fluid (UOP) PA-300	1975	Developers moved to Development Dept	1975	Selected cellulose acetate-based materials		
								1976	Delivered for starch wastewater treatment (Hokkaido)
		1977	FilmTec was founded						
		1978	FilmTec FT-30			1978	RO Business Development Dept	1978	H. Yamamoto appointed to develop the membrane business Began developing composite membranes
						1979	Iwakuni plant	1979	Integrated research groups
		1980	Dow Chemical low-pressure cellulose triacetate hollow fiber membrane	1980	PEC Group moved to Development Dept	1980	Iwakuni Plant began operations	1980	Membrane Module Development Department NTR-7197 and NTR-7199

(continued)

(continued)

The United States				Japan					
Public sector		Private company		Toray		Toyobo		Nitto Denko	
		1981	J. E. Cadotte (FilmTec) US4277344 patent granted			1981	Demonstrations in Saudi Arabia	1981	Membrane Module Business Promotion Dept
				1982	PEC2000 developers moved to Development Dept	1982	Announced Three-year Plan to raise its membrane sales to 2 billion yen in 1984	1982	Sales reached 600 million yen
				1983	Sales reached approx. 2 billion yen	1983	Functional Membrane Division	1983	Sales reached 1 billion yen NTR-7250
				1984	Announced the plan to increase membrane sales to 10 billion yen in 3 years	1984	Received orders from Saudi Arabia	1984	Announced plan to increase membrane sales to 6 billion yen in 1987
		1985	Dow Chemical acquired FilmTec	1985	RO Production Dept				
						1986	Received orders for Jiddah 1 Phase 1	1986	Membrane Division Shiga Plant
				1987	K. Maeda appointed president			1987	Acquired Hydranautics
								1988	NTR-759

(continued)

(continued)

The United States				Japan					
Public sector		Private company		Toray		Toyobo		Nitto Denko	
						1989	Jeddah 1 Phase 1 operation started		
								1990	Patent dispute began
				1991	SU-800	1991	Jeddah 1 Phase 1 troubled	1991	Emergency project
						1992	Jeddah 1 Phase 1 trouble resolved Received orders for the Phase 2		
						1993	Polyamide hollow fiber membranes for semiconductor applications	1993	344 patent was confirmed to belong to the U.S. government
		1994	J. E. Cadotte retired	1994	K. Maeda ordered developed to build an efficient system	1994	Phase 2 operation started	1994	Turned a single-year profit
								1995	ES-10
								1996	H. Yamamoto appointed president
				1997	BCS system			1997	LF-10
		1998	Koch acquired Fluid Systems			1998	Turned a single-year profit		
		1999	DuPont announced withdrawal			1999	J. Tsumura appointed president		

(continued)

(continued)

The United States				Japan					
Public sector		Private company		Toray		Toyobo		Nitto Denko	
						2000	Received an order for seawater desalination plant in Fukuoka		
				2002	S. Sakakibara appointed president	2002	HB9155	2002	Membrane sales reached 7.9 billion yen
				2004	Suido kiko became Subsidiary	2004	Eliminated accumulated deficits	2004	Membrane sales exceeded 13 billion yen
				2005	Global sales team	2005	R. Sakamoto appointed president HJ9155	2005	Membrane sales reached 14 billion yen
				2006	Prioritized the water treatment business	2006	Iwakuni Plant capacity increased 1.5 times	2006	Membrane sales reached 17.7 billion yen
				2010	A. Nikkaku appointed president				
						2012	Saudi Arabia plant started operation		
				2014	Re-prioritized the water treatment business				

Source Compiled from the authors' interviews, newspaper, and magazine articles
Note The year of development fluctuates depending on the delimitation method, such as patent application year, publication year, paper publication year, news release year, product realization year, and market launch year. Therefore, there may be discrepancies with other papers. In fact, even the developers themselves may note different development years in different papers

Bibliography

Abell, D. F. (1980). *Defining the business: The starting point of strategic planning.* Prentice-Hall.

Abernathy, W. J., & Utterback, J. M. (1978). Patterns of industrial innovation. *Technology Review, 80*, 40–47.

Akgul, D., Çakmakcı, M., Kayaalp, N., & Koyuncu, I. (2008). Cost analysis of seawater desalination with reverse osmosis in Turkey. *Desalination, 220*, 123–131.

American Water Works Association 2007. *Reverse Osmosis and Nanofiltration (2nd edition).* American Water Works Association.

Aoki, M., & Takeyama, T. (1970). (2) Sangyō shizai ((2) Industrial Materials) in *Sen-i Kōgaku (Fiber Engineering), 23*, 21–26.

Applegate, L. E. (1981). Effect of seven years RO service on aramid hollow fiber permeators, technical proceedings. In *9th Annual Conference and International Trade Fair of the National Water Supply Improvement Association, 2.*

Boehiert, C. R. (Annual editions). *Legislative History: Saline Water Conversion Act.* Washington D. C.: Office of Saline Water, U.S. Department of Interior.

Breton, E. J. (1957a). Water and ion flow through imperfect osmotic membranes. *Ph.D dissertation at the University of Florida.* http://www.ufdc.ufl.edu/AA00003986/00001/.

Cooper, R. G. (2011). *Winning at new products: Creating value through innovation.* Basic Books.

Cushnie, G. (2009). *Pollution prevention and control technologies for plating operations*, 2nd edn. Ann Arbor: National Center for Manufacturing Sciences (NCMS).

Denryoku Chuo Kenkyusho Otemachi Kenkyusho (Otemachi Research Laboratory, Central Research Institute of Electric Power Industry). (1968). *Kenkyu hokoku dai III shu (Research report No. III)* (pp. 152–167).

Department of Finance, State of California. http://www.dof.ca.gov. Accessed 6 May 2018.

Dow Jones & Co. (1995). *The dow jones guide to the world stock market*. Prentice Hall.

Drioli, E., & Giorno, L. (Eds.). (2010). *Comprehensive membrane science and engineering* (Vol. 1). Elsevier Science.

Food and Agriculture Organization of the United Nations AQUASTAT website. http://www.fao.org/nr/water/aquastat/data/query/index.html?lang=en. Accessed 9 Apr 2018.

Francis, P. S. (1966). *Fabrication and Evaluation of New Ultrathin Reverse Osmosis Membranes. Research and Development Progress Report*, No. 177. Washington, D.C.: U.S. Dept. of the Interior.

Fuji Keizai (2016b). *Mizushigen kanrenshijō no genjō to shōraitembō 2016 (Current status and future outlook of water resources related markets 2016).*

M. Fujiwara and Y. Aoshima, *Mechanisms for Long-Term Innovation*, Advances in Japanese Business and Economics 31,
https://doi.org/10.1007/978-981-19-4896-1

Fujiwara, M., & Aoshima, Y. (2014b). Nitto Denko kabushiki gaisha: Gyakushintō maku no kaihatsu to yōto kaitaku (Nitto Denko Corporation: Development and application of reverse osmosis membranes). *IIR Case Study*, CASE#14–02.

Fujiwara, M., & Aoshima, Y. (2014a). Toyobo: Development and business expansion of reverse osmosis membrane. *Hitotsubashi Business Review, 62*, 102–119.

Fujiwara, M., Aoshima, Y., & Miki, T. (2011). Toray: Gyakushintō maku Jigyō no sōzō process (Toray: Creation process of reverse osmosis membrane business). *Hitotsubashi Business Review*, *59*, 150–167.

Fujiyama, K. (2015b). Influences of business development approaches: A case of reverse osmosis industry. *Journal of Business Management, 35*, 28–40.

Fujiyama, K. (2015a). Nitto Denko: Gyakushintō maku shijō ni okeru share gyakuten no process (Nitto Denko: The process of market share reversal in the reverse osmosis membrane market). *Hitotsubashi Business Review*, *62*, 126–141.

Funaki, H. (2009a). Mizu no sekai o lead suru nihon no 'maku' gijutsu (1): Kōdatsuen/Kōtōka o jitsugen shita Nitto Denko no 'gyakushintō maku' (Japan's 'membrane' technology leading the water world (1): Nitto Denko's 'reverse osmosis membrane' realizes high desalination and permeation. *Foresight*, February 2009a, 68–71.

Funaki, H. (2009b). Mizu no sekai o lead suru Nihon no 'maku' gijutsu (2): Kyū seichō no 'osui sairiyō' de kisou Toray to Mitsubishi Rayon (Japan's 'membrane' technology leading the water world (2): Toray and Mitsubishi Rayon competing in fast-growing 'sewage reuse'). *Foresight*, March 2009b, 60–62.

Global Water Intelligence. (2008). *Water Desalination Report*, *44*, 2008. http://www.nmbworks.com/other_pdf/WDR-5.pdf.

Gotō, F. ((1979). Kaisui tansuika no gijutsu kaihatsu wa ikani okonawaretaka (How was the technological development of seawater desalination carried out?). *Denki gakkai zasshi (Journal of the Institute of Electrical Engineers of Japan)*, *99*, 40–44.

Hassler, G. L., & McCutchan J. W. (1960). *Saline Water Conversion. Advances in Chemistry Series*, *No.27*. Washington DC: American Chemistry Society.

Hightower, S., Price, K., & Henthorne, L. (1994). The US bureau of reclamation's research programs in water treatment and desalting technologies. *Desalination, 99*, 201–210.

Hoffmann, V. H., Trautmann, T., & Hamprecht, J. (2009). Regulatory uncertainty: A reason to postpone investments? Not necessarily. *Journal of Management Studies, 46*, 1227–1253.

Holum, K. (1970). *Recorded interview by William W. Moss, John F. Kennedy Library Oral History Program.*

http://www.id.loc.gov/authorities/names/n50057680. Accessed 23 Feb 2015.

http://www.carlsbad-desal.com/news.aspx?id=99

http://www.caselaw.findlaw.com/us-federal-circuit/1287586.html

http://www.fivecitieswater.com/Well_Water_Treatment/0803Executive_Insight.pdf. Accessed 3 June 2018.

http://www.ftp.resource.org/courts.gov/c/F2/982/982.F2d.1546.92-1091.html

http://www.main.spsj.or.jp/nenpyo/1966-1967.htm. Accessed 18 Feb 2015.

http://www.waterworld.com/index/display/article-display/340334/articles/water-wastewater-international/volume-23/issue-4/features/the-reverse-osmosis-membrane-evolution.html

http://www.workingwithwater.net/view/3230/water-purification-ultrafiltration-for-portable-water-purification/

https://www.nap.edu/read/12473/chapter/36#208. Accessed 7 June 2018.

https://www.sandiego.gov/sites/default/files/legacy/water/pdf/purewater/060501.pdf. Accessed 10 Feb 2017.

https://www.uop.com/about-us/uop-history/a-friendly-acquisition/. Accessed 6 June 2018.

IDA. (2009). *Desalination Yearbook 2009–2010.*

Ikeda, K. (1991). Development and practical uses of low pressure type reverse osmosis membranes. *Membrane,16*, 223–232.

Inoue, T. (2004). Reverse osmosis membrane. *Sen'i Gakkaishi (journal of Fiber Science and Technology),60,* 163–165.

International Atomic Energy Agency. (1964). Desalination of water using conventional and nuclear energy. *Technical Reports Series,* No. 24.

Iwahori, H., & Funayama, K. (2004). Maku maeshori 2 dan system (Membrane pretreatment two-stage system). *Maku o mochiita mizushori gijutsu (Water treatment technology using membrane)* (pp. 260–276). Tokyo: CMC Publishing.

Izumoji, Y., & Matsubara T. (1970). Kaisui tansuika no kaigaino jōkyō (Overseas status of seawater desalination). *Mizushori Gijutsu (Water Purification and Liquid Wastes Treatment), 11,* 1–8.

Japan Water Works Association. (2008). *WHO guidelines for drinking water quality* (3rd ed.). Japan Water Works Association.

Japan Water Works Association. (2012). *WHO guidelines for drinking water quality* (4th ed.). Japan Water Works Association.

Joint Committee on Atomic Energy. (1964). *Hearings on Use of Nuclear Power for the Production of Fresh Water from Salt Water.* 88th Congress, 2nd Session.

Jōsui maku (Dai 2 han) Henshū iinkai hen (Water Purification Membranes (2nd Edition) Editorial Committee). (2008). *Jōsui maku (Water purification membranes).* Tokyo: Gihodo Publishing Co.

Kagaku Kōgyō Nippō (Chemical Daily) (August 18, 1995).

Kamiyama, Y., Yoshioka, N., Matsui, K., & Nanakgome, K. (1984). New thin-film composite reverse osmosis membranes and spiral wound modules. *Desalination,51,* 79–92.

Kijima, J. (1976). Kaisui oyobi kansui no tansuika (Desalination of seawater and brine). *Kankyō Gijutsu (Journal of Environmental Conservation Engineering), 5,* 30–37.

Kōgyō Gijutsuin (Agency of Industrial Science and Technology), Ministry of International Trade and Industry. (1970). Kaisui tansuika to fukusanbutsu riyō no kenkyū kaihatsu jokyō (Status of research and development of seawater desalination and by-product utilization). *Ōgata Project News (Large Project News),* 20–23.

Kōgyō Gijutsuin (Agency of Industrial Science and Technology), Ministry of International Trade and Industry. (1958). *Showa 33 nendo shiken kenkyūjo kenkyū keikaku (FY1958 research plan for the experiment laboratories).*

Kōgyō Gijutsuin (Agency of Industrial Science and Technology), Ministry of International Trade and Industry. (1961). *Showa 36 nendo shiken kenkyūjo kenkyū keikaku (FY1961 research plan for the experiment laboratories).*

Kōgyō Gijutsuin (Agency of Industrial Science and Technology), Ministry of International Trade and Industry. (1970). *Ōgata project news (Major project news).*

Kubota, S. (1992). Production methods of ultra pure water and its applications. *Sen'i Gakkaishi (journal of Fiber Science and Technology),48,* 20–27.

Kumano, A. (2000). Seawater desalination technology by hollow fiber reverse osmosis module. *Sen'i Gakkaishi (journal of Fiber Science and Technology),56,* 7–11.

Kumano, A. (2007). Chūtō chiiki deno kaisui tansuika RO plant no unten jōkyō (Operation of Seawater Desalination RO Plants in the Middle East). In *New Membrane Technology Symposium 2007.* Presentation Material.

Kurihara, M., Uemura, T., Himeshima, Y., Ueno, K., & Bairinji, R. (1994). Development of crosslinked aromatic polyamide composite reverse osmosis membrane. *Journal of the Chemical Society of Japan,2,* 97–107.

Kurihara, M., Yamamura, H., & Nakanishi, T. (1999). High recovery/high pressure membranes for brine conversion SWRO process development and its performance data. *Desalination,125,* 9–15.

Kurihara, M., & Takeuchi, H. (2008). Mizu mondai! Nihon no kōken wa? (Water shortage! What is Japan's contribution?). *Keiei Sensor (Management Sensor), Toray Business Research Institute,* 31–41.

Kurita Water Industries. (2002). *Kore de wakaru mizushori gijutsu* (Now you understand water treatment technology). Tokyo: Kōgyō Chōsakai.

Lee, K. P., Arnot, T. C., & Mattia, D. (2011). A review of reverse osmosis membrane materials for desalination-development to date and future potential. *Journal of Membrane Science,370*, 1–22.

Lewis, W. M. (Ed.). (1980). *Developments in water treatment-2.* Applied Science.

Lonsdale, H. K. (1982). The growth of membrane technology. *Journal of Membrane Science,10*, 81–181.

Merton, U. (1970). "Makuhō ni yoru datsuen" (Desalination by the membrane method). *Bulletin of the Society of Sea Water Science Japan,23*, 206–209.

Ministry of Economy, Trade and Industry. (2011). *Heisei 22-nendo sangyō gijutsu chōsa: Wagakuni kigyō no kenkyū kaihatsu toshi kōritsu ni kakaru open innovation no teiryōteki hyōkatō ni kansuru chōsa hōkokusho (Shiryo-hen) (FY2010 Industrial technology survey: Survey report on quantitative evaluation, etc. of open innovation on R&D investment efficiency of Japanese firms (Data version)).*

Ministry of Economy, Trade and Industry. (2008). Dai 3 shō dai 4 setsu mizu mondai to wagakuni no torikumi (Chapter 3, Section 4: Water issues and Japan's efforts) in *White Paper on International Economy and Trade* (p. 351).

Morita, H. (1999). Ultrapure water (Production, analysis and functional water). *Journal of the Surface Finishing Society of Japan,50*, 873–878.

Nakagome, K. (1984). Chōjunsui seizō no tame no maku module (Membrane module for ultrapure water production). *Kagaku sōchi (Plant and Process)* (pp. 1–8).

National Development and Reform Commission of China. (2016). *The "13th Five-Year" seawater use plan.* http://www.ndrc.gov.cn/zcfb/zcfbghwb/201612/W020161230519130996138.pdf. Accessed 9 Apr 2018.

National Institute of Resources, Science and Technology Agency Committee. (1967). *Kaisui tansuika no gijutsu kaihatsu ni kansuru hōkoku/ Kagaku gijutsuchō Shigen chōsakai hōkoku dai 41 gō (Report on technological development of seawater desalination/ National institute of resources, science and technology agency committee report No. 41).* Tokyo.

Nihon Keizai Shimbun. July 31, 2010; September 27, 2013.

Nihon Keizai Shimbun. May 9, 1978; June 23, 2018.

Nikkei Business. March 19, 2007. p. 134.

Nikkei Sangyō Shimbun (Nikkei Business Daily). November 12, 1983; July 24, 1984b.

Nikkei Sangyō Shimbun (Nikkei Business Daily). July 29, 1983; March 3, 1984a; April 12, 1984a.

Nikkei Sangyō Shimbun (Nikkei Business Daily). November 25, 1977; April 5, 1980; July 10, 1982; March 10, 1984; August 10, 1985; August 26, 1986.

Nikkei Sangyō Shimbun (Nikkei Business Daily) (Eds.). (2011). *Nikkei share chōsa 2012-nemban (Nikkei share survey* (2012th ed.). Nikkei Inc.

Nitto Denko. *Nitto group report kakunemban (Nitto Group report annual editions).*

Nitto Denko. *Securities Report.*

Nitto Denko. *Nitto gihō kakunemban (Nitto Technical Report annual editions).*

Office of Saline Water. (1974). *Saline Water Conversion Summary Report.*

Okazaki, M., Taniguchi, Y., & Suzuki, H. (2006). *Water treatment integrated membrane system.* Tokyo: Nikkan Kogyo Shimbun.

Oshio, K. (2003). *Environmental history of water: Southern California in the twentieth century.* Tamagawa University Press.

Parnell, J. A., Lester, D. L., & Menefee, M. L. (2000). Strategy as a response to organizational uncertainty: An alternative perspective on the strategy-performance relationship. *Management Decision,38*, 520–530.

Petersen, R. J., & Cadotte, J. E. (1988). Thin film composite reverse osmosis membranes. In M.C. Porter (Ed.), *Handbook of Industrial Membrane Technology* (pp. 307–348). Noyes Publications.

Ranson, W. W., & Godfrey, W. J., Jr. (1974). Kōnōdo ensui ni mochiiru Gyakushintō yō shinchūkū sen-i (New hollow fiber membranes for reverse osmosis for highly concentrated saline water). *Bulletin of the Society of Sea Water Science Japan,28*, 52–56.

Redd, H. E. (1974). Gyakushintō hō no kogyō oyobi toshi haisuishori gijutsu e no ōyō (Application of the reverse osmosis method to industrial and municipal wastewater treatment technology). *Bulletin of the Society of Sea Water Science, Japan,28*, 57–62.

Ryūkyū Shimpō. April 10, 1997.

Sawada, D. (2010). 2: Mizu shigen mondai no kaiketsu ni torikumu Nihon no maku gijutsu (2: Japanese membrane technology for solving water resource problems) in *Jizoku kanō na Shakai no Kōchiku (Building a Sustainable Society)* (pp. 135–144).

Sekino, M., Kumano, A., & Fujiwara, N. (2006). Expanding technology of seawater desalination membrane. *Bulletin of the Society of Sea Water Science Japan,60*, 408–414.

Shokuhin sangyō maku Riyō gijutsu Kenkyū kumiai (Membrane Research Association in Food Industry). (1987). *Saikin no maku module kaihatsu to shokuhin kogyō e no ōyōrei (Recent membrane module developments and applications in the food industry).*

Shūkan Diamond (Weekly Diamond). January 27, 2007. p. 82.

Strathmann, H., Giorno, L., & Drioli, E. (2006). *An Introduction to Membrane Science and Technology*. Roma: Consiglio Nazionale delle Ricerche.

Tanaka, K., Matsui, K., Hori, T., Iwahashi, H., Takeuchi, K., & Ito, Y. (2009). World's first large-scale 3-stage series reverse osmosis (rO) method seawater desalination plant. *Mitsubishi Heavy Industries Technical Review,46*, 13–15.

Taniguchi, M. (2009). Present state and prospective of seawater desalination technologies using reverse osmosis membranes. *Nihon Kaisui Gakkaishi (bulletin of the Society of Sea Water Science, Japan),63*, 214–220.

Taylor, J. S. (2000). *Flat Sheet, Bench and Pilot Testing for Pesticide Removal Using Reverse Osmosis*. Denver: American Water Works Association.

Tokyo Kōgyō Shikenjo (The Government Chemical Industrial Research Institute, Tokyo). (1971). *Tokyo kōgyō shikenjo 70-nen shi (Tokyo National Industrial Research Institute: 70 Years of History)*. Tokyo.

Toray Industries, Inc. (2003). Development of polyamide composite reverse osmosis membranes and their membrane. *Reports of achievements in industrialization Awarded the Okochi Memorial Prize*, 49, 51–60

Toray Industries, Inc. *Securities Report*.

Toray Industries, Inc. (1997). *70-nen shi (Seventy Years of History)*. Tokyo.

Toray Research Center. (2008). Dai 2 shō gyakushintō (RO) hō (Chapter 2 Reverse Osmosis (RO) Method) in *Ekitai no Maku bunri (Membrane Separation of Liquid)*. Tokyo.

Toyobo Corporation. *Securities Report*.

Toyobo Corporation. (2015). *Toyobo 130-nen shi (130-Year History of Toyobo)*. Osaka.

Ukai, T., & Yanaga, Y. (1994). Gyakushintō maku hō ni yoru kaisuitansuika gijutsu (Seawater Desalination Technology by the Reverse Osmosis Membrane Method). *Kagaku kōgaku (Chemical Engineering), 58*, 20–23. http://www.toyobo.co.jp/news/2014/release_4747.html. Accessed 17 Jan 2019.

United States Department of the Interior. (Annual editions) *Saline Water Conversion Report*. Washington, D.C.

Utterback, J. M. (1994). *Mastering the dynamics of innovation: How companies can seize opportunities in the face of technological change*. Harvard Business School Press.

Water Reuse Promotion Center. *Kaisui tansuika Gijutsu kaihatsu Chōsa hōkokusho Kaku nemban (Seawater Desalination Technology Development Research Report. Annual editions).* Tokyo.

Water Reuse Promotion Center. (1987). *Kaisui tansuika gijutsu kaihatsu ni tsuite: Chigasaki kenkyū shisetsu ni okeru gijutsu kaihatsu no gaiyō (On the Development of Seawater Desalination Technology: An Overview of Technological Development at the Chigasaki Research Facility).* Tokyo.

World Water Vision River and Water Committee. (2001). *World water vision.* Tokyo: Sankaido. http://www.toray.co.jp/ir/pdf/lib/lib_a385.pdf. Accessed 1 Dec 2015.

Yamano, A. (1987). Cleaning and estimation technology of semiconductor manufacturing process. *Jitsumu Hyōmen Gijutsu (practical Surface Technology),34*, 44–50.

Yarisawa, T. (1970a). Kaisui tansuika to fukusanbutsu riyō no kenkyū kaihatsu ni tsuite (On research and development of seawater desalination and by-product utilization) in *Japanese scientific monthly*, *22*, 27–35.